MOLECULAR ORBITAL THEORY for organic chemists

MOLECULAR ORBITAL THEORY for organic chemists

ANDREW STREITWIESER, JR., Associate Professor of Chemistry, University of California, Berkeley, California

John Wiley & Sons, Inc., New York · London

SECOND PRINTING, DECEMBER, 1962

Library of Congress Catalog Card Number: 61—17363
Printed in the United States of America

Preface

This book concerns, principally, the simple molecular orbital methods and their application to organic chemistry. It is a commonplace that the chemical literature has grown apace in recent decades. The organic chemical molecular orbital literature dramatically exemplifies this trend. A mere handful of papers (approximately 20) in the thirties was followed by approximately 70 papers in the forties, whereas the decade of the fifties just completed has witnessed some 600 papers on this subject. This developing interest in the molecular orbital method is accented by the increased use of molecular orbital concepts in a number of recent elementary organic chemical courses and textbooks. Yet, despite this evidence of the popularity and importance of the molecular orbital method, few reviews and no adequate modern textbook for organic chemists exist in this vast body of literature.

The present book is a threefold attempt to annul this void. As a textbook, it offers instruction in the simple molecular orbital method with a minimum background in quantum mechanics. The introductory chapter is a review of elements of quantum chemistry which should be largely familiar to the experienced student but which are actually not necessary to an understanding of the simple molecular orbital methods described in the succeeding chapters. Exercises and examples are provided to help the student to learn enough to make his own calculations and applications. For the serious student the study of group theory, as described in Chapter 3, may be rewarding; however, such study is not essential to an understanding of the material that follows. A number of the applications involves topics not normally included in an organic chemical curriculum; for example, electron spin resonance, electronic transitions, electron affinity, and ionization potential. For such subjects an introductory exposition of the topic accompanies the molecular orbital treatment.

I have attempted in this book to provide an essentially complete review of the literature. It would be foolhardy to claim success in this attempt, and I shall simply apologize in advance for neglecting those contributions

v

that I have undoubtedly, but inadvertently, overlooked. The literature citations should be relatively complete through mid-1960, although a few important references have been mentioned up to early 1961.

Finally, the present volume is in part a research paper. Established correlations between theory and experiment have been extended and new ones are documented. Some of the qualitative interpretations of chemistry in terms of theory are paraphrased from the literature; others are original.

In an effort to be all things to all chemists there is danger in not succeeding in any single phase; however, the present need for both a critical evaluation and a textbook is so great that the attempt had to be made.

In any work of this sort necessary innovations and uniformity of symbolism require apologia, particularly when reference is made to so many fields apart from my own areas of competency. Indeed, on rereading the manuscript, I am reminded of the story of the suspect, who, after long and diligent interrogation by experienced police officers, finally blurted out, "But fellows, I've told you more than I know already!" An explanation is demanded of my adoption of the term HMO, or Hückel molecular orbital. The foundations of molecular orbital theory were erected notably by Hund and Mulliken. The simple molecular orbital theory of organic chemistry was developed extensively by Lennard-Jones, Longuet-Higgins, Coulson, Wheland, and a host of others. Yet it remains that Hückel applied a certain set of approximations used previously in a simple theory of crystals to the π-electronic systems of conjugated compounds and showed how the results provided significant interpretations of the chemistry of such systems. These approximations, which are fully detailed in Chapter 2, are referred to frequently and are so important to the further development of the theory that they require a simple label; it seems singularly appropriate to call them the Hückel molecular orbital, or HMO method. Some of the approximations may be removed by the use of additional parameters within the framework of the basic theory. These methods constitute the "simple molecular orbital" methods. They, and their applications, make up the bulk of this book. Introduction of additional elaborations such as electronic repulsion integrals requires modification of the basic framework of the simple theory. Adaptation of these elaborations constitutes the "advanced molecular orbital methods" which are discussed in Chapter 16.

I must apologize also for the use of the "natural order" for the direction of arrows to represent absorption transitions in Chapter 8. The application of $\rightarrow$ for emission and $\leftarrow$ for absorption transitions has been advocated. The extensive employment of absorption spectra by organic chemists would result in almost exclusive use of the "unnatural" sequence, $A^* \leftarrow A$, according to this convention. In considering this convention, I have been advised that I would be damned if I did and damned if I didn't. With such unhealthy alternatives, I have followed my instincts and have adopted the "natural order."

This book had its origin in a series of seminars that I presented in the spring of 1958 at the Shell Development Company at the suggestion of David P. Stevenson. The extensive transformation from lecture notes to manuscript was largely accomplished at Berkeley while I was on sabbatical leave as a National Science Foundation Science Faculty Fellow with additional assistance from the Alfred P. Sloan Foundation. The research and calculations described were supported in part by the Air Force Office of Scientific Research and in part by the National Science Foundation. I am indebted to George S. Hammond and John R. Platt for valued criticisms of the entire manuscript. Individual chapters were scrutinized by a number of persons, including W. H. Saunders, Jr., W. B. Schaap, K. S. Pitzer, W. D. Gwinn, J. Howe, R. L. Ward, and J. B. Bush. Last, but certainly not least, I am indebted to Miss Jayne Kravig for her superb typing of most of the manuscript and to my wife for her patience and her other specialized contributions.

ANDREW STREITWIESER, JR.

Berkeley, California
August 1961

Contents

Table of symbols and abbreviations

a_{0r} coefficient of rth AO in NBMO

a a vector

A_{rs} cofactor of element rs in a determinant

A electron affinity; annelation energy

$\mathscr{A}$ a matrix

AH   alternant hydrocarbon

AIP   auxiliary inductive parameter

AO   atomic orbital

ASMO   antisymmetrized molecular orbital

c_{ir} coefficient of rth AO in ith MO

C_j j-fold rotation axis

D dissociation energy; units of dipole moment in Debye

DE delocalization energy

E total energy; identity operation

E_π π-Energy

EI   electron impact

ESR   electron spin resonance

F free energy

$\Delta F^\ddagger$ free energy of activation

F_r free valence at position r

FE   free electron

h order of a group

h_r Coulomb integral increment of rth AO in units of β_0

H_0 Hammett's acidity function

$\Delta H^{\ddagger}$ enthalpy of activation

H magnetic field strength

$\mathbf{H}_{rs}$ Hamiltonian integral between orbitals r and s

$\mathscr{H}$ Hermitian matrix

H Hamiltonian operator

hfs   hyperfine splitting

HMO   Hückel molecular orbital

i center of inversion

i, j indices of molecular orbitals

I ionization potential

k_i a rate constant

k_{rs} bond integral for r—s bond in units of β_0

K_i an equilibrium constant

l angular momentum quantum number; dimension of a group representation

L_b bond localization energy

L_p para-localization energy

L_r localization energy of atom r

L_r^{ω} ω-technique localization energy

LCAO   linear combination of atomic orbitals

m mass; magnetic quantum number; index of highest occupied MO

m_i bonding energy coefficient in ith MO

M bonding energy coefficient in total π-energy

MO   molecular orbital

n quantum number, principal quantum number; total number of AO's in MO

n index of MO of highest energy

N_r Dewar's reactivity number

NBMO   nonbonding molecular orbital

non-AH   nonalternant hydrocarbon

p atomic orbital with $l = 1$

p_{rs} π-bond order between atoms r and s

PI   photoionization

q_r electron density at atom r

Q hyperfine splitting proportionality constant

r, s indices of atoms or AO's

R_{RS} conjugation energy between π-systems R and S

RE resonance energy

s atomic orbital with $l = 0$

S overlap integral

$\Delta S^{\ddagger}$ entropy of activation

S_r super-delocalizability at position r

S_{rs} overlap integral between AO's r and s

SCF   self-consistent field

T kinetic energy

$\mathcal{U}$ unitary matrix

V potential energy

VB   valence-bond

x diagonal element in HMO determinant equivalent to $(\alpha - \epsilon)/\beta$

Z_r Brown's reactivity index

α Coulomb integral

α_0 standard Coulomb integral

α_r Coulomb integral of rth AO or atom

β bond integral

β_0 standard bond integral

β_{rs} bond integral between AO's r and s

γ bond integral with inclusion of overlap

Γ group representation

δ_{jk} Kronecker delta

ϵ_i energy of ith MO

$\epsilon_{1/2}$ half-wave potential

ζ_r charge density of position r

μ dipole moment

$\pi_{r,s}$ atom-atom polarizability

$\pi_{rs,t}$ bond-atom polarizability

$\pi_{t,rs}$ atom-bond polarizability

$\pi_{rs,tu}$ bond-bond polarizability

ρ reaction constant

ρ_r net spin density at atom r

σ plane of symmetry

σ_r substituent reactivity constant at position r

χ character

ψ a wave function, molecular orbital

ψ_j jth molecular orbital

Ψ total product wave function

ω parameter in ω-technique

I SIMPLE MOLECULAR ORBITAL THEORY

1 Introduction

1.1 Quantum Mechanics

The development of the dual wave-particle concepts of light and matter only a few decades ago has altered along with much of the philosophical and intellectual world the thought and language of the organic chemist. The brilliant triumph of structure theory and the tetrahedral carbon was based virtually on a concept of bonds formed from miniscule hooks reminiscent of Lucretius' prickly atoms. The various intellectual constructions of ingenious but largely meaningless symbols for benzene represent an example of the frustration existent prior to the wave electron.

A major impetus for this change comes from Schrödinger's postulate, for the Schrödinger equation is a wave equation. With its development Dirac[1] was led to state, "The underlying physical laws necessary for the mathematical theory of a large part of physics and the whole of chemistry are thus completely known"—fortunately the quote does not end here, or we should miss the fun of doing an experiment—"and the difficulty is only that the application of these laws leads to equations much too complicated to be soluble."

In one dimension for one electron the Schrödinger equation for a conservative system, one which does not interact with its surroundings, may be written as

$$\frac{d^2\psi}{dx^2} + \frac{8\pi^2 m}{h^2}(E - V)\psi = 0 \tag{1}$$

in which E is the total energy, a constant, and V is the potential energy, which in general is a function of x. Solution of this equation involves finding a *wave function*, ψ, a function of x, which will satisfy (1). This equation can be solved for some important simple cases such as the harmonic oscillator and the electron in a box (p. 27).

A simple and instructive case is that in which an electron is constrained to move in a circle of radius r. We let $V = 0$ at the circumference of the

[1] P. A. M. Dirac, *Proc. Roy. Soc.*, **A123**, 714 (1929).

circle and $V = \infty$ elsewhere; x is then a distance parameter along the edge of the circle. Since $x = r\theta$, conversion to polar coordinates gives

$$\frac{1}{r^2} \frac{d^2\psi}{d\theta^2} + \frac{8\pi^2 m}{h^2} E\psi = 0 \tag{2}$$

A solution to this equation is readily found to be

$$\psi = a \cos \lambda\theta \tag{3}$$

in which

$$\lambda^2 = \frac{8\pi^2 mr^2}{h^2} E \tag{4}$$

One of the integration constants, a, is determined by the *normalization condition* which, in our case, is

$$\int_0^{2\pi} \psi^2 \, d\theta = 1 \tag{5}$$

We find that

$$a = \frac{1}{\sqrt{\pi}} \tag{6}$$

As is generally the case with differential equations, we are left with an infinite number of possible solutions. However, the only physically acceptable solutions are those that are single-valued and continuous. In our case this condition implies

$$\psi(\theta = 0) = \psi(\theta = 2\pi) \tag{7}$$

that is, in traversing the circle the wave must exactly superimpose in any region of overlap. Equations 3 and 7 give

$$\cos 2\pi\lambda = 1 \tag{8}$$

but this equation can hold only for integral values of λ. From (4), the condition that λ is an integer means that

$$E = \frac{n^2 h^2}{8\pi^2 mr^2} \qquad n = 0, 1, 2, \cdots \tag{9}$$

We still have an infinite number of acceptable solutions, but our energy values are constrained to a set of discrete quantities determined by a *quantum number*, n, which must be integral. Hence for each value of n there corresponds an energy E_n and a wave function ψ_n. Wave functions for some values of n are illustrated schematically in Fig. 1.1. Note that the different wave functions are characterized by nodes along the circle, that is, points at which the wave function has the value zero. If the wave function is defined to be positive above the plane of the circle and negative below, the nodes are points that mark a change in sign of the wave function. The nodes occur in pairs and may be marked by nodal lines which are illustrated as dotted lines in Fig. 1.1. Note especially that the *number of nodal lines is equal to the quantum number, n*; an increase in the number of nodes is accompanied by an increase (more positive-less stable) in the

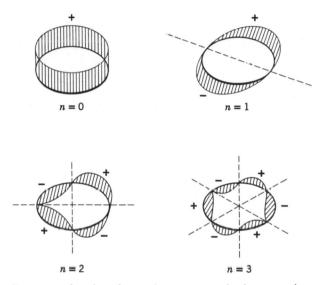

$n = 0$ $n = 1$

$n = 2$ $n = 3$

Fig. 1.1   Some wave functions for an electron constrained to move in a circle.

energy. We shall find that the parallelisms among quantum numbers, numbers of nodes in the wave functions, and energies are rather common in quantum mechanics.

In three dimensions the second derivative in (1) is replaced by the Laplacian operator,

$$\nabla^2 \equiv \frac{\partial^2}{\partial x^2} + \frac{\partial^2}{\partial y^2} + \frac{\partial^2}{\partial z^2} \tag{10}$$

Hence we write

$$\nabla^2 \psi + \frac{8\pi^2 m}{h^2}(E - V)\psi = 0 \tag{11}$$

This familiar expression for the Schrödinger equation may be rearranged to

$$\left(-\frac{h^2}{4\pi^2} \cdot \frac{1}{2m} \nabla^2 + V\right)\psi = E\psi \tag{12}$$

The quantity within the parentheses is the sum of the kinetic and potential energies of the electron written in operator form. For more complicated systems it is usually convenient to work with the *Hamiltonian*, **H**, of classical mechanics, a function of the momenta and coordinates which embodies the equations of motion.[2] In the cases of interest to us the potential energy is independent of velocity; for such cases the Hamiltonian is equal to the sum of the kinetic (T) and potential energies

$$H = T + V \qquad (13)$$

The Schrödinger postulate can be set down in terms of a restatement of the Hamiltonian in which the latter is converted to an operator. We shall not concern ourselves with the details of this process,[3] since it will develop that we shall rarely need to write out the Hamiltonian explicitly. For a single electron, the Hamiltonian is simply the expression in parentheses in (12). Equation 12 may then be written as

$$H\psi = E\psi \qquad (14)$$

a common expression of the Schrödinger equation. In equations of this type, in which **H** is an operator, the constant E is called an *eigenvalue* and the function ψ is called an *eigenfunction*. As an example of this equation, for the operator d/dx the constant c is an eigenvalue of the eigenfunction e^{cx} because

$$\frac{de^{cx}}{dx} = ce^{cx}$$

Similarly, one is the eigenvalue and any function is an eigenfunction of the operator "divide by one."

The Hamiltonian can be set up for a given case of interest, representing, perhaps, an atom or a molecule. The problem in quantum mechanics is then to find the eigenfunctions and the eigenvalues of the resulting operator. Such complete solution, however, has been achieved only for relatively few systems such as those mentioned above and the hydrogenlike atom. In practice, one finds exact solutions for an approximate Hamiltonian or approximate solutions to the actual or complete Hamiltonian. Indeed, in systems of interest to organic chemists the most common procedure has been to find approximate solutions for approximate Hamiltonians. For organic chemists the importance of quantum mechanics lies not at all in exact calculations from first principles (*ab initio* calculations) but rather

---

[2] *H* in general is also a function of time; however, we are concerned primarily with *stationary states* in which time does not enter.

[3] For these details see the texts listed under Supplemental Reading at the end of the chapter.

in providing heuristic concepts and insights in establishing qualitative and quantitative semiempirical correlations of experimental data and, especially, in facilitating the application of what has long been the organic chemist's most important tool: reasoning by analogy.

1.2 Atomic Orbitals

For the hydrogen atom the Schrödinger equation becomes

$$-\left(\frac{h^2}{8\pi^2 m}\nabla^2 + \frac{e^2}{r}\right)\psi - E\psi \tag{15}$$

in which e is the electronic charge, r is the distance of the electron from the nucleus, and m is the mass of the electron or, more properly, the reduced mass of the proton and electron. This case is clearly the more general three-dimensional version of (2) in which r is now a variable. Solution of (15) can be accomplished but not without a considerable amount of complex mathematical skullduggery. Physically acceptable solutions again require the introduction of quantum numbers, which come into the problem as constants of integration limited to integral values in order to yield single-valued, continuous wave functions. Because of the additional degrees of freedom in this case, the quantum numbers are now three in number and are listed in Table 1.1. Assignment of appropriate

TABLE 1.1

QUANTUM NUMBERS

Symbol	Name	Possible Values
n	Principal quantum number	$1 \leqslant n \leqslant \infty$
l	Angular momentum quantum number	$0 \leqslant l \leqslant n - 1$
m	Magnetic quantum number	$-l \leqslant m \leqslant l$

numbers gives specific wave functions which are solutions to (15). Such one-electron wave functions are called *atomic orbitals*. For such three-dimensional wave functions the nodes occur as nodal surfaces, which, for systems having the spherical symmetry of the hydrogen atom, take the form of spheres, planes, and cones.

The principal quantum number, n, is one more than the total number of nodal surfaces. The angular momentum quantum number, l, is equal to the number of nonspherical nodal surfaces. The magnetic quantum number, m, indicates the character and orientation of the nonspherical nodes.

In the common symbolic representation for the atomic orbitals the numerical value of n is followed by a letter representing l in which s, p, d, and f refer to $l = 0, 1, 2, 3$, respectively. Thus $1s$ refers to a spherically symmetrical wave function having no nodes; $2s$ is a spherically symmetrical orbital having one spherical nodal surface. Frequently a subscript is used to indicate the orientation of the node and represents m. Thus $2p_x$ has one nodal surface, the yz-plane.

The "intensity" or "density" of the wave in a given volume element, $d\tau$, is given by $\psi^2\, d\tau$ or, for the case of a wave function formulated in the complex domain, $\psi\psi^*\, d\tau$, in which the asterisk marks the complex conjugate.[4] The normalization condition is given as

$$\int \psi\psi^*\, d\tau = 1 \tag{16}$$

the integration being taken over all space. This equation is the more general form of (5). The normalized intensity of a given volume element is postulated to have the physical significance of the probability of finding an electron therein; that is, $\psi\psi^*$ is the electron density at a point.

The orbitals are commonly represented pictorially in two dimensions by a locus of points of equal electron density such that some given fraction, say 90%, of the total electronic charge is contained within the figure. Such loci for some orbitals important in organic chemistry are pictured symbolically in Fig. 1.2. The signs are the algebraic signs of the wave functions in different regions. Note again that the wave function changes sign in crossing a node.

As n increases, the region of maximum electron density shifts farther from the nucleus and the corresponding eigenvalue is that of increasing energy. If the zero energy level is that of the proton and electron separated by a vast difference ($n = \infty$), the energies for finite n are negative and refer to the energy liberated when an electron initially far away drops into an orbital characteristic of n.

As the nuclear charge is increased (hydrogenlike atoms), the atomic orbitals are qualitatively similar but are shifted closer to the nucleus because of the increased Coulombic attraction. The ground state is still the $1s$ wave function.

The concept of electronic spin introduces a fourth quantum number, the spin quantum number, which can have one of two values, usually written as the spin functions α or β. By the Pauli principle, no two electrons can have all four quantum numbers the same; that is, two electrons of opposite spin can simultaneously occupy the same orbital. By successively increasing the nuclear charge, the matching electrons may be put

---

[4] For the complex number, $a + ib$, the complex conjugate is $a - ib$.

pairwise into orbitals of successively higher energy to build up the neutral atoms of the periodic table (Aufbau principle), the order of orbital energies being $1s$, $2s$, $2p$, etc. The three $2p$-orbitals ($m = -1, 0, +1$) have identical energies and are said to be *threefold degenerate*. In such a case, because of Coulombic repulsion between electrons, none of the orbitals can have two electrons in the ground state if any of the degenerate orbitals

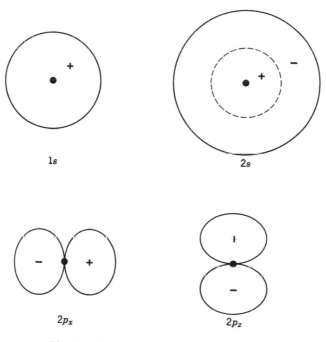

Fig. 1.2   Some atomic orbitals (schematic).

is unoccupied (Hund's rule). Hence the ground state of carbon, for example, is $1s^2 2s^2 2p_x 2p_y$, in which the superscripts total the number of electrons in each orbital. The first row of the periodic table built up in this way is given in Table 1.2.

It is important to remember that in this process of inserting successive electrons into what started as one-electron wave functions we have assumed that the additional electrons do not distort the relative energies of the various orbitals. Increased electronic repulsion will affect orbital energies. In the excited hydrogen atom, for example, the energies are independent of l; that is, $2s$- and $2p$-orbitals have the same energy. Yet, in the Aufbau process, the $2p$-orbitals in the ground state of a neutral atom are of higher energy than the $2s$-orbital.

The concept of orbitals as entities even in polyelectronic systems is justified by more detailed self-consistent field calculations of atomic systems. In this procedure the wave function is determined for an electron in the field of the nucleus and the average field of the other electrons; this wave function is used to find the average field for a second electron,

TABLE 1.2

Element	Electronic Structure
H	$1s$
He	$1s^2$
Li	$1s^2 2s$
Be	$1s^2 2s^2$
B	$1s^2 2s^2 2p_x$
C	$1s^2 2s^2 2p_x 2p_y$
N	$1s^2 2s^2 2p_x 2p_y 2p_z$
O	$1s^2 2s^2 2p_x^2 2p_y 2p_z$
F	$1s^2 2s^2 2p_x^2 2p_y^2 2p_z$
Ne	$1s^2 2s^2 2p_x^2 2p_y^2 2p_z^2$

and the iterative process is continued until the orbitals no longer change within the desired degree of accuracy. Such calculations show clearly the successive shell character of atomic structure. The resulting orbitals are called self-consistent field (SCF) orbitals. Unfortunately, these orbitals are expressed tabularly and not as analytical functions. Approximate analytic wave functions have been devised, the most common of which are those resulting from Slater's rules.[5] Slater orbitals are sufficiently good for many purposes for the atoms of interest to us. Some Slater orbitals for carbon in its ground state are

$$\psi(1s) = 7.66e^{-10.8r}$$
$$\psi(2s) = 2.06re^{-3.07r}$$
$$\psi(2p_x) = 3.58r \cos \theta e^{-3.07r}$$

r is the distance from the nucleus in angstroms and θ is the angle from the x-coordinate. Figure 1.3 shows plots of these functions. The approximate nature of Slater orbitals is demonstrated by the absence of a nodal surface in the 2s-orbital. Figure 1.4 shows a contour diagram of a carbon $2p_x$-orbital. A different type of representation of a Slater carbon $2p_z$-orbital is shown in Fig. 1.5. For each point along the x-axis of such an orbital there corresponds a value of z at which the wave function has a

---

[5] These rules are detailed in C. A. Coulson, *Valence*, Oxford University Press, London (1952), p. 40.

maximum value. Figure 1.5 shows the locus of points of maximum ψ as a function of x. Actually, two such plots, corresponding to the positive and negative lobes of the p-orbital, could be given, but only one is actually depicted. We shall frequently find this type of representation very convenient.

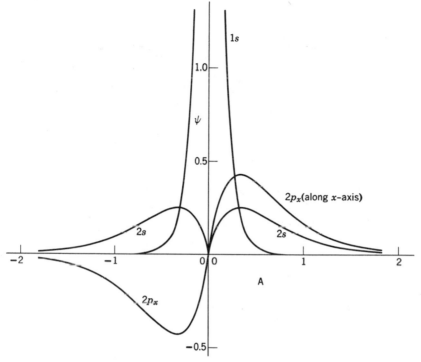

Fig. 1.3   Slater atomic orbital functions for carbon along the x-axis.

1.3   Bonds

The union of a proton and a hydrogen atom produces the hydrogen molecule-ion, H_2^+, with the liberation of 64 kcal./mole of energy. The two protons are held in close proximity (equilibrium distance, 1.06 A) by the single electron; we say that a bond exists between the two protons. The solution of the Schrödinger equation can be accomplished to a high degree of approximation for this molecule with the assumption of stationary nuclei (Born-Oppenheimer approximation).[6] The resulting

[6] Numerical integration was carried out by Ø. Burrau, *Det. Kgl. Danske Vid. Selskab.*, **7**, 1 (1927). Analytical solutions involve infinite series which can be evaluated to any desired degree of approximation. Rather accurate treatments have been presented by E. A. Hylleraas, *Z. Physik*, **71**, 739 (1931); G. Jaffé, *Z. Physik.*, **87**, 535 (1934).

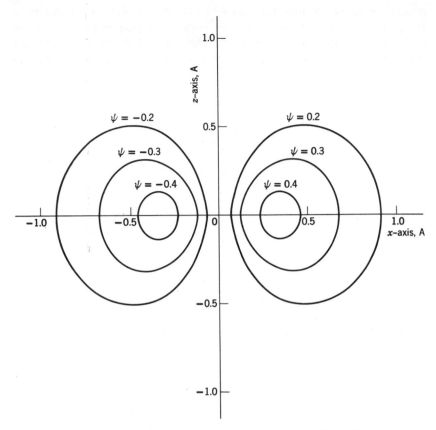

Fig. 1.4   Slater wave function contours for a carbon $2p_x$-orbital.

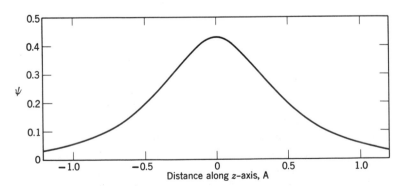

Fig. 1.5   Maximum magnitude of ψ along the x-axis of one lobe of a Slater carbon $2p_x$-orbital as a function of the distance along the z-axis.

ground state wave function shows a concentration of electron density in the region between the nuclei. The hydrogen molecule, on the other hand, with two electrons cannot be solved exactly, although extremely good approximate solutions have been obtained which give excellent agreement with experiment.[7] For diatomic molecules containing more electrons and for polyatomic molecules the necessary additional approximations give results in only qualitative or semiquantitative agreement with experiment.

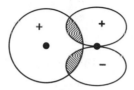

Fig. 1.6   Pi overlap between an s- and a p_z-orbital.

A useful concept is that bonding results from the overlap of two atomic orbitals. This concept may be given quantitative expression through the overlap integral S, defined by (17), in which ψ_a and ψ_b are normalized atomic orbitals. The asterisk indicates the complex conjugate and the integration of the space element $d\tau$ is taken over all space.

$$S = \int \psi_a^* \psi_b \, d\tau \tag{17}$$

S is clearly a function of the distance separating a and b. For $a = b$ we have simply the normalization condition and $S = 1$. When a and b are separated by infinity, the atomic orbitals will have no space elements in common with significant values for the wave functions and $S = 0$.

In the *principle of maximum overlapping*[8] we presume a direct relationship between the overlap integral and the bond strength. We should add the phrase, "other things being equal." Clearly, from this principle alone we anticipate that the bond strength should increase as the two nuclei approach each other. With appropriate orbitals it is true that the electronic contribution to bond strength increases with decreasing internuclear distance; on the other hand, electronic and nuclear repulsion energies also contribute to the total bond energy. The equilibrium bond distance is the distance for which the sum of the repulsion and attraction energies is a minimum.

Although an s-orbital and a p_z-orbital may overlap as in Fig. 1.6, the positive value of the integral over one lobe of the p-orbital is matched by

[7] H. M. James and A. S. Coolidge, *J. Chem. Phys.*, **1**, 825 (1933).

[8] J. C. Slater, *Phys. Rev.*, **37**, 481 (1931); **38**, 325, 1109 (1931); R. S. Mulliken, *Phys. Rev.*, **41**, 67 (1932); L. C. Pauling, *J. Am. Chem. Soc.*, **53**, 1367 (1937).

the negative value of the integral over the other lobe, and the sum which is the total overlap integral vanishes.

Such a combination does not give rise to net bonding. Two orbitals with zero overlap are said to be *orthogonal*; hence the 1s-, 2s-, 2p-, etc., orbitals of an atom are mutually orthogonal. For a p_x-orbital overlapping with

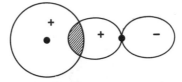

Fig. 1.7   Sigma overlap between an s- and a p_x-orbital.

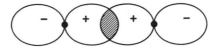

Fig. 1.8 $p_\sigma p_\sigma$ bond.

an s-orbital (Fig. 1.7) the integral over the (+) lobe has a far greater value than the negative integral over the (−) lobe; the overlap integral is non-zero and bonding can occur. Note that we can conceive of the internuclear distance as being the same in these two cases.

Overlapping of orbitals as in Fig. 1.7 produces a bond that is cylindrically symmetrical about the bond axis (x-axis). Such bonds are called *sigma* (σ)

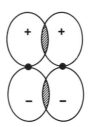

Fig. 1.9 $p_\pi p_\pi$ bond.

bonds. Two p-orbitals can overlap to form a sigma bond ($p_\sigma p_\sigma$), as in Fig. 1.8, or they can overlap as in Fig. 1.9. With the wave functions oriented as shown, the overlap integral is greater than zero; the corresponding bond is not cylindrically symmetrical about the bond axis but has a node—the xy-plane. Such a bond is called a *pi* (π) bond.[9]

[9] We can refer to an angular momentum quantum number about the bond axis; by analogy with the atomic quantum number l, the values 0, 1, 2, etc., are assigned the letters σ, π, δ, etc., respectively. This symbolism was introduced by F. Hund, *Z. Physik*, **51**, 759 (1928).

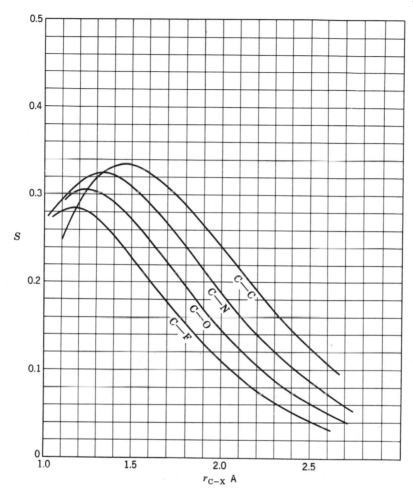

Fig. 1.10   Slater overlap integrals for $p_\sigma p_\sigma$ bonds.

Overlap integrals have been tabulated for a variety of bonds. The extensive tables of Mulliken, Rieke, Orloff, and Orloff[10] provide overlap integrals for Slater orbitals. In Figs. 1.10 and 1.11 overlap integrals are plotted as a function of internuclear distance for σ and π 2p-2p bonds between carbon and carbon, oxygen, nitrogen, and fluorine.

Consider two adjacent p-orbitals twisted through an angle ϕ, as in Fig. 1.12. The twisted orbital can be resolved into x- and z-components

[10] R. S. Mulliken, C. A. Rieke, D. Orloff, and H. Orloff, *J. Chem. Phys.*, **17**, 1248 (1949).

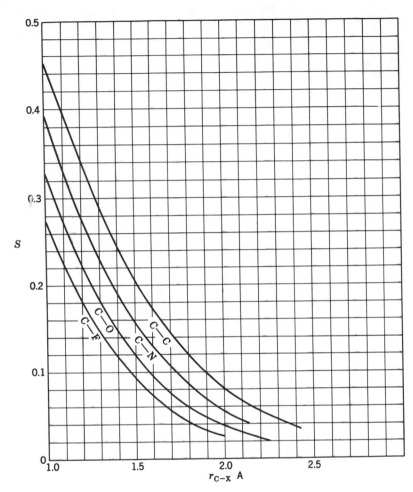

Fig. 1.11   Slater overlap integrals for $p_\pi p_\pi$ bonds.

by (18). Note in this connection the correct implication that the orbitals can be treated as *vectors*.

$$\psi_p = \psi_{p_x} \sin \phi + \psi_{p_z} \cos \phi \tag{18}$$

Since the π-overlap of x- and z-orbitals is zero, the overlap integral in this case is simply $S(p_z p_z) \cos \phi$. In a similar manner, any two p-orbitals, however oriented, may be resolved into appropriate p_x-, p_y-, and p_z-orbitals for calculation of S. For this purpose it is usually convenient to set up coordinates such that one axis passes through both nuclei.

For example, in Fig. 1.13

$$p_a = p_{x_a} \cos \phi + p_{z_a} \sin \phi$$

and
$$p_b = p_{x_b} \cos \theta + p_{z_b} \sin \theta$$

hence
$$S_{ab} = S(\pi, \pi) \sin \phi \sin \theta + S(\sigma, \sigma) \cos \phi \cos \theta$$

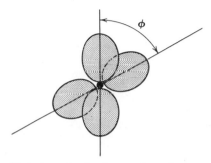

Fig. 1.12   End view of twisted π-overlap of two p-orbitals.

These orbitals partly overlap σ and partly π, and the total overlap integral is given as the sum of the component S's.

Consider now an excited carbon atom with one electron in each of the $2s$-, $2p_x$-, $2p_y$-, and $2p_z$-orbitals. This situation is exactly equivalent to

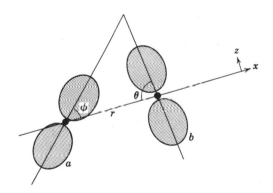

Fig. 1.13 p-Orbitals overlapping partly σ and partly π.

having one electron in each of four orbitals formed as linear combinations of the original four provided only that the four new orbitals are also mutually orthogonal. Such *hybrid* orbitals usually have improved directional properties for bond formation. By judicious choice of linear combinations, four equivalent hybrid orbitals can be formed. One such set of four orbitals is

$$\varphi_1 = \tfrac{1}{2}(s + p_x + p_y + p_z)$$
$$\varphi_2 = \tfrac{1}{2}(s - p_x - p_y + p_z)$$
$$\varphi_3 = \tfrac{1}{2}(s + p_x - p_y - p_z)$$
$$\varphi_4 = \tfrac{1}{2}(s - p_x + p_y - p_z)$$

(19)

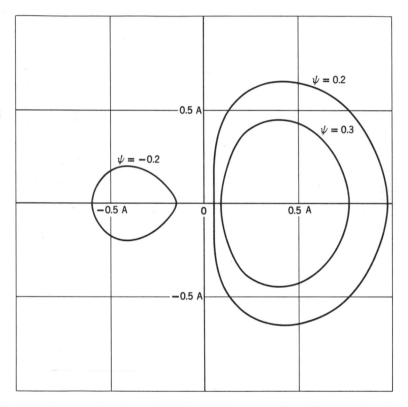

Fig. 1.14   Contours for carbon sp^3 hybrid orbital constructed from Slater wave functions.

Exercise. Show that these four hybrid orbitals are mutually orthogonal.

Each of these orbitals is $\tfrac{1}{4}s$ and $\tfrac{3}{4}p$ and is called a hybrid sp^3-orbital. The improved directional properties are illustrated by the Slater sp^3-wave function shown in Fig. 1.14.[11]

[11] Diagrams of SCF hybrid orbitals may be found in W. E. Moffitt and C. A. Coulson, *Phil. Mag.*, Ser. 7, **38**, 634 (1947).

A second set of four sp^3-orbitals is

$$\varphi_1' = \tfrac{1}{2}(s + p_x + \sqrt{2p_z})$$
$$\varphi_2' = \tfrac{1}{2}(s + p_x - \sqrt{2p_z})$$
$$\varphi_3' = \tfrac{1}{2}(s - p_x + \sqrt{2p_y})$$
$$\varphi_4' = \tfrac{1}{2}(s - p_x - \sqrt{2p_y})$$

(20)

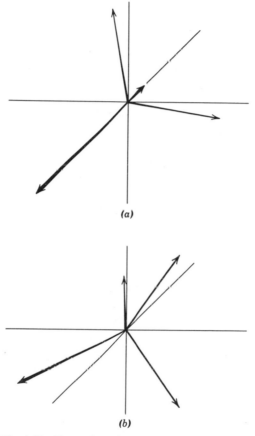

(a)

(b)

Fig. 1.15   Two orientations of a set of sp^3-orbitals.

If the orbitals are represented simply as vectors, (19) are four orbitals directed towards the corners of a tetrahedron oriented as in Fig. 1.15a, whereas (20) corresponds to a tetrahedron oriented as in Fig. 1.15b. Infinite sets of four orthogonal linear combinations can be written to represent four equivalent orbitals differing only in their spacial orientation.

Exercise. Given $\varphi_1'' = \tfrac{1}{2}(s + \sqrt{3}p_x)$. Find three other sp^3-orbitals in this set.

Note: For any $\varphi = \frac{1}{2}(s + ap_x + bp_y + cp_z)$ normalization requires $a^2 + b^2 + c^2 = 3$, whereas the requirement that the set of orbitals be mutually orthogonal means that $1 + aa' + bb' + cc' = 0$, in which the primed letters refer to any other hybrid in the set, φ''.

The molecule of methane may clearly be represented by overlapping a hydrogen $1s$-orbital with each of the sp^3-orbitals. In a familiar manner other hybrids are conceived. Linear combinations of s- and two of the p-orbitals result in three *trigonal* sp^2-orbitals directed toward the corners of an equilateral triangle in the plane defined by the two original p-orbitals. The planar structure of methyl radical can be represented by overlapping each of these sp^2-orbitals with a hydrogen orbital forming three two-electron bonds. The odd electron is retained in the remaining unhybridized p-orbital.

Hybridization of an s- and a p-orbital forms two *digonal* sp-orbitals directed toward opposite ends of the line defined by the component p-orbital. The matter is too familiar to demand further elaboration.

1.4 Localized Bonds—Valence-Bond (VB) Theory

In the Heitler-London[12] treatment of the hydrogen molecule an approximate wave function was suggested of the form

$$\psi = \varphi_r(1)\varphi_s(2) + \varphi_r(2)\varphi_s(1) \tag{21}$$

in which φ_r and φ_s are hydrogen $1s$-orbitals of atoms r and s and the numbers (1) and (2) refer to two electrons of opposite spin. This wave function is a sum of functions such that one electron is associated with one atom and the second electron is associated with the other atom. In this function the electrons are never associated together at one atom; the motion of the two electrons is completely correlated. This simple function is fairly decent for hydrogen, giving two-thirds of the total observed binding energy of the molecule. Because of this success and its embodiment of the electron-pair concept of bonding, the Heitler-London treatment has been extensively developed as the Heitler-London-Slater-Pauling (HLSP), or *valence-bond* (VB), theory. In its simplest form a bond, R—S, formed by overlapping two identical atomic orbitals (either pure or hybrid), φ_r and φ_s, is represented by the wave function in (21).[13] Calculation of the

---

[12] W. Heitler and F. London, *Z. Physik.*, **44**, 455 (1927).

[13] In this function we have neglected the spin part. One statement of the Pauli exclusion principle is that the total wave functions must be antisymmetric; i.e., the wave function must change sign on interchanging the labels of two electrons. The anti-symmetrized function corresponding to (21) is

$$\psi = [\varphi_r(1)\varphi_s(2) + \varphi_r(2)\varphi_s(1)] \cdot [\alpha(1)\beta(2) - \alpha(2)\beta(1)]$$

α and β are spin functions for the electrons. Note that this equation is not normalized. See Chap. 16 for further discussion of antisymmetrization.

binding energy E for the ground state of the bond gives

$$E = \frac{Q + J}{1 + S^2} \tag{22}$$

The expression is actually considerably more complicated, since the integrals involved have been replaced by the symbols Q, J, and S. S is simply the overlap integral discussed before (17).

$$Q = \int \varphi_r(1)\varphi_s(2)H\varphi_r(1)\varphi_s(2)\, d\tau \tag{23}$$

$$J = \int \varphi_r(1)\varphi_s(2)H\varphi_r(2)\varphi_s(1)\, d\tau \tag{24}$$

Q represents the interaction energy between each electron and its respective nucleus and is called the *Coulomb integral*. It is commonly taken to have 10 to 15% the magnitude of J which represents the energy associated with exchanging electrons between the nuclei; it is called the *exchange integral*.[14] The dominant term is J, which must be negative to have a significant bonding energy. J can be negative only if $S \neq 0$.[15] This feature provides justification for the maximum overlapping criterion for bonding.

The foregoing applies to most ordinary single bonds in organic compounds. In particular, the energy of methane treated as independent or localized C—H electron-pair bonds does not differ greatly from the energy calculated by considering the electron pairing within the entire molecule. Different pairing schemes can be represented by *structures* in which a line represents an electron pair with a substantial negative (bonding) value for J and a dotted line represents an electron pair with $J \cong 0$. For example, the localized bond electron-pair representation corresponds to structure I in which an electron in each sp^3-orbital of carbon is paired with an electron in the appropriate hydrogen $1s$-orbital. An alternative pairing scheme is structure II in which an electron in one carbon sp^3-orbital is paired with that in another and an electron in a hydrogen $1s$-orbital is paired with that in another.

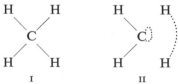

[14] K is sometimes used instead of Q, and J is sometimes replaced by α. The definitions of J and K have also sometimes been reversed.

[15] J. H. Van Vleck and A. Sherman, *Rev. Modern Phys.*, **7**, 167 (1935).

Each structure, I and II, corresponds to a wave function, ψ_I and ψ_{II}, respectively. According to the variation method (Chap. 2), a linear combination (25) of these wave functions should be a better approximation than ψ_I alone. The coefficients c_I and c_{II} can be found to make the energy of Ψ a minimum, this energy being lower than that of E_I alone.

$$\Psi = c_I\psi_I + c_{II}\psi_{II} \qquad (25)$$

In fact, the energy of Ψ is close to that of ψ_I and c_{II} is small compared to c_I. Consequently, the localized bond approximation is valid for many purposes. Our ability to treat such bonds as independent or localized is reflected experimentally in constancy of bond energies, bond dipole moments, characteristic infrared frequencies, etc. In conjugated systems and in aromatic compounds this independence is lost; some of the bonds must be described as *nonlocalized* or *delocalized*.

1.5   Localized Bonds—Molecular Orbital (MO) Theory

In the molecular orbital (Hund-Mulliken)[16] treatment of the hydrogen molecule wave functions are constructed to encompass the entire molecule as a unit. Each wave function can house two electrons of opposite spin. With polyatomic molecules, in principle, wave functions can still be set up to encompass all of the constituent atoms. However, if the molecule consists entirely of relatively localized bonds, for many purposes it is sufficient to consider only sets of wave functions that encompass the two atoms of individual bonds; such wave functions may be called *two-center molecular orbitals*. One simple approach is to construct the wave function as a *linear combination of atomic orbitals (LCAO)*.[17] For a homonuclear bond, R—S, constructed from identical atomic orbitals, φ_r and φ_s, the molecular orbital of lowest energy is

$$\psi = \varphi_r + \varphi_s \qquad (26)$$

and the energy of an electron in this orbital is given by

$$E = \frac{\alpha + \gamma}{1 + S} \qquad (27)$$

[16] The molecular orbital method was developed and extensively applied particularly to diatomic molecules independently by F. Hund and R. S. Mulliken in a long series of papers: F. Hund, *Z. Physik.*, **36**, 657 (1926); **37**, 742 (1927); **40**, 742 (1927); **42**, 93 (1927); **43**, 805 (1927); **51**, 759 (1928); **73**, 1, 565 (1931); **63**, 719 (1930); **74**, 1, 429 (1932); R. S. Mulliken, *Phys. Rev.*, **26**, 561 (1925); **29**, 648 (1927); **32**, 186, 761 (1928); **33**, 730 (1929); **40**, 55 (1932); **41**, 49, 751 (1932); **43**, 279 (1933).

[17] R. S. Mulliken, *J. Chem. Phys.*, **3**, 375 (1935).

α and γ are again symbols for integrals. α represents the energy of an electron in one of the isolated atomic orbitals alone and is again called a *Coulomb integral*; γ represents the energy of an electron in the field of both nuclei and is called the *bond* or *resonance integral*.[18] These integrals are defined by

$$\alpha = \int \varphi_r^* H \varphi_r \, d\tau \tag{28}$$

$$\gamma = \int \varphi_r^* H \varphi_s \, d\tau \tag{29}$$

Since two electrons are housed in this molecular orbital, the total energy is $2E$. For cases in which S is small compared to one, the binding energy is then given approximately as 2γ. Even this simple treatment yields more than half the observed bond energy of hydrogen.[19] The bond integral depends on S. γ is zero between truly orthogonal atomic orbitals and for appropriate orbitals becomes greater as S becomes larger.

An alternative linear combination is

$$\psi' = \varphi_r - \varphi_s \tag{30}$$

which has energy

$$E' = \frac{\alpha - \gamma}{1 - S} \tag{31}$$

Because of the coefficient of γ, this energy is higher than that of an electron in the isolated atomic orbital; hence ψ' is an unstable or *antibonding* molecular orbital. After two electrons have been placed in ψ, an additional one or two electrons may be housed in ψ' but only with a sacrifice of net binding energy.

The VB and MO approaches may be compared as follows: in the VB method we consider a bond to be formed by bringing together two initially distant atomic orbitals, each containing one electron of opposite spin. As overlap becomes appreciable, each electron feels the influence of the opposite nucleus and spends an increasing amount of time in the neighborhood of the other nucleus. In a sense it has achieved an increasing degree of motional freedom.

In the MO approach we start with two nuclei at the equilibrium bond distance stripped of the bonding electrons. We place the electrons one by one into orbitals centered on both nuclei. The situation is exactly analogous to the Aufbau procedure of placing electrons successively into atomic orbitals (Sec. 1.2).

An important deficiency of the HLSP(VB) method is that the exchange of the two electrons is perfectly correlated. The two electrons are not permitted to enjoy simultaneously the company of a single nucleus. In the LCAO method no such restriction is placed on the two electrons; an important deficiency of this method is that such *ionic* configurations are made much more important than

---

[18] The symbol Q is sometimes used in place of α; β sometimes replaces γ.
[19] C. A. Coulson, *Trans. Faraday Soc.*, **33**, 1479 (1937).

they should be. The difference is demonstrated by an expansion of the two electron wave functions:

$$\psi_{HLSP} = \varphi_r(1)\varphi_s(2) + \varphi_r(2)\varphi_s(1) \tag{32}$$

$$\psi_{LCAO} = (\varphi_r + \varphi_s)(1)(\varphi_r + \varphi_s)(2)$$

$$= \varphi_r(1)\varphi_s(2) + \varphi_r(2)\varphi_s(1) + \varphi_r(1)\varphi_r(2) + \varphi_s(1)\varphi_s(2) \tag{33}$$

The difference is seen to be the inclusion of the two final ionic terms in (33). Clearly, the HLSP wave function can be improved by the addition of ionic terms:

$$\psi_{HLSP}(\text{improved}) = \varphi_r(1)\varphi_s(2) + \varphi_r(2)\varphi_s(1) + \lambda[\varphi_r(1)\varphi_r(2) + \varphi_s(1)\varphi_s(2)] \tag{34}$$

λ is chosen to give the best value for the energy. An equivalent change may be made in (33) by conversion to

$$\psi_{LCAO}(\text{improved}) = (\varphi_r + \varphi_s)(1)(\varphi_r + \varphi_s)(2) + \mu(\varphi_r - \varphi_s)(1)(\varphi_r - \varphi_s)(2) \tag{35}$$

This change amounts to including some antibonding character in the original wave function and is a simple example of *configuration interaction* (see Chap. 16). The reader may show that, for $\lambda = -(1 + \mu)/(1 - \mu)$, (34) and (35) become identical.[20] Hence both methods are completely equivalent when each is taken to the next higher approximation.

1.6   Nonlocalized Bonds—Valence-Bond (VB) Theory

The localized-bond approximation breaks down for systems with conjugated multiple bonds. Even in these cases, however, it is usually possible to dissect the molecule into sets of localized and delocalized bonds and to treat each set independently. For example, the complete σ-bond framework of benzene can be treated as a system of twelve localized bonds—six C—H bonds formed from carbon sp^2- and hydrogen $1s$-orbitals and six C—C bonds formed from carbon sp^2-orbitals. The six π-electrons may be paired as in structure III in which the sigma skeleton has been omitted. Each electron-pair bond in this structure is formed from

III

π-overlap of two carbon $2p$-orbitals.

If, for convenience, we neglect the overlap integrals, the energy of III is

$$E_{III} = Q + 1.5J_\pi \tag{36}$$

[20] C. A. Coulson, *Valence*, Oxford University Press, London (1952), p. 148.

The energy is not $3Q + 3J$, as may be thought from (22). Q is now defined for six electrons instead of two; that is, it should be recalled that Q is a Coulomb integral for a structure of several electrons and not for just a single electron as in LCAO theory (28). The coefficient of J_π includes repulsions between bonded but unpaired orbitals as well as the attraction between bonded and paired orbitals.

An alternative and equivalent pairing scheme is that represented by IV.

IV

A better wave function than either ψ_{III} or ψ_{IV} alone is the linear combination

$$\psi = c_{III}\psi_{III} + c_{IV}\psi_{IV} \tag{37}$$

The energy calculated from ψ is $Q + 2.4J_\pi$. This energy is significantly lower than that for III alone; the limitation of the localization treatment is apparent for this case.

Additional structures which could be included are V, VI, and VII, the so-called Dewar structures. Inclusion of these structures lowers the energy

V   VI   VII

to $Q + 2.61J_\pi$. Their contribution to the total wave function is significant, but they are clearly not so important as the two equivalent Kekulé structures, III and IV.[21]

Valence-bond calculations become rapidly more difficult as the size of the molecule increases. Although calculations have been made for naphthalene, anthracene, and phenanthrene,[22] they are incomparably more difficult than calculations by LCAO method (*vide infra*). The mathematical difficulties of the VB method preclude application to molecules containing heteroatoms (atoms other than carbon) in the π-bond system. The relatively few early VB calculations[23] made have led to generalizations

---

[21] VB solutions for benzene were reported by E. Hückel, *Z. Physik*, **70**, 204 (1931) and by L. Pauling and G. W. Wheland, *J. Chem. Phys.*, **1**, 362 (1933). For details of obtaining VB energies and wave functions, consult Supplemental Reading at the end of the chapter.

[22] J. Sherman, *J. Chem. Phys.*, **2**, 488 (1934); C. Vroelant and R. Daudel, *Bull. soc. chim. France*, **1949**, 36; M. B. Oakley and G. E. Kimball, *J. Chem. Phys.*, **17**, 706 (1949).

[23] The total number of VB papers is nevertheless large. Active interest has continued particularly among the French workers. For further details and applications, cf. B. Pullman and A. Pullman, *Les Théories électroniques de la chimie organique*, Masson et Cie, Paris (1952).

that were seized upon by several pioneering organic chemists and developed as the now familiar resonance theory of organic chemistry. The transition was undoubtedly the more facile because the formalism of VB theory embodied the symbolism of the structural concept long known to organic chemists. There can be no question but that the development of resonance theory had much heuristic value in stimulating thought, understanding, and research in organic chemistry. Nevertheless, the theory is essentially qualitative and intuitive and consequently has been misused frequently. Neither resonance theory nor its parent VB theory is suited for general quantitative or semiquantitative calculations and correlations.

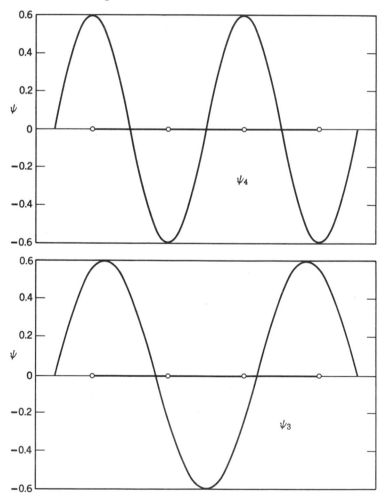

Fig. 1.16   Free electron (FE) wave functions for 1,3-butadiene.

1.7   Nonlocalized Bonds—Free Electron (FE) Theory[24]

In a manner similar to the handling of benzene (*vide supra*), all of our approximate theories of polyenes and other conjugated systems involve separation and independent treatment of the σ-bond framework considered as a set of localized two-center orbitals and the π-bond network considered to be occupied by a set of delocalized electrons. Consider a conjugated polyene stripped of its π-electrons. The result can be represented as a set of points of unit positive charge arranged as a zigzag line in a plane. In one approach we make the assumption that the positive charge is spread out uniformly along this line such that the original periodic potential along the line is replaced by a constant potential. If we now allow interaction to occur with one π-electron, the problem is that of the *electron in a one-dimensional box*, for which complete solution of the Schrödinger

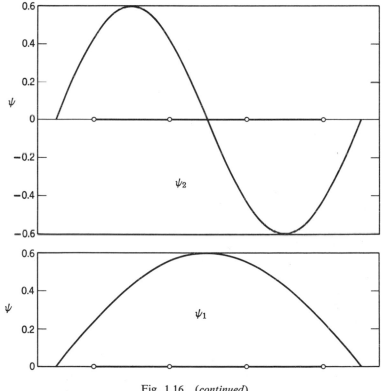

Fig. 1.16   (*continued*)

[24] For reviews see N. S. Bayliss, *Quart. Revs.*, **6**, 319 (1952); H. Kuhn, *Experientia*, **9**, 41 (1953).

equation is possible. This is an example of finding exact solutions for an approximate Hamiltonian. For an electron in a one-dimensional box such that the potential energy is zero everywhere within the box and infinite outside the box, the energy levels are given by

$$\epsilon_l = \frac{l^2 h^2}{8ma^2} \tag{38}$$

in which m is the electronic mass, a is the length of the box, and l is a quantum number. The corresponding wave functions are given by

$$\psi_l = \sqrt{2/a} \sin \frac{l\pi x}{a} \tag{39}$$

As a reasonable simplifying approximation, we may consider that our box stretches for half a bond length on either side of the ends of the conjugated

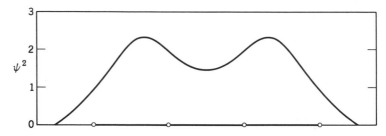

Fig. 1.17   FE electron density function for butadiene.

chain. Since the average bond length in a conjugated polyene is about 1.4 A, $a = 1.4n$, n being the number of atoms in the conjugated chain. Four wave functions for butadiene are plotted in Fig. 1.16. Note that as in the case of atomic orbitals the number of nodes increases as l increases. Each of the FE wave functions encompasses the entire π-system and could be called a FE molecular orbital. As the number of nodes increases, the net bonding of the orbital decreases; the orbital corresponds to progressively higher energy, as indicated in (38). If we neglect the effects of electronic interaction, we can put the total available π-electrons pairwise into these orbitals. The total electron density at each point is then given by

$$\psi^2 = \frac{2n}{a} \sum_{l=1}^{n/2} \sin^2 \frac{l\pi x}{a} \tag{40}$$

This function is plotted for butadiene in Fig. 1.17. Note the concentration of electron density between carbons 1 and 2 and between 3 and 4, corresponding to the usual diene structure. Except for the assumed length of the one-dimensional box, the FE method involves no empirical parameters.

Nevertheless, it gives a reasonably good account of the electronic spectra of polyenes.[25]

In a similar manner, a FE model of benzene would restrict movement of the electrons to the edge of a circle having the radius of a benzene molecule. Some other aromatic hydrocarbons can be treated by considering that the electrons peregrinate the periphery only; satisfactory correlations with electronic spectra are obtained nevertheless.[26]

The FE method has been applied successfully to simple models of dyes.[27] The theory has been elaborated to include extensive π-bond networks;[28] numerical application to polycyclic aromatic hydrocarbons gave satisfactory agreement with electronic spectra, bond lengths, and relative reactivities.[29] Nonetheless, compared to the research activity in LCAO MO theory, relatively little has been done with FE theory. Indeed, it seems that many of the correct predictions of the FE theory are the result of a basic similarity to LCAO MO theory;[30] both correctly treat important invariants of many-electron problems such as nodal lines and orthogonality of wave functions.

1.8 Nonlocalized Bonds—Molecular Orbital (MO) Theory

Molecular orbital theories are of various types and levels of complexity, although only relatively simple examples, such as FE and LCAO methods, have been applied at all extensively to polyatomic molecules. The LCAO method is based on the expectation that close to one atom of a molecule the influence of that atom will predominate and the wave function will resemble that of an atomic orbital. Consequently, it seems natural to approximate the molecular orbital as a *linear combination of atomic orbitals*. This presumption notwithstanding, of undoubtedly far greater importance is the fact that this approximation is mathematically tractable and, furthermore, that the solutions have a particularly convenient form. We shall see that even the most elementary form of this method, the *simple, naive,* or *Hückel*[31] MO theory, gives a successful account of many aspects of organic chemistry.

[25] N. S. Bayliss, *J. Chem. Phys.*, **16**, 287 (1948); cf., H. Kuhn, **16**, 840.

[26] J. R. Platt, *J. Chem. Phys.*, **17**, 484 (1949).

[27] Examples: H. Kuhn, *Helv. Chim. Acta*, **31**, 1441 (1948); W. T. Simpson, *J. Chem. Phys.*, **16**, 1124 (1948).

[28] K. Ruedenberg and C. W. Scherr, *J. Chem. Phys.*, **21**, 1565 (1953).

[29] C. W. Scherr, *J. Chem. Phys.*, **21**, 1583 (1953).

[30] K. Ruedenberg, *J. Chem. Phys.*, **22**, 1878 (1954); A. A. Frost, **23**, 310 (1955); N. S. Ham and K. Ruedenberg, **29**, 1199 (1958).

[31] The first application in organic chemistry was the treatment of benzene by E. Hückel, *Z. Physik*, **70**, 204 (1931).

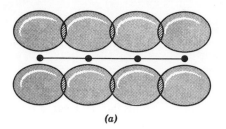

(a)

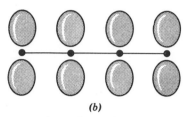

(b)

Fig. 1.18 Symbolic representations of constituent $2p_z$-orbitals in the π-network of butadiene.

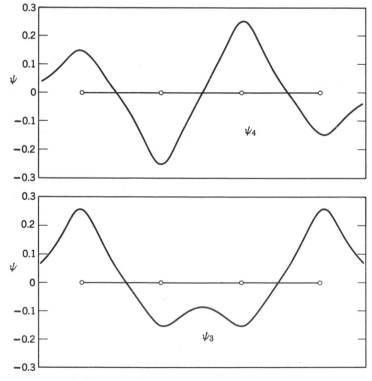

Fig. 1.19 Hückel molecular orbitals for butadiene.

We begin as usual by separating the σ-bonds from the π-orbitals and treat only the latter. In the treatment we make a number of simplifying assumptions, such as coplanarity of the π-orbitals (i.e., the π-orbitals share the same nodal plane), equality of all bond distances, and neglect of all non-neighbor interactions. In the case of butadiene, for example, the system of atomic orbitals to be used in our linear combination is represented in Fig. 1.18a in which the contours of the constituent carbon $2p$-orbitals are symbolic representations taken from Fig. 1.4. Because non-neighbor interactions are neglected, *syn-* and *anti-*butadiene are not distinguished, and the constituent carbon atoms can be represented as lying on a straight line. A set of molecular orbitals is frequently symbolized in terms of the constituent atomic orbitals. Aspects of stereochemistry, particularly, are frequently emphasized in this manner. For clarity, the atomic orbitals are sometimes altered and isolated as in the symbol for butadiene shown in Fig. 1.18b. The π-network in such cases is sometimes further symbolized by dotted lines joining the overlapping atomic orbitals.

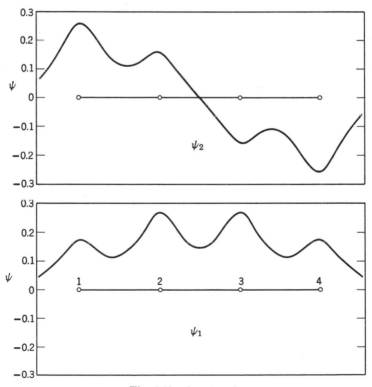

Fig. 1.19   (*continued*)

The set of linear combinations contains as many molecular orbitals as the number of constituent atomic orbitals. By the procedure detailed in Chap. 2, the four molecular orbitals (ψ_i) for butadiene comprise the following linear combinations of carbon $2p$ atomic orbitals (φ_r):

$$\psi_1 = 0.371\varphi_1 + 0.600\varphi_2 + 0.600\varphi_3 + 0.371\varphi_4$$
$$\psi_2 = 0.600\varphi_1 + 0.371\varphi_2 - 0.371\varphi_3 - 0.600\varphi_4$$
$$\psi_3 = 0.600\varphi_1 - 0.371\varphi_2 - 0.371\varphi_3 + 0.600\varphi_4$$
$$\psi_4 = 0.371\varphi_1 - 0.600\varphi_2 + 0.600\varphi_3 - 0.371\varphi_4$$

These orbitals are represented in Fig. 1.19. In these diagrams constructed from Slater orbitals the bond, or x-axis, is the abscissa and the maximum z-value of the wave function for each point x is plotted as the ordinate. The LCAO MO's may be compared with the corresponding FE MO's (Fig. 1.16). The LCAO MO's look somewhat different, particularly because of the "peaking" at atoms due to the use of atomic orbitals, but the nodal properties and general form of the MO's are very much the same in the two methods. As before, the four available π-electrons are accommodated pairwise in ψ_1 and ψ_2, the orbitals of lowest energy.

Supplemental Reading

Each of the following books presents an introduction to quantum mechanics, and each includes discussions in various degrees of detail of the material presented in this chapter.

C. A. Coulson, *Valence*, Oxford University Press, London (1952).

H. Eyring, J. Walter, and G. E. Kimball, *Quantum Chemistry*, John Wiley and Sons, New York (1944).

H. Hartmann, *Theorie der chemischen Bindung auf quantentheoretischer Grundlage*, Springer-Verlag, Berlin (1954).

W. Kauzmann, *Quantum Chemistry*, Academic Press, New York (1957).

K. S. Pitzer, *Quantum Chemistry*, Prentice-Hall, Englewood Cliffs, N.J. (1953).

G. W. Wheland, *Resonance in Organic Chemistry*, John Wiley and Sons, New York (1955).

H. A. Staab, *Einführung in die theoretische organische Chemie*, Verlag Chemie, Weinheim (1959).

J. R. Platt, "The Chemical Bond," Vol. 37 of *Handbuch der Physik*, Springer-Verlag, Heidelberg (1960).

R. Daudel, R. Lefebvre, and C. Moser, *Quantum Chemistry*, Interscience Publishers, New York (1959).

2 The Hückel molecular orbital (HMO) method

2.1 Variation Method

We have seen that an ideal objective of quantum mechanics is the finding of wave functions, Ψ, which are solutions to the Schrödinger equation, $H\Psi = E\Psi$, in which the Hamiltonian operator, H, includes all interaction terms for m electrons and n nuclei. We have seen further the thwarting of this ideal by the lack of exact analytical expressions for Ψ for polyelectronic systems. In principle, numerical expression would be possible, but this alternative is an inefficient way of storing information. A listing of the value of the wave function at each of a values for all $3(m + n)$ coordinates would require $a^{3(m+n)}$ entries. Moreover, it would be difficult in the extreme for the organic chemist to glean useful patterns from such tabulations.

We resort instead to the approximation that Ψ may be factored into a set of independent or noninteracting Φ, each of which describes, in effect, a separate set of electrons. At a later stage we may return and attempt to correct for those interactions that were assumed to be nonexistent in the first approximation. An example was the dissection of saturated compounds into component localized bonds which in turn were approximated in terms of constituent atomic or hybrid orbitals (Secs. 1.4 and 1.5).

Unsaturated molecules were assumed to factor into sets of σ-bonds and π-bonds. The set of σ-bonds is assumed to be further divisible into individual relatively noninteracting localized bonds as in the saturated compounds. A similar dissection of π-bonds seems to be too gross an approximation. At this point we have $\Psi = \Phi_\sigma \Phi_\pi$. In the MO method the properties of Φ_σ are assumed to be predictable from its component bonds; Φ_σ is assumed to reduce to a product of two-center molecular orbitals. Φ_π, on the other hand, is approximated as a product of molecular orbitals, each of which in the LCAO method is taken to be a linear combination of $2p_z$-orbitals. For unsaturated hydrocarbons carbon $2p_z$-orbitals are involved, and we assume that they share the same nodal plane. Each LCAO MO is of the form

$$\psi_j = c_{j1}\varphi_1 + c_{j2}\varphi_2 + \cdots + c_{jn}\varphi_n \qquad (1)$$

or

$$\psi_j = \sum_{r=1}^{n} c_{jr}\varphi_r \qquad (2)$$

ψ_j is the jth molecular orbital, φ_r is the atomic orbital (taken as a carbon $2p_z$-orbital in the present case) for the rth atom, and c_{jr} is the coefficient of the rth atomic orbital in the jth molecular orbital. The subscripts r, $s, \cdots$, are generally used to represent atoms or atomic orbitals and i, $j, \cdots$, are generally used to represent molecular orbitals. These molecular orbitals are eigenfunctions of a Hamiltonian operator which is considered for the π-system alone. In principle, this Hamiltonian can be set up explicitly (Sec. 16.1), although it may develop that we shall rarely need to consider the explicit form; initially, at least, it may also be taken as a one-electron Hamiltonian.

Our problem now is to find the best set of values for the coefficients, that is, to find that set of coefficients that gives the best value for the energy of the molecular orbital. Solution of this problem makes use of the variation principle, which states that

$$\epsilon = \frac{\displaystyle\int \psi H \psi \, d\tau}{\displaystyle\int \psi^2 \, d\tau} \geqslant E_0 \qquad (3)$$

Any wave function other than the correct one will yield a value for the ground-state energy algebraically higher than the true value. Only the following intuitive proof is given here.[1] The ground state of a molecule is that of the most stable distribution of electrons and is described by a wave function. Any other distribution—hence any other wave function—represents a less stable arrangement and consequently corresponds to a higher energy.

---

[1] The variation method is discussed in almost all books on quantum mechanics. Some of the texts in which proofs are presented are H. Eyring, J. Walter, and G. E. Kimball, *Quantum Chemistry*, John Wiley and Sons, New York (1944), p. 99; H. Hartmann, *Theory der chemischen Bindung auf quantentheoretischer Grundlage*, Springer-Verlag, Berlin (1954), p. 22; W. Kauzmann, *Quantum Chemistry*, Academic Press, New York (1957), p. 119; L. Pauling and E. B. Wilson, *Introduction to Quantum Mechanics*, McGraw-Hill Book Co., New York (1935), p. 180; K. S. Pitzer, *Quantum Chemistry*, Prentice-Hall, Englewood Cliffs, N.J. (1953), p. 66; G. W. Wheland, *Resonance in Organic Chemistry*, John Wiley and Sons, New York (1955), p. 570.

Our problem is reduced to finding the set of coefficients that yields the lowest energy when put into (3). This is accomplished by minimizing the function in (3) with respect to each of the coefficients:

$$\frac{\partial \epsilon}{\partial c_r} = 0 \tag{4}$$

Substituting (2) into (3) and omitting the MO indices for the present,

$$\epsilon = \frac{\int \left(\sum_r c_r \varphi_r \right) H \left(\sum_r c_r \varphi_r \right) d\tau}{\int \left(\sum_r c_r \varphi_r \right)^2 d\tau} \tag{5}$$

$$= \frac{\sum_r \sum_s c_r c_s \int \varphi_r H \varphi_s \, d\tau}{\sum_r \sum_s c_r c_s \int \varphi_r \varphi_s \, d\tau} \tag{6}$$

For convenience we introduce the symbols

$$\mathbf{H}_{rs} \equiv \int \varphi_r H \varphi_s \, d\tau \tag{7}$$

$$\mathbf{S}_{rs} \equiv \int \varphi_r \varphi_s \, d\tau \tag{8}$$

With these symbols, (6) becomes

$$\epsilon = \frac{\sum_r \sum_s c_r c_s \mathbf{H}_{rs}}{\sum_r \sum_s c_r c_s \mathbf{S}_{rs}} \tag{9}$$

Since the denominator consists of a series of overlap integrals whose sum cannot vanish, we may write

$$\epsilon \sum_r \sum_s c_r c_s \mathbf{S}_{rs} = \sum_r \sum_s c_r c_s \mathbf{H}_{rs} \tag{10}$$

We differentiate with respect to a particular coefficient, c_t, and use (4)

$$\epsilon \sum_s c_s \mathbf{S}_{ts} + \epsilon \sum_r c_r \mathbf{S}_{rt} = \sum_s c_s \mathbf{H}_{ts} + \sum_r c_r \mathbf{H}_{rt} \tag{11}$$

It may be shown that

$$\mathbf{S}_{rs} = \mathbf{S}_{sr} \tag{12}$$

$$\mathbf{H}_{rs} = \mathbf{H}_{sr} \tag{13}$$

Hence

$$\epsilon \sum_r c_r \mathbf{S}_{rt} = \sum_r c_r \mathbf{H}_{rt} \tag{14}$$

or

$$\sum_r c_r (\mathbf{H}_{rt} - \epsilon \mathbf{S}_{rt}) = 0 \tag{15}$$

Since the minimization is carried out with each coefficient, we end up with n equations:

$$c_1(\mathbf{H}_{11} - \mathbf{S}_{11}\epsilon) + c_2(\mathbf{H}_{12} - \mathbf{S}_{12}\epsilon) + \cdots + c_n(\mathbf{H}_{1n} - \mathbf{S}_{1n}\epsilon) = 0$$
$$c_1(\mathbf{H}_{21} - \mathbf{S}_{21}\epsilon) + c_2(\mathbf{H}_{22} - \mathbf{S}_{22}\epsilon) + \cdots + c_n(\mathbf{H}_{2n} - \mathbf{S}_{2n}\epsilon) = 0 \qquad (16)$$
$$\cdots\cdots\cdots\cdots\cdots\cdots\cdots\cdots\cdots\cdots\cdots\cdots\cdots\cdots$$
$$c_1(\mathbf{H}_{n1} - \mathbf{S}_{n1}\epsilon) + c_2(\mathbf{H}_{n2} - \mathbf{S}_{n2}\epsilon) + \cdots + c_n(\mathbf{H}_{nn} - \mathbf{S}_{nn}\epsilon) = 0$$

Exercise. The student unfamiliar with the handling of summation signs should repeat the derivation explicitly for a π-network of three atoms.

Thus far we have introduced no approximations beyond the original use of a linear combination function as a solution to a one-electron Hamiltonian. At this point we now introduce some dillies. It is these approximations that constitute the simple LCAO or Hückel MO (HMO)[2] method.

The terms, $\mathbf{H}_{rr}$, are called *Coulomb integrals*.[3] From the definition, $\mathbf{H}_{rr} = \int \varphi_r H \varphi_r \, d\tau$, the Coulomb integral represents approximately the energy of an electron in a carbon $2p$-orbital. Since our π-lattice consists entirely of carbons, we assume that all such integrals are equal and we replace them by the symbol α.

The terms $\mathbf{H}_{rs}$ for $r \neq s$ are called the *resonance*[3] or *bond integrals*. From the definition $\mathbf{H}_{rs} = \int \varphi_r H \varphi_s \, d\tau$, these integrals represent the energy of interaction of two atomic orbitals. This interaction energy clearly depends on the distance of separation of the two orbitals. Hence the following assumptions are not too unreasonable: when atoms r and s are not bonded in a classical structural expression, the interaction energy is likely to be small and we take $\mathbf{H}_{rs} = 0$. For atoms r and s bonded, $\mathbf{H}_{rs}$ is finite, but if all the bond distances are equal and if the atomic orbitals share the same nodal plane the values of the various $\mathbf{H}_{rs}$ will be of closely comparable magnitude. We assume all $\mathbf{H}_{rs}$ for atoms r and s bonded are equal, and we replace them by the symbol β. Relative to the energy of an electron at infinity, both α and β are *negative* energy quantities.

The integrals $\mathbf{S}_{rs}$ are the *overlap integrals* which we have mentioned previously. If we have taken normalized atomic orbitals, $\mathbf{S}_{rr} = 1$. For atoms r and s separated by a large distance the overlap integral is vanishing small. Indeed, we make the assumption that $\mathbf{S}_{rs} = 0$ for $r \neq s$. This assumption greatly simplifies the mathematics, although it does seem to be a rather drastic measure. $\mathbf{S}_{rs}$ for π-bonded carbons r and s has a value

[2] These approximations were first introduced by E. Hückel, *Z. Physik*, **70**, 204 (1931). The method was extensively exploited by him; cf. *ibid.*, **72**, 310 (1931); **76**, 628 (1932); **83**, 632 (1933); International Conference on Physics, London, 1934, Vol. II, The Physical Society, London (1935), p. 9.

[3] E. Hückel, *Z. Physik*, **76**, 628 (1932).

close to 0.25 (Fig. 1.11), which is far from zero. We shall see later (Sec. 4.2), however, that this assumption is not so drastic as it seems. The assumption is equivalent to treating the constituent atomic orbitals as orthogonal. Hence the assumption is frequently referred to as the "assumption of orthogonality," and S_{rs} is sometimes called the *nonorthogonality integral*.

With the introduction of these additional approximations, (16) now reduce to

$$c_1(\alpha - \epsilon) + c_2\beta_{12} + \cdots + c_n\beta_{1n} = 0$$
$$c_1\beta_{21} + c_2(\alpha - \epsilon) + \cdots + c_n\beta_{2n} = 0 \qquad (17)$$
$$\cdots\cdots\cdots\cdots\cdots\cdots\cdots\cdots\cdots\cdots\cdots\cdots$$
$$c_1\beta_{n1} + c_2\beta_{n2} + \cdots + c_n(\alpha - \epsilon) = 0$$

in which $\beta_{rs} = \beta_{sr} = \beta$ or 0, depending on whether atoms r and s are bonded or not bonded, respectively, in a conventional structure. These equations form a set of simultaneous homogeneous linear equations in n unknowns. Such a set of equations is well known to have a nontrivial solution (all coefficients not all zero) only if the corresponding *secular determinant* vanishes:

$$\begin{vmatrix} \alpha - \epsilon & \beta_{12} & \beta_{13} \cdots \beta_{1n} \\ \beta_{21} & \alpha - \epsilon & \beta_{23} \cdots \beta_{2n} \\ \cdots\cdots\cdots\cdots\cdots\cdots\cdots\cdots \\ \beta_{n1} & \beta_{n2} & \beta_{n3} \cdots \alpha - \epsilon \end{vmatrix} = 0 \qquad (18)$$

Expansion of this determinant yields a polynomial equation of the type

$$(\alpha - \epsilon)^n + a_1\beta(\alpha - \epsilon)^{n-1} + a_2\beta^2(\alpha - \epsilon)^{n-2} +$$
$$\cdots + a_{n-1}\beta^{n-1}(\alpha - \epsilon) + a_n\beta^n = 0 \quad (19)$$

This *characteristic equation* has n real roots of the form

$$(\alpha - \epsilon) = -m_j\beta \qquad j = 1, \cdots, n \qquad (20)$$

or
$$\epsilon_j = \alpha + m_j\beta \qquad (21)$$

This procedure yields n values for the energy given as an algebraic sum of a Coulomb integral and some fraction of a bond integral. Hence the energies can be represented as a series of energy levels above and below an energy zero taken as α itself. Since β is negative, negative values of the roots (i.e., positive m_j) represent energy levels more negative (more stable) than the energy of an electron in a single carbon $2p$-orbital and are *bonding levels* which correspond to *bonding MO's*. For $m_j = 0$ the energy of the MO is the same as that of any constituent carbon $2p$-orbital, and such an MO is said to be *nonbonding* (NBMO). Negative values of m_j

represent higher energies (lower stability) than an isolated carbon $2p$-orbital; the corresponding MO's are said to be *antibonding*. An energy-level diagram can be set up as in Fig. 2.1.

With n values of the energies at hand, we can now reinsert them individually into (17) to produce n sets of n simultaneous equations. Each

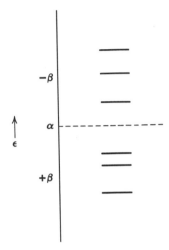

Fig. 2.1   Energy level diagram for HMO's.

set represents a separate problem which can be solved to yield all of the values of the coefficients.

At first sight it appears that we are determining $n + 1$ unknowns (n number of c's and one ϵ) with only n simultaneous equations. Actually, these equations will yield only the ratios between the coefficients; for example,

$$c_2/c_1, \, c_3/c_1, \cdots, c_n/c_1.$$

The $n + 1$th equation which gives all of our final values is simply the normalization condition

$$\int \psi^2 \, d\tau = 1 \tag{22}$$

Substituting (2) into (22) gives

$$\sum_r \sum_s c_r c_s \int \varphi_r \varphi_s \, d\tau = \sum_r \sum_s c_r c_s S_{rs} = 1 \tag{23}$$

But our prior assumption of orthogonality, $S_{rs} = 0$ for $r \neq s$, causes all cross products in our double summation to vanish and leaves

$$\sum c_r^2 = 1$$

This is our extra equation—the sum of the squares of the coefficients in an MO must equal unity.

Application of the HMO procedure to specific compounds involves the following operations:

1. Setting up the secular determinant
2. Expanding the secular determinant
3. Finding the roots of the characteristic equation[4]
4. Determining the coefficients

The first three operations determine the energy levels of the MO's. We shall apply these operations to several examples and return later to the matter of finding the coefficients.

2.2 Ethylene. $\overset{1}{\bigcirc}\!\!-\!\!-\!\!\overset{2}{\bigcirc}$

The determinant is simply

$$\begin{vmatrix} \alpha - \epsilon & \beta \\ \beta & \alpha - \epsilon \end{vmatrix} = 0 \tag{24}$$

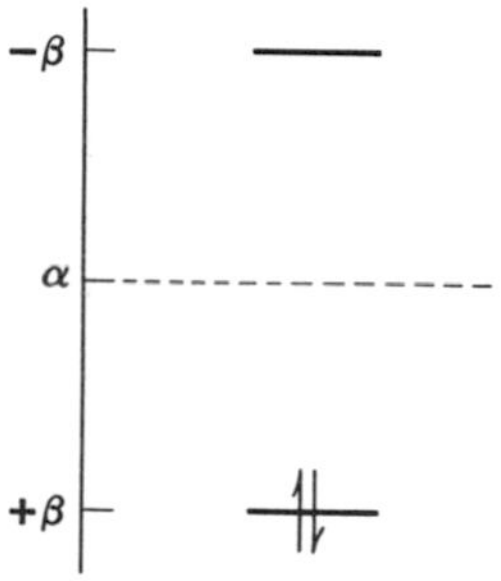

Fig. 2.2   HMO π-energy level diagram for ethylene.

We shall find it convenient to divide through each element by β and substitute $x \equiv (\alpha - \epsilon)/\beta$.

$$\begin{vmatrix} x & 1 \\ 1 & x \end{vmatrix} = x^2 - 1 = 0 \tag{25}$$

It follows that $x = \pm 1$ or $\epsilon_1 = \alpha + \beta$ and $\epsilon_2 = \alpha - \beta$. Note that the characteristic equation takes a simple dimensionless form that we may call the *secular polynomial*. The two available π-electrons are accommodated with paired spins in the MO of energy $\alpha + \beta$. The total π-energy, $E_\pi = 2\alpha + 2\beta$ (Fig. 2.2). Since the energy of two electrons in isolated carbon $2p$-orbitals is 2α, the π-bond energy is 2β.

---

[4] D. S. Urch, *J. Chem. Soc.*, **1958**, 4767, suggests deriving the characteristic equation directly from the simultaneous equations without going through the determinant.

2.3 Allyl. $\overset{1}{\circ}\!\!-\!\!\overset{2}{\circ}\!\!-\!\!\overset{3}{\circ}$

With the new symbolism, we see that the secular determinant consists of x's along the diagonal and either 1 or 0 in the off-diagonal elements, depending on the bonding or nonbonding relation, respectively, of the

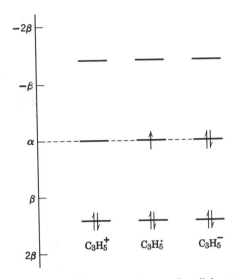

Fig. 2.3   HMO π-energy level diagram for allyl systems.

atoms indicated by the corresponding row and column indices. The determinant for the allyl system is, consequently,

$$\begin{vmatrix} x & 1 & 0 \\ 1 & x & 1 \\ 0 & 1 & x \end{vmatrix} = 0 \tag{26}$$

According to the normal expansion of a third-order determinant,

$$\begin{vmatrix} a_1 & b_1 & c_1 \\ a_2 & b_2 & c_2 \\ a_3 & b_3 & c_3 \end{vmatrix} = \begin{matrix} a_1b_2c_3 + b_1c_2a_3 + c_1a_2b_3 \\ -a_3b_2c_1 - b_3c_2a_1 - c_3a_2b_1 \end{matrix} \tag{27}$$

The secular polynomial for allyl is

$$x^3 - 2x = 0$$

and $x = 0$, $\pm\sqrt{2}$, or $\epsilon_1 = \alpha + \sqrt{2}\beta$, $\epsilon_2 = \alpha$, and $\epsilon_3 = \alpha - \sqrt{2}\beta$. The allyl cation has but two π-electrons which can be placed in ψ_1. The total

π-energy of the cation $E_\pi^+ = 2\alpha + 2\sqrt{2}\beta$. Allyl radical contains a third electron which must be placed in the next higher MO giving $E_\pi^\cdot = 3\alpha + 2\sqrt{2}\beta$. The additional electron in the allyl anion can also be placed in this MO giving $E_\pi^- = 4\alpha + 2\sqrt{2}\beta$. These relationships are diagrammed in Fig. 2.3. Note that the bonding energy in all three species is the same ($2\sqrt{2}\beta$) in the HMO method, even though more electrons should mean greater repulsion. This result is undoubtedly a gross oversimplification; it arises from our specific neglect of electron repulsions in the Aufbau process. This limitation of the HMO method, which must also be kept in mind, is discussed in greater detail in Chap. 4.

2.4 Butadiene. $\overset{1}{\text{O}}\!\!-\!\!-\!\!\overset{2}{\text{O}}\!\!-\!\!-\!\!\overset{3}{\text{O}}\!\!-\!\!-\!\!\overset{4}{\text{O}}$

The secular determinant is rapidly set up as

$$\begin{vmatrix} x & 1 & 0 & 0 \\ 1 & x & 1 & 0 \\ 0 & 1 & x & 1 \\ 0 & 0 & 1 & x \end{vmatrix} = 0 \tag{28}$$

In principle, all determinants can be expanded as a total of $n!$ products formed by taking one element from each row and column. Each product is assigned a $+$ or $-$ sign determined as follows: for each element considered as a_{rs} the products are first arranged in the natural order with respect to r. If an even number of interchanges is required to restore the s-indexes to natural order, the product is $+$; if odd, the product is $-$. *Example:* $a_{13}a_{21}a_{34}a_{42}$ is $-$. However, expansion in terms of *cofactors* or *signed minors* is far more satisfactory in practice for determinants of fourth or higher orders. The author's experience has been that a systematic expansion along the top of the determinant reduces mistakes to a minimum.

It will be recalled that the *minor* of the element a_{rs} is the determinant of next lower order formed by striking out row r and column s of the original determinant. The *cofactor*, A_{rs}, of a_{rs} is $(-1)^{r+s}$ times the minor. The determinant is reduced to a sum of determinants of next lower order by taking the product of each element and its cofactor along the entire row.

Expanding along the top of the butadiene determinant gives

$$x\begin{vmatrix} x & 1 & 0 \\ 1 & x & 1 \\ 0 & 1 & x \end{vmatrix} - \begin{vmatrix} 1 & 1 & 0 \\ 0 & x & 1 \\ 0 & 1 & x \end{vmatrix} = 0 \tag{29}$$

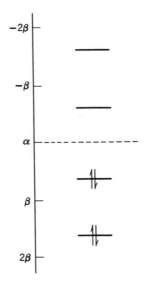

Fig. 2.4   HMO π-energy level diagram for butadiene.

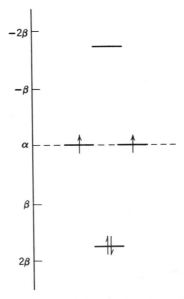

Fig. 2.5   HMO π-energy level diagram for trimethylenemethane.

Expansion of the third-order determinants follows directly to give the secular polynomial

$$x(x^3 - 2x) - (x^2 - 1) = x^4 - 3x^2 + 1 = 0 \tag{30}$$

Since this equation is in quadratic form, the roots are found easily. The resulting energy levels are $\epsilon_1 = \alpha + [(\sqrt{5} + 1)/2]\beta$, $\epsilon_2 = \alpha + [(\sqrt{5} - 1)/2]\beta$, $\epsilon_3 = \alpha - [(\sqrt{5} - 1)/2]\beta$, $\epsilon_4 = \alpha - [(\sqrt{5} + 1)/2]\beta$. The four π-electrons are put into the first two MO's, giving $E_\pi = 4\alpha + 2\sqrt{5}\beta$ (Fig. 2.4). Note that the bonding energy comes out to be greater than that of two isolated ethylenes. The difference is the calculated energy released when two pairs of electrons originally localized in two isolated double bonds are allowed to delocalize. This energy, called the *delocalization energy*, or *DE*, amounts to 0.472β for butadiene and seems clearly to correspond to what we think of as a resonance energy. The *DE* is a useful *defined* quantity for which we frequently quote values; the relationship to a resonance energy is discussed in Chap. 9.

2.5   Trimethylenemethane.

The secular determinant is

$$\begin{vmatrix} x & 1 & 1 & 1 \\ 1 & x & 0 & 0 \\ 1 & 0 & x & 0 \\ 1 & 0 & 0 & x \end{vmatrix} = 0 \tag{31}$$

which expands to

$$x^4 - 3x^2 = 0 \tag{32}$$

Thus $\epsilon_1 = \alpha + \sqrt{3}\beta$, $\epsilon_2 = \epsilon_3 = \alpha$, and $\epsilon_4 = \alpha - \sqrt{3}\beta$. Two of the four available π-electrons are put as usual in the first energy level. The next level is two-fold degenerate. In such a case Hund's rule applies, and one electron is placed in each level with parallel spins (Fig. 2.5). Since $E_\pi = 4\alpha + 2\sqrt{3}\beta$, *DE* is negative, and this π-lattice is less stable than that of two isolated double bonds. Note that the parallel spins of two electrons implies that this compound is a diradical. This result is expected from the structure, I, since two double bonds cannot be written for this structure without invoking "long bonds."

I

Exercises. Determine the π-energy levels for each of the following π-networks and determine the total π-energy and the *DE*

1. $\triangle$

   Ans. $x = -2, 1, 1$
   $E_\pi^+ = 2\alpha + 4\beta; \quad DE = 2\beta$
   $E_\pi^{\cdot} = 3\alpha + 3\beta; \quad DE = \beta$
   $E_\pi^- = 4\alpha + 2\beta; \quad DE = 0$

2. $\triangleright\!-$

   Ans. $x = -2.170, -0.311, +1.000, +1.481$
   $E_\pi = 4\alpha + 4.962\beta; \quad DE = 0.962\beta$

3. $\square$

   Ans. $x = 0, 0, \pm 2$
   $E_\pi = 4\alpha + 4\beta; \quad DE = 0$

4. $\boxtimes$

   Ans. $x = -2.562, 0, 1, 1.562$
   $E_\pi = 4\alpha + 5.123\beta; \quad DE = 1.123\beta$

5. $\wedge\!\!\wedge$

   Ans. $x = 0, \pm\sqrt{3}, \pm 1$
   $E_\pi^+ = 4\alpha + 5.464\beta; \quad DE = 1.464\beta$
   $E_\pi^{\cdot} = 5\alpha + 5.464\beta; \quad DE = 1.464\beta$
   $E_\pi^- = 6\alpha + 5.464\beta; \quad DE = 1.464\beta$

6. $\wedge\!\!\wedge\!\!\vee$

   Ans. $x = \pm 1.802, \pm 1.247, \pm 0.445$
   $E_\pi = 6\alpha + 6.988\beta; \quad DE = 0.988\beta$

7.

   Ans. $x = \pm 1, \pm(2 \pm \sqrt{3})^{\frac{1}{2}}$
   $E_\pi = 6\alpha + 6.899\beta; \quad DE = 0.899\beta$

(Note that this cross-conjugated triene turns out to be less stable than the preceding hexatriene.)

2.6 Roots of Polynomial Equations

A vital particular of the HMO method is finding the roots of the secular polynomial. For a second-order equation, of course, the quadratic formula is easily applicable. Although the general solution for cubic equations is well known, the author has found that for most cases the application of the formula is more work than it is worth. For fourth and higher polynomials other methods must be employed.

The obvious first step is to test the several possible integral values for x, 0, ± 1, ± 2. Note that since no carbon in a π-lattice can be bound to more than three other carbons no root can equal or exceed three in magnitude; that is, all $|m_j| < 3$. A rapid procedure for testing these values is the following: starting at the far left, the first coefficient of the secular polynomial, which is generally unity for our cases, is multiplied by the assumed value for the argument and the product is added algebraically to the next lower coefficient. The sum is multiplied by the assumed argument and the product is added to the next lower coefficient, etc. If the final sum vanishes, the assumed value for the argument is a root, and, moreover, the set of sums constitutes the coefficients for the next lower secular polynomial obtained by factoring the found root. An example follows in which $x = -2$ is found to be a root:

$$
\begin{array}{cccc}
x^3 + x^2 & -3x & -2 \\
-2 & 2 & 2 \\
\hline
-1 & -1 & 0
\end{array}
$$

The polynomial obtained by factoring this root is $x^2 - x - 1 = 0$.

With a calculating machine this method affords a rapid (with practice) means of testing assumed values of the argument. When an approximate value is found, a method such as Newton's may be used to obtain the root to any desired precision. Alternatively, a plot of the function may be constructed for the first approximation to the roots.[5]

An additional help: the reader will recall from his algebra days that the number of alternations of sign between successive coefficients equals the number of negative roots.

Some generalizations are also helpful. An important theorem which is not proved here (see Chap. 3) is

$$\sum_j m_j = 0 \tag{33}$$

that is, the algebraic sum of all the roots vanishes. The form of the roots takes an even more restricted pattern for special types of systems.

An especially important distinction is the classification into alternant hydrocarbons[6] (AH) and nonalternant hydrocarbons (non-AH). Alternant

[5] Other methods for finding roots of polynomials are known. See H. Margenau and G. M. Murphy, *The Mathematics of Physics and Chemistry*, Van Nostrand, New York (1943), p. 477; and J. W. Mellor, *Higher Mathematics for Students of Chemistry and Physics*, Dover Publications, New York (1955), Chap. VI.

[6] C. A. Coulson and H. C. Longuet-Higgins, *Proc. Roy. Soc.*, **A192**, 16 (1947).

hydrocarbons are planar conjugated hydrocarbons, having no odd-membered rings, in which the carbons can be divided into two sets, s (starred), and u (unstarred), such that each s-carbon has only u neighbors and vice-versa. In an even-AH the numbers of starred and unstarred positions are usually equal ($n_s = n_u$); for such cases the roots take the form

$$x_j = \pm\mu, \pm\mu', \pm\mu,'' \text{ etc.} \tag{34}$$

that is, the roots occur in pairs of opposite sign.[7] An example of such a system is II. In some cases the starred set is larger than the unstarred set ($n_s > n_u$); such cases have ($n_s - n_u$) MO's with no bonding energy[8] $[(n_s - n_u)x's = 0]$ and may normally be described as polyradicals. An example is III. Note that for such systems one cannot write normal Kekulé structures having more than n_u double bonds.

In most odd-AHs the starred set exceeds the unstarred set by one; the starred carbons are referred to as *active positions*.[8] For these systems, the roots also occur in pairs and the extra root has the value zero. An example is IV. For some systems n_s exceeds n_u by an odd number greater than one. For these cases $2n_u$ roots occur as $\pm$ pairs and the remainder are zero. For these systems, also, Kekulé structures can be written only with n_u double bonds. An example is V.

II   III   IV

V

For alternant systems the energy levels are symmetrically disposed about the zero level.

For some simple systems the energies may be written in closed form.[3,6,9] For straight chains of n atoms

$$x_j = -2\cos\frac{j\pi}{n+1}; \qquad j = 1, 2, 3, \cdots, n \tag{35}$$

[7] C. A. Coulson and G. S. Rushbrooke, *Proc. Cambridge Phil. Soc.*, **36**, 193 (1940).
[8] H. C. Longuet-Higgins, *J. Chem. Phys.*, **18**, 265 (1950).
[9] C. A. Coulson, *Proc. Roy. Soc.*, A**164**, 383 (1938); F. G. Fumi, *Nuovo cimento*, **8**, 1 (1951); A. A. Frost and B. Musulin, *J. Chem. Phys.*, **21**, 572 (1953). The last reference puts the formulas in a convenient mnemonic form.

For rings of n atoms

$$x_k = -2\cos\frac{2k\pi}{n}; \quad k = 0, \pm 1, \pm 2, \cdots, \quad \begin{array}{l} \pm\dfrac{n-1}{2} \text{ for } n \text{ odd} \\[2mm] \pm\dfrac{n}{2} \text{ for } n \text{ even} \end{array} \tag{36}$$

Heilbronner[10] has shown how the secular polynomials for a complex system may be derived from those of component parts. If K_n, K_{n-1}, and K_{n-2} are the secular polynomials for chains of n, $n-1$, and $n-2$ atoms, respectively,

$$K_n = xK_{n-1} - K_{n-2} \tag{37}$$

If R_n is the polynomial for a ring of n atoms,

$$R_n = K_n - K_{n-2} - 2(-1)^n \tag{38}$$

Let P_n be any system of n atoms having the secular polynomial, $\mathbf{P}_n$; let $\mathbf{P}_{n-1}^r$ be the polynomial for the system P_{n-1}^r which is P_n with atom r removed from the π-lattice. Then, if Q_{n+1} (having polynomial $\mathbf{Q}_{n+1}$) is derived from P_n by attachment of a carbon atom at atom r,

$$\mathbf{Q}_{n+1} = x\mathbf{P}_n - \mathbf{P}_{n-1}^r \tag{39}$$

Example. Let $Q_{n+1} = \underset{\text{(structure)}}{}$, $P_n = \underset{\text{(structure)}}{}$, $P_{n-1}^r = \underset{\text{(structure)}}{}$, $\mathbf{P}_n = x^7 - 7x^5 + 13x^3 - 7x$; $\mathbf{P}_{n-1}^r = x^6 - 5x^4 + 6x^2 - 1$. Therefore, $\mathbf{Q}_{n+1} = x^8 - 8x^6 + 18x^4 - 13x^2 - 1$.

Consider the system Q_{n+m} formed by joining atom r of system R_n to atom s of system S_m. We find that

$$\mathbf{Q}_{n+m} = \mathbf{R}_n\mathbf{S}_m - \mathbf{R}_{n-1}^r\mathbf{S}_{m-1}^s \tag{40}$$

Example. Let $R_n = \underset{\text{(structure)}}{}$, $S_m = \underset{\text{(structure)}}{}$, $Q_{n+m} = \underset{\text{(structure)}}{}$. $\mathbf{R}_n = x^5 - 5x^3 + 5x + 2$; $\mathbf{R}_{n-1}^r = x^4 - 3x^2 + 1$; $\mathbf{S}_m = x^3 - 3x + 2$; $\mathbf{S}_{m-1}^s = x^2 - 1$.

Therefore,

$$\mathbf{Q}_{n+m} = x^8 - 9x^6 + 2x^5 + 24x^4 - 8x^3 - 19x^2 + 4x + 5.$$

These relations are of general use for simplifying computational problems.[11]

[10] E. Heilbronner, *Helv. Chim. Acta*, **36**, 170 (1953).

[11] For additional properties of some use see I. Samuel, *Compt. rend.*, **238**, 2422 (1954); **229**, 1236 (1949); R. Gouarné, *Compt. rend.*, **230**, 844 (1950); H. O. Pritchard and F. H. Sumner, *Phil. Mag.*, **45**, 466 (1954); T. H. Goodwin and V. Vand, *J. Chem. Soc.*, **1955**, 1683; H. H. Günthard and H. Primas, *Helv. Chim. Acta*, **39**, 1645 (1956); A. L. Chistyavkov, *Izvest. Akad. Nauk S.S.S.R., Otdel Khim. Nauk*, **1959**, 1349.

2.7   Calculation of Coefficients

To find the coefficients for the molecular orbital that corresponds to each energy, we return to the original set of simultaneous equations, insert each energy in turn, and solve for the set of coefficients.

The case of ethylene is trivial. The secular determinant (25) corresponds to the equations

$$c_1 x + c_2 = 0$$
$$c_1 + c_2 x = 0$$

For $x = -1$ (lowest level)

$$-c_1 + c_2 = 0 \qquad \therefore c_1 = c_2$$

Applying the normalization condition,

$$c_1^2 + c_2^2 = 1$$

we find

$$c_1 = c_2 = 1/\sqrt{2}$$

Thus

$$\psi_1 = \frac{1}{\sqrt{2}}(\varphi_1 + \varphi_2); \qquad \epsilon_1 = \alpha + \beta$$

Using $x = 1$ in the same manner, we find

$$\psi_2 = \frac{1}{\sqrt{2}}(\varphi_1 - \varphi_2); \qquad \epsilon_2 = \alpha - \beta$$

The two MO's for ethylene are now completely characterized.

Butadiene is somewhat more laborious by this method. The secular determinant (28) corresponds to

$$
\begin{aligned}
c_1 x + c_2 &= 0 \\
c_1 + c_2 x + c_3 &= 0 \\
c_2 + c_3 x + c_4 &= 0 \\
c_3 + c_4 x &= 0
\end{aligned}
$$

For the first (lowest-lying) MO, $x = -[(1 + \sqrt{5})/2]$. Working down, we find

$$c_2 = \left(\frac{1 + \sqrt{5}}{2}\right)c_1$$

$$c_3 = -c_1 + c_2\left(\frac{1 + \sqrt{5}}{2}\right) = \left(\frac{1 + \sqrt{5}}{2}\right)c_1$$

$$c_3 - \left(\frac{1 + \sqrt{5}}{2}\right)c_4 = 0$$

$$c_4 = \left(\frac{2}{1 + \sqrt{5}}\right)c_3 = \left(\frac{2}{1 + \sqrt{5}}\right)\left(\frac{1 + \sqrt{5}}{2}\right)c_1 = c_1$$

From the normalization condition,

$$c_1^2 + c_2^2 + c_3^2 + c_4^2 = 1$$

$$c_1^2 + \left(\frac{1 + \sqrt{5}}{2}\right)^2 c_1^2 + \left(\frac{1 + \sqrt{5}}{2}\right)^2 c_1^2 + c_1^2 = 1$$

$$\therefore\ c_1 = \left(\frac{1}{5 + \sqrt{5}}\right)^{\frac{1}{2}} = c_4$$

and $$c_2 = c_3 = \left(\frac{1 + \sqrt{5}}{2}\right)\left(\frac{1}{(5 + \sqrt{5})^{\frac{1}{2}}}\right)$$

Hence $$\psi_1 = \frac{1}{\sqrt{5 + \sqrt{5}}}\left(\varphi_1 + \frac{1 + \sqrt{5}}{2}(\varphi_2 + \varphi_3) + \varphi_4\right)$$

This procedure can be repeated for each energy to compute each MO. However, the equivalent method of cofactors is much more systematic and straightforward. In principle, the cofactors along any row can be used, although there is generally little advantage to using any row other than the first. For cofactors, A_r, along the first row, the coefficient of the rth atom (and the rth column of the secular determinant)

$$c_r \sim A_r = (-1)^{r+1}(\text{minor of } a_{1r}) \qquad (41)$$

Example. Butadiene:

$$A_1 = x^3 - 2x$$
$$A_2 = -(x^2 - 1)$$
$$A_3 = x$$
$$A_4 = -1$$

For ψ_1, $x = -\left(\dfrac{1 + \sqrt{5}}{2}\right)$

r	A_r	A_r^2	$c_r = A_r/(\sum A^2)^{\frac{1}{2}}$	
1	-1	1	$\dfrac{1/\sqrt{5 + \sqrt{5}}}{}$	$= 0.371$
2	$-\left(\dfrac{1 + \sqrt{5}}{2}\right)$	$\dfrac{3 + \sqrt{5}}{2}$	$\dfrac{1 + \sqrt{5}}{2\sqrt{5 + \sqrt{5}}}$	$= 0.600$
3	$-\left(\dfrac{1 + \sqrt{5}}{2}\right)$	$\dfrac{3 + \sqrt{5}}{2}$	$\dfrac{1 + \sqrt{5}}{2\sqrt{5 + \sqrt{5}}}$	$= 0.600$
4	-1	1	$\dfrac{1}{\sqrt{5 + \sqrt{5}}}$	$= 0.371$

$$\sum A_r^2 = \sqrt{5} + 5$$

$$\therefore\ (\sum A_r^2)^{\frac{1}{2}} = (5 + \sqrt{5})^{\frac{1}{2}}$$

By this procedure normalized coefficients are obtained directly. Furthermore, the work is kept neat and orderly and can be inspected for errors rather conveniently. Each value of x is substituted in turn to determine each MO. It should be noted that occasionally the cofactors along a particular row vanish; in such cases cofactors along a different row must be used.

For some special systems several theorems are useful for the determination of the coefficients. These are stated without proof. For an AH, in which the energies occur in pairs, $\pm\mu$, the coefficients for the MO of $m = \mu$ are readily found from the MO of $m = -\mu$ by changing the sign of the coefficients of the unstarred atoms.[8] In the nonbonding MO ($m = 0$) of an odd-AH coefficients of unstarred positions are zero and each sum of the coefficients of starred atoms attached to any one unstarred position vanishes[6,7] (see p. 54).

For straight chains of n carbons the coefficients take the form[3,6,9]

$$c_{jr} = \sqrt{\frac{2}{n+1}} \sin \frac{rj\pi}{n+1} \tag{42}$$

whereas for rings of n atoms

$$c_{jr} = \frac{1}{\sqrt{n}} e^{\frac{2\pi i r(j-1)}{n}} \tag{43}$$

in which $j = 1, 2, 3, \cdots, n$. In the latter case note that the coefficients are given in complex form. When using such coefficients in other equations, attention must be given to the use of complex conjugates where appropriate.[12]

Exercise. Determine the coefficients of all of the MO's for each of the following compounds (see Exercise 2.5):

Ans.

Allyl,

ψ_i	x_i	c_1	c_2	c_3
1	$-\sqrt{2}$	$1/2$	$1/\sqrt{2}$	$1/2$
2	0	$1/\sqrt{2}$	0	$-1/\sqrt{2}$
3	$\sqrt{2}$	$1/2$	$-1/\sqrt{2}$	$1/2$

Cyclopropenyl,

ψ_i	x_i	c_1	c_2	c_3
1	-2	$1/\sqrt{3}$	$1/\sqrt{3}$	$1/\sqrt{3}$
2	1	$1/\sqrt{6}$	$1/\sqrt{6}$	$-2/\sqrt{6}$
3	1	$1/\sqrt{2}$	$-1/\sqrt{2}$	0

[12] The use of complex exponents is discussed in Sec. 3.10.

Note: In this case of degenerate energy levels the MO's cannot be uniquely defined. We shall find that the requirement that MO's be orthogonal means that the two MO's in this case have nodes at right angles to each other. Any linear combinations of the degenerate MO's which satisfy the orthogonality and normalization conditions are equally valid MO's. The MO's are related by these equations:

$$\psi_2' = \lambda\psi_2 + (1 - \lambda^2)^{1/2}\psi_3, \qquad 0 \leqslant \lambda \leqslant 1$$
$$\psi_3' = (1 - \lambda^2)^{1/2}\psi_2 - \lambda\psi_3$$

Butadiene,

ψ_i	x_i	c_1	c_2	c_3	c_4
1	-1.618	0.371	0.600	0.600	0.371
2	-0.618	0.600	0.371	-0.371	-0.600
3	0.618	0.600	-0.371	-0.371	0.600
4	1.618	0.371	-0.600	0.600	-0.371

Pentadienyl,

ψ_i	x_i	c_1	c_2	c_3	c_4	c_5
1	-1.732	0.288	0.500	0.576	0.500	0.288
2	-1	0.500	0.500	0	-0.500	-0.500
3	0	0.576	0	-0.576	0	0.576
4	1	0.500	-0.500	0	0.500	-0.500
5	1.732	0.288	-0.500	0.576	-0.500	0.288

3-Methylene-1,4-pentadiene

ψ_i	x_i	c_1	c_2	c_3	c_4	c_5	c_6
1	-1.932	0.230	0.444	0.628	0.325	0.444	0.230
2	-1.000	0.500	0.500	0	0	-0.500	-0.500
3	-0.518	0.444	0.230	-0.325	-0.628	0.230	0.444
4	$+0.518$	-0.444	0.230	0.325	-0.628	0.230	-0.444
5	1.000	0.500	-0.500	0	0	0.500	-0.500
6	1.932	0.230	-0.444	0.628	-0.325	-0.444	0.230

2.8 Electron Densities

We recall that the probability that an electron will be found in a small region of space, $\delta\tau$, is given by $\psi_\tau\psi_\tau^*\,\delta\tau$, in which ψ_τ is the value of the normalized wave function in the small region τ. Since our wave functions are generally real, we can omit the complex conjugate notation. For our linear combination MO's

$$\psi = \sum_r c_r\varphi_r$$

$$\int \psi^2 \, d\tau = \int \left(\sum_r c_r\varphi_r\right)^2 d\tau \tag{44}$$

$$= \int \sum_r c_r^2\varphi_r^2 \, d\tau + \int \sum_r \sum_{s \neq r} c_r c_s \varphi_r \varphi_s \, d\tau$$

Our previous orthogonality assumption was

$$\int \varphi_r \varphi_s \, d\tau = 0 \quad \text{for } r \neq s \tag{45}$$

Hence

$$\int \psi^2 \, d\tau = \int \sum_r c_r^2 \varphi_r^2 \, d\tau = \sum_r c_r^2 \int \varphi_r^2 \, d\tau \tag{46}$$

which we recognize as the normalization condition; however, $\int \varphi_r^2 \, d\tau$ represents the probability of finding an electron in the region of space associated with the atomic orbital φ_r. Because the integration is made over

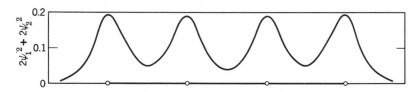

Fig. 2.6   Electron density function, $2\psi_1^2 + 2\psi_2^2$, for butadiene, using Slater orbitals.

all space, the value of each integral is unity, but each term is multiplied by a factor, c_r^2. In the simple LCAO (HMO) approximation c_r^2 consequently has the physical significance of the probability that an electron in an MO is associated with atomic orbital φ_r or, more simply, that c_r^2 is the electron density at atom r in the MO.[13]

The total electron density at an atom, q_r, is the sum of electron densities contributed by each electron in each MO:

$$q_r = \sum_j n_j c_{jr}^2 \tag{47}$$

c_{jr} is the coefficient of atom r in the jth MO, which is occupied by n_j electrons. The sum is taken over all of the MO's.

Examples. Butadiene: ψ_1 and ψ_2 are each occupied by two electrons. Hence $q_1 = 2c_{11}^2 + 2c_{21}^2 = 2(0.140) + 2(0.360) = 1.000$ and $q_2 = 2c_{12}^2 + 2c_{22}^2 = 2(0.360) + 2(0.140) = 1.000$. By symmetry, atoms 3 and 4 give the same answer; that is, the electron density is unity at each atom.

We should emphasize that the electron density as we *define* it is a convenient fiction which has only approximate physical significance. This point is demonstrated by an inspection of the function $2\psi_1^2 + 2\psi_2^2$ in Fig. 2.6, in which the electron density is the area beneath the curve. Although this function clearly peaks at the nuclei, it has substantial value between the atoms where one clearly cannot discern a boundary between the domains of individual atoms.

[13] G. W. Wheland and L. Pauling, *J. Am. Chem. Soc.*, **57**, 2086 (1935).

Allyl: For allyl cation, ψ_1 is doubly occupied.

r	c_r	c_r^2	q_r	ζ_r
1	$\frac{1}{2}$	$\frac{1}{4}$	0.5	+0.5
2	$\sqrt{2}/2$	$\frac{1}{2}$	1	0
3	$\frac{1}{2}$	$\frac{1}{4}$	0.5	+0.5
			$\Sigma q_r = 2$	$\Sigma \zeta_r = +1$

Since the neutral carbon would contain one electron, both carbons 1 and 3 in allyl cation have half a positive charge. We say that the *charge density* $\zeta_r = 1 - q_r$ is $+0.5$. The central carbon is neutral, that is, $\zeta_2 = 0$.

This result corresponds to a resonance hybrid of the two structures:

$$\overset{\oplus}{C} - C = C \leftrightarrow C = C - \overset{\oplus}{C}$$

Note that the sum of the electron densities over all of the atoms must equal the total number of π-electrons. If it does not, an error has been made in the computations.

Allyl radical contains, in addition, one electron in ψ_2 which has $c_1 = \sqrt{2}/2 = -c_3$ and $c_2 = 0$. For the total π-electron densities the square of each coefficient is added to the electron densities of the cation. We find $q_1 = q_2 = q_3 = 1$ for allyl radical.

In allyl anion ψ_2 houses two electrons; hence each square above is doubled and added to the electron densities for allyl cation, yielding $q_1 = q_2 = 1.5$; $q_2 = 1$. Both end carbons in allyl anion have a half unit of additional negative charge; that is, $\zeta_1 = \zeta_3 = -0.5$, $\zeta_2 = 0$.

Exercise. Find the electron densities for each of the following systems:

Methylenecyclopropene,

ψ_i	x	c_1	c_2	c_3	c_4
1	-2.170	0.278	0.612	0.524	0.524
2	-0.311	0.814	0.253	-0.368	-0.368
q_r		1.478	0.882	0.820	0.820
ζ_r		-0.478	$+0.118$	$+0.180$	$+0.180$

Cyclopropenyl,

Cation: $q_1 = q_2 = q_3 = 0.677$

Radical: Degeneracy, see note.

Anion: $q_1 = q_2 = q_3 = 1.333$

Note: In cyclopropenyl radical the third π-electron may be placed in either of two degenerate orbitals. A molecule cannot exist in one of several degenerate states (Jahn-Teller theorem) and will distort slightly to destroy the symmetry which gives rise to the degeneracy (Jahn-Teller distortion). In some cases a time-average electron density is useful. For such cases the Jahn-Teller effect may be ignored, and the electron density is calculated by assuming that $1/n$th of the available electrons is in each of the n degenerate MO's. The cyclopropenyl radical case would then be treated as if half an electron were in each of the degenerate MO's ψ_2 and ψ_3. This procedure yields $q_1 = q_2 = q_3 = 1$. The cyclopropenyl anion is straightforward—the third electron is put in ψ_2 and the fourth, having the same spin as the third, is put into ψ_3. The anion, consequently, is a diradical, but the Jahn-Teller effect does not apply.

Cyclobutadiene,   *Ans.* $q_1 = q_2 = q_3 = q_4 = 1$

Coulson and Rushbrooke[7] have shown that for even alternant hydrocarbons (AH) and for odd-AH radicals the electron density at each position is unity. In general, the electron densities in nonalternant systems differ from unity. An odd-AH cation or anion is formed by removing one electron from or adding one electron to the singly occupied nonbonding MO of an odd-AH radical having uniform electron distribution. The charge densities in such cases, consequently, are determined entirely by the coefficients of the NBMO which may be found without solving secular equations. The procedure has been discussed by Longuet-Higgins.[14] We use the fact that in the NBMO the coefficients of active atoms sum to zero about each inactive position. In the examples of benzyl and α-naphthylmethyl which follow the active positions are circled. In benzyl c_4 is taken as a. About inactive position 3, $c_2 + c_4 = 0$; therefore, $c_2 = -a$. The other coefficients are determined in the same manner.

$$(2a)^2 + 3a^2 = 1$$
$$\therefore \quad a^2 = \tfrac{1}{7}$$
$$a = 1/\sqrt{7}$$
$$\zeta_2 = \zeta_4 = \zeta_6 = \tfrac{1}{7}$$
$$\zeta_7 = \tfrac{4}{7}$$

$$(3a)^2 + 2(2a)^2 + 3a^2 = 1$$
$$\therefore \quad a^2 = \tfrac{1}{20}$$
$$a = 1/2\sqrt{5}$$
$$\zeta_2 = \zeta_4 = \tfrac{1}{5}$$
$$\zeta_5 = \zeta_7 = \zeta_9 = \tfrac{1}{20}$$
$$\zeta_{11} = \tfrac{9}{20}$$

In these examples ζ_r is positive or negative, depending on whether we are considering the cation or anion, respectively.

[14] H. C. Longuet-Higgins, *J. Chem. Phys.*, **18**, 275 (1950); cf. also M. J. S. Dewar, *J. Am. Chem. Soc.*, **74**, 3345, 3357 (1952); *Ann. Repts. Prog. Chem., Chem. Soc. London*, **48**, 112 (1951).

Exercise. Find the charge densities for the heptatrienyl, cinnamyl, and β-naphthylmethyl cations.

Ans.

2.9   Bond Orders and Free Valence

In 1939 Coulson[15] introduced the *partial mobile bond order* p_{rs}^{j} for the r—s bond in the *j*th MO, defined as

$$p_{rs}^{j} = c_{jr}c_{js} \tag{48}$$

The corresponding *total mobile bond order* or, more simply, the *bond order* is

$$p_{rs} = \sum_{\substack{\text{all} \\ \text{electrons}}} p_{rs}^{j} = \sum_{j} n_{j}c_{jr}c_{js} \tag{49}$$

in which n_j is the number of electrons in the *j*th MO.

The bond order is merely a convenient defined quantity, although this particular definition has some important mathematical advantages (*vide infra*). Nevertheless, as so defined, this quantity has a rough physical significance that we might associate with the binding power of a bond, since the product of the coefficients of adjacent bonded atoms may be construed as a bond electron density. When both coefficients are large and of like sign, the product is large and corresponds to substantial electronic cement binding the atoms. When one of the coefficients is zero, indicative of a node at an atom, the partial mobile bond order vanishes in agreement with our expectations for a nonbinding situation. Finally, if the coefficients are of opposite sign, indicative of a node between the atoms, we have a negative bond order in agreement with what we would demand for an antibonding situation.

Examples.
Ethylene: $p_{12} = 2c_1c_2 = 1$
Butadiene: $p_{12} = 2(0.371)(0.600) + 2(0.600)(0.371) = 0.894$

$$p_{23} = 2(0.600)(0.600) + 2(0.371)(-0.371) = 0.447$$

Note: In this case the 1–2 bond, which is a double bond in the conventional structural symbolism, has a large *p*-value, whereas the "single" bond, 2–3, has a rather small *p*-value.

[15] C. A. Coulson, *Proc. Roy. Soc.*, **A169**, 413 (1939).

Allyl: Cation: $p_{12} = 2(0.5)(0.707) = 0.707$

Radical: $p_{12} = 0.707 + (0)(0.707) = 0.707$

Anion: $p_{12} = 0.707 + 2(0)(0.707) = 0.707$

Bond orders in allyl are the same in the cation, radical, and anion because the upper level is a nonbonding MO (NBMO). This equality is true of all odd-AH systems.

Cyclopropenyl:

$$\text{Cation: } p_{12} = 2\left(\frac{1}{\sqrt{3}}\right)\left(\frac{1}{\sqrt{3}}\right) = 0.667$$

$$\text{Radical (see note on p. 53): } p_{12} = 0.667 + \frac{1}{2}\left(\frac{1}{\sqrt{6}}\right)\left(\frac{1}{\sqrt{6}}\right) + \frac{1}{2}\left(\frac{1}{\sqrt{2}}\right)\left(-\frac{1}{\sqrt{2}}\right)$$
$$= 0.500$$

$$\text{Anion: } p_{12} = 0.667 + \left(\frac{1}{\sqrt{6}}\right)\left(\frac{1}{\sqrt{6}}\right) + \left(\frac{1}{\sqrt{2}}\right)\left(-\frac{1}{\sqrt{2}}\right)$$
$$= 0.333$$

In this case a variation in electron population occurs in antibonding MO's and a corresponding variation in bond order results.

Exercise. Calculate the bond orders for methylenecyclopropene, pentadienyl, and trimethylenemethane.

Ans.

Coulson and Longuet-Higgins[16] have shown that the total π-electron energy may be expressed as a function of electron densities and bond orders:

$$E_\pi = \sum_r q_r \alpha + 2\sum\sum_{r<s} p_{rs}\beta \tag{50}$$

This theorem provides a valuable check on one's computations, for, after obtaining the coefficients, these may be used to calculate the bond orders, the sum of which over all the bonds should equal the total binding energy.

The free valence index, F_r, has been defined by Coulson[17] as

$$F_r = N_{max} - N_r \tag{51}$$

in which N_r is the sum of the orders of all bonds joining atom r and N_{max} is the maximum value, which is usually taken as $\sqrt{3}$.[18]

[16] C. A. Coulson and H. C. Longuet-Higgins, *Proc. Roy. Soc.*, **191**, 39 (1947).

[17] C. A. Coulson, *Discussions Faraday Soc.*, **2**, 9 (1947); *J. Chim. Phys.*, **45**, 243 (1948).

[18] J. D. Roberts, A. Streitwieser, Jr., and C. M. Regan, *J. Am. Chem. Soc.*, **74**, 4579 (1952); H. H. Greenwood, *Trans. Faraday Soc.*, **48**, 677 (1952).

For butadiene, $F_1 = 1.732 - 0.894 = 0.838$ and $F_2 = 1.732 - 0.894 - 0.447 = 0.391$. The free valence is a modern version of Thielé's concept of residual affinity. As required by such a concept, the free valence is higher at the terminal position of butadiene.

The bond order for a bond in trimethylenemethane is $\sqrt{3}/3$; hence F for a terminal position is $\sqrt{3} - (\sqrt{3}/3) = \frac{2}{3}\sqrt{3} = 1.16$, whereas F for the central position is $\sqrt{3} - 3(\sqrt{3}/3) = 0$. The value for N_{max} comes from this system.[18] For aromatic carbons F is typically about 0.4, and in free radicals some positions have values of about unity. F has been used as a measure of ease of attack by free radicals (see Chaps. 10 and 13).[19]

2.10   Nodal Properties of MO's

The MO wave functions for butadiene were diagrammed in Fig. 1.18. The four wave functions, ψ_1, ψ_2, ψ_3, and ψ_4, had 0, 1, 2, and 3 nodes, respectively, which, in this representation occur as points along a line. All straight chain π-systems can be represented along a straight line because of our neglect of non-neighbor interactions. For such systems each MO, ψ_j, can be represented in a similar manner with $j - 1$ nodal points. In straight-chain even systems the nodal points occur only between atoms and divide the segments between nuclei into *bonds* and *antibonds*: ψ_1 of butadiene can be described as having a net of three bonds; ψ_2, with two bonds and one antibond, has a net of only one bond; ψ_3 has one bond and two antibonds, hence is net antibonding; ψ_4 has three antibonds. The order of orbital energies clearly parallels the bonding and antibonding characteristics of the MO's.

A more schematic representation of the MO's of pentadienyl is given in Fig. 2.7. For this odd-membered system the nodal points in some of the MO's coincide with nuclear positions giving regions that we may call nonbonds; ψ_3 is totally nonbonding in agreement with its energy and with our prior classification as a NBMO.

As soon as we introduce branches or rings, the nuclei of the π-network must be described as points in a plane. The nodes now become lines, and further possibilities for nodes open up. We now have two MO's with one node, the nodal lines in these cases being at right angles. The orientation of the nodal lines is simply that which leaves the most net bonding and fulfills symmetry requirements. If both orthogonal nodal lines give MO's with identical energy, we have a degeneracy, and any orientation of the

[19] For a recent review of the free-valence concept, see B. Pullman and A. Pullman, "Free Valence in Conjugated Organic Molecules," Chap. 2 in J. W. Cook, *Progress in Organic Chemistry*, Vol. 4, Butterworths Scientific Publications, London (1958).

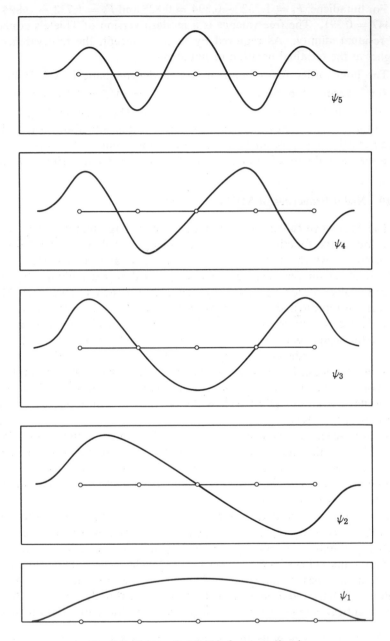

Fig. 2.7 Schematic HMO's for pentadienyl.

nodal lines corresponding to rotation about a point is equally valid. These features are illustrated below.

The six MO's of 3-methylene-1,4-pentadiene are represented schematically in Fig. 2.8: ψ_1 is totally bonding; ψ_2 has two bonds and three nonbonds; ψ_3, with a node at right angles to that in ψ_2, has three bonds and two antibonds and, consequently, has less net bonding than ψ_2; ψ_4, with two parallel nodes, has two bonds and three antibonds, hence has net antibonding; ψ_5 has two orthogonal nodes with two antibonds and three nonbonds; ψ_6 is totally antibonding. Note that in this case we do not get two parallel nodes along the long axis of the molecule. The reason has to do with symmetry considerations which become evident in Chap. 3.

An interesting case is that of cyclobutadiene, whose MO's are shown in Fig. 2.9; ψ_1 is completely bonding; ψ_2 has one node which forms two bonds and two antibonds or a net of zero bonding; ψ_3 has an orthogonal node and is clearly identical with ψ_2, except for orientation in space, and must have identical energy. Consequently, any pair of MO's obtained by a rotation of the nodes about the center of the molecule are likewise perfectly acceptable MO's and have the same energy. Examples are ψ_2' and ψ_3' formed by rotating the nodes counterclockwise by 45°.

Supplemental Reading

Although descriptions of the HMO method have been presented frequently, the following references treat the method in some detail.

Dictionary of Values of Molecular Constants, edited by C. A. Coulson and R. Daudel, Vol. I, Centre de Chemie Théorique de France, Paris.

G. W. Wheland, *Resonance in Organic Chemistry,* John Wiley and Sons, New York (1955), p. 660.

B. Pullman and A. Pullman, *Les theories electroniques de la chimie organique,* Masson et Cie, Paris (1952), p. 178.

H. Hartmann, *Theorie der chemischen Bindung,* Springer-Verlag, Berlin (1954), p. 250.

C. A. Coulson, *Valence,* Oxford University Press, London (1952), p. 238.

C. A. Coulson and H. C. Longuet-Higgins, *Proc. Roy. Soc.,* **A191,** 39 (1947).

R. Daudel, R. Lefebvre, and C. Moser, *Quantum Chemistry,* Interscience Publishers, New York (1959).

A. Streitwieser, Jr., and J. I. Brauman, *Tables of Molecular Orbital Calculations,* Pergamon Press, New York (in press).

This book gives an extensive list of HMO energies and coefficients.

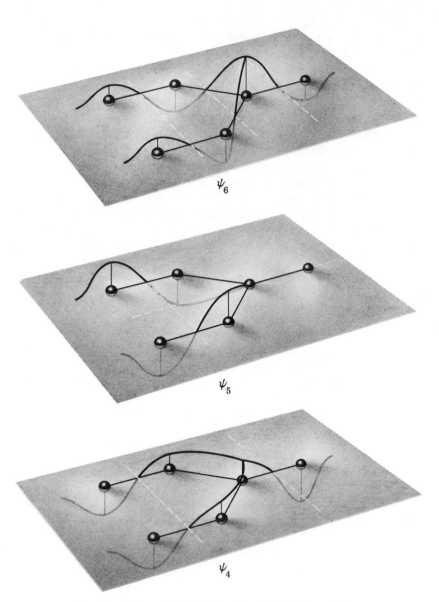

Fig. 2.8 Schematic HMO's for 3-methylene-1,4-pentadiene.

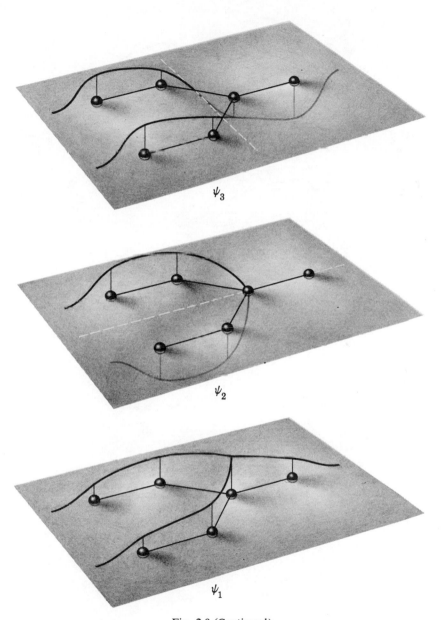

Fig. 2.8 (Continued)

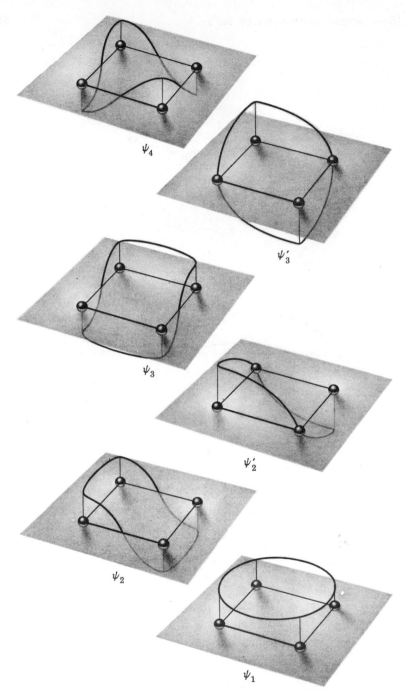

Fig. 2.9   Schematic representation of cyclobutadiene HMO's.

3 Matrix formulation and group theory

3.1 Vectors and Matrices

The techniques discussed in Chap. 2 suffice for complete HMO calculations on systems of any size and complexity; however, solution of the secular determinant involves a great deal of labor for even medium-sized systems. By using techniques founded in group theory, this labor can be substantially reduced when the system contains elements of symmetry. Before a discussion of group theory, an introduction to matrix concepts is desirable. A formulation of some of the LCAO operations in matrix terms can provide further insight to their significance. Furthermore, matrix notation is frequently used in molecular orbital literature.

A matrix is simply an array of numbers or symbols which is usually written in the form

$$\begin{pmatrix} a_{11} & a_{12} & a_{13} \\ a_{21} & a_{22} & a_{23} \end{pmatrix}$$

The number of rows and columns in a matrix may be different (*rectangular matrix*) or they may be equal (*square matrix*). An aggregate of numbers $(a_1, a_2, a_3, \cdots, a_n)$ is an n-dimensional *vector*, or a vector in n-dimensional space. Vectors may be represented as matrices with a single row (*row vector*) or a single column (*column vector*); hence each row or column of a matrix may be regarded as a vector. We are concerned only with vectors and square matrices.

Certain operations are defined for matrices. Addition is defined only when the two matrices have identical dimensions. For the matrix equation $\mathscr{A} + \mathscr{B} = \mathscr{C}$ each element, c_{kl} (k is the row index, l the column index), of $\mathscr{C}$ is simply the sum of corresponding elements of $\mathscr{A}$ and $\mathscr{B}$; that is,

$$c_{kl} = a_{kl} + b_{kl} \tag{1}$$

Two matrices, $\mathscr{A}$ and $\mathscr{B}$, are equal only if all the elements are equal; that is, all $a_{kl} = b_{kl}$.

Multiplication is defined as follows: for the product $\mathscr{A}\mathscr{B} = \mathscr{C}$: if $\mathscr{A}$ has m rows and n columns, $\mathscr{B}$ must have n rows and m columns and $\mathscr{C}$ is a square matrix with m rows and m columns. Each element, c_{kl}, of $\mathscr{C}$ is obtained as

$$c_{kl} = \sum_{j=1}^{n} a_{kj} b_{jl} \tag{2}$$

That is, we take a sum of products along the kth *row* of $\mathscr{A}$ and the lth *column* of $\mathscr{B}$. The following examples illustrate the procedure (remember: row × column):

$$\begin{pmatrix} a_{11} & a_{12} \\ a_{21} & a_{22} \end{pmatrix} \begin{pmatrix} b_{11} & b_{12} \\ b_{21} & b_{22} \end{pmatrix} = \begin{pmatrix} a_{11}b_{11} + a_{12}b_{21} & a_{11}b_{12} + a_{12}b_{22} \\ a_{21}b_{11} + a_{22}b_{21} & a_{21}b_{12} + a_{22}b_{22} \end{pmatrix}$$

$$(3 \quad 1 \quad 0) \begin{pmatrix} 4 \\ 2 \\ -1 \end{pmatrix} = (14) \qquad \begin{pmatrix} 4 \\ 2 \\ -1 \end{pmatrix} (3 \quad 1 \quad 0) = \begin{pmatrix} 12 & 4 & 0 \\ 6 & 2 & 0 \\ -3 & -1 & 0 \end{pmatrix}$$

The last two examples demonstrate that the product of two matrices depends in general on the order of the factors; that is, matrices do not necessarily commute.

The associative law holds, however:

$$\mathscr{A}(\mathscr{B}\mathscr{C}) = (\mathscr{A}\mathscr{B})\mathscr{C} \tag{3}$$

In the transformation of a set of variables, $x_1, x_2, \cdots, x_n$, to a new set, $x_1', x_2', \cdots, x_n'$, as in a change of coordinates, one uses a set of equations such as

$$x_1' = a_{11}x_1 + a_{12}x_2 + \cdots + a_{1n}x_n$$
$$x_2' = a_{21}x_1 + a_{22}x_2 + \cdots + a_{2n}x_n \tag{4}$$
$$\cdots\cdots\cdots\cdots\cdots\cdots\cdots\cdots\cdots$$
$$x_n' = a_{n1}x_1 + a_{n2}x_2 + \cdots + a_{nn}x_n$$

Such a *linear transformation* can be written conveniently in matrix notation as

$$\mathscr{X}' = \mathscr{A}\mathscr{X} \tag{5}$$

in which $\mathscr{X}'$ and $\mathscr{X}$ are column vectors and $\mathscr{A}$ is a square matrix:

$$\mathscr{X}' = \begin{pmatrix} x_1' \\ x_2' \\ \cdot \\ \cdot \\ \cdot \\ x_n' \end{pmatrix} \qquad \mathscr{X} = \begin{pmatrix} x_1 \\ x_2 \\ \cdot \\ \cdot \\ \cdot \\ x_n \end{pmatrix} \qquad \mathscr{A} = \begin{pmatrix} a_{11} & a_{12} & \cdots & a_{1n} \\ a_{21} & a_{22} & \cdots & a_{2n} \\ \cdot & & & \cdot \\ \cdot & & & \cdot \\ \cdot & & & \cdot \\ a_{n1} & a_{n2} & \cdots & a_{nn} \end{pmatrix}$$

This same notation may be used for our linear combinations of atomic orbitals (Eq. 2.1):

$$\Psi = \mathscr{C}\Phi \tag{6}$$

The atomic orbitals may be considered to define the dimensions of an n-dimensional space. Each MO is a vector in this *Hilbert space* and is transformed by the Hamiltonian operator into a colinear vector whose length corresponds to the energy. In a commonly used nomenclature, the atomic orbitals are *basis functions*, the MO's are *eigenfunctions*, the sets of coefficients are *eigenvectors*, and the energies are *eigenvalues*.

Some special products of vectors are important. The *scalar product* of two vectors, a and b, is simply the product of a as a row vector times b as a column vector, or $a_1 b_1 + a_2 b_2 + \cdots + a_n b_n$. When this product vanishes, the vectors are said to be *orthogonal*. When the scalar product, $aa = 1$, vector a is said to be a unit vector or to be normalized.

These statements are correct only if the vectors are real. If the vectors have imaginary elements, these statements refer to the *Hermitian scalar product*, defined as $a_1^* b_1 + a_2^* b_2 + \cdots + a_n^* b_n$, in which a_1^* is the complex conjugate of a_1.

According to these properties, we may describe the matrix of LCAO coefficients as a set of unit orthogonal row vectors.

Matrix division is not defined, although for every matrix $\mathscr{A}$, with which we are concerned, there corresponds an inverse $\mathscr{A}^{-1}$, such that

$$\mathscr{A}\mathscr{A}^{-1} = \mathscr{A}^{-1}\mathscr{A} = \mathscr{E} \tag{7}$$

Note that a matrix commutes with its inverse. The matrix $\mathscr{E}$ is a square matrix with the same number of rows and columns as $\mathscr{A}$ and having the number one as each diagonal element, $a_{11}, a_{22}, \cdots, a_{nn}$, and zero elsewhere. This *unit matrix* is an example of a diagonal matrix, $\mathscr{D}$:

$$\mathscr{D} = \begin{pmatrix} d_1 & 0 & \cdots & 0 \\ 0 & d_2 & \cdots & 0 \\ \cdot & \cdot & & \cdot \\ \cdot & \cdot & & \cdot \\ \cdot & \cdot & & \cdot \\ 0 & 0 & \cdots & d_n \end{pmatrix}$$

3.2 Special Matrices

The *transpose* $\mathcal{A}'$ is formed by interchanging rows and columns of the square matrix $\mathcal{A}$. If, in addition to interchanging rows and columns, we take the complex conjugate of each element, we obtain the *adjoint*, $\mathcal{A}^\dagger = \mathcal{A}^{*\prime}$, of $\mathcal{A}$. If each element of a matrix is real, the matrix is said to be real; that is, for a real matrix $\mathcal{A}^* = \mathcal{A}$.

If the transpose of a matrix leaves the matrix unchanged, the matrix is said to be *symmetric*; that is, a symmetric matrix is symmetrical about the principal diagonal. The adjoint of a *Hermitian* matrix, $\mathcal{H}$, is the same as the original matrix. A Hermitian matrix has a type of symmetry along the diagonal—each element a_{kl} is the complex conjugate of a_{lk}. A *real symmetric* matrix is also Hermitian. Examples of these matrices follow.

$$\mathcal{A} = \begin{pmatrix} a_{11} & a_{12} & a_{13} \\ a_{21} & a_{22} & a_{23} \\ a_{31} & a_{32} & a_{33} \end{pmatrix} \qquad \mathcal{A}' = \begin{pmatrix} a_{11} & a_{21} & a_{31} \\ a_{12} & a_{22} & a_{32} \\ a_{13} & a_{23} & a_{33} \end{pmatrix}$$

$$\mathcal{A}^\dagger = \begin{pmatrix} a_{11}^* & a_{21}^* & a_{31}^* \\ a_{12}^* & a_{22}^* & a_{32}^* \\ a_{13}^* & a_{23}^* & a_{33}^* \end{pmatrix}$$

$$\begin{pmatrix} a_{11} & a_{12} & a_{13} \\ a_{12} & a_{22} & a_{23} \\ a_{13} & a_{23} & a_{33} \end{pmatrix} \qquad \begin{pmatrix} a_{11} & a_{12} & a_{13} \\ a_{12}^* & a_{22} & a_{23} \\ a_{13}^* & a_{23}^* & a_{33} \end{pmatrix}$$
$$\text{Symmetric} \qquad\qquad \text{Hermitian}$$

A matrix, $\mathcal{U}$, for which $\mathcal{U}^\dagger = \mathcal{U}^{-1}$, is called a *unitary* matrix; that is, for a unitary matrix the adjoint is also the inverse. If we take the kth row to be a vector, α_k, the definition of a unitary matrix implies

$$\sum_k a_k^* a_l = 0 \tag{8}$$

$$\sum_k a_k^* a_k = 1 \tag{9}$$

since
$$\mathcal{U}\mathcal{U}^{-1} = \mathcal{E} \tag{10}$$

Note that (8) and (9) mean that the component vectors of a unitary matrix are unit and orthogonal. It follows that the matrix of the LCAO coefficients [$\mathcal{C}$ in (6)] is a unitary matrix.

The *null matrix* $\mathcal{O}$ is a square matrix, all of whose elements are zero.

3.3   Similarity Transformations

Consider the linear transformation

$$\mathscr{H}x = \lambda x \tag{11}$$

Multiplication of vector x by the square matrix $\mathscr{H}$ has the effect of multi-plying x by a scalar. In this equation, rewritten as (12), the matrix $\mathscr{H} - \lambda \cdot \mathscr{E}$ is the characteristic matrix of $\mathscr{H}$.

$$(\mathscr{H} - \lambda \cdot \mathscr{E})x = \mathscr{O} \tag{12}$$

The scalars λ are the *eigenvalues* of $\mathscr{H}$, and the vectors x are the corre-sponding *eigenvectors* of $\mathscr{H}$. To the characteristic matrix of $\mathscr{H}$ there corresponds a determinant which, on expansion, yields a *characteristic equation*, the roots of which are the eigenvalues.

In our present application $\mathscr{H} - \lambda \cdot \mathscr{E}$ is clearly our *secular matrix* and the corresponding *eigenvectors* are the coefficients of the atomic orbitals within each MO. The importance of this formulation lies in the fact that our secular matrix is Hermitian. For Hermitian matrices the so-called *similarity transformations* have important properties.

Matrices $\mathscr{A}$ and $\mathscr{B}$ are said to be related by a similarity transformation if a matrix $\mathscr{Q}$ exists such that

$$\mathscr{Q}\mathscr{A}\mathscr{Q}^{-1} = \mathscr{B} \tag{13}$$

It can be shown that the eigenvalues of a matrix are not changed by a similarity transformation; that is, $\mathscr{A}$ and $\mathscr{B}$ have the same eigenvalues. The *trace* of a matrix is defined as the sum of the diagonal elements:

$$\text{Tr } \mathscr{A} = \sum_{k=1}^{n} a_{kk} \tag{14}$$

The trace is also unchanged by a similarity transformation.

We state without proof that every Hermitian matrix can be converted to diagonal form by a similarity transformation with a unitary matrix. A similarity transformation with a unitary matrix is also called a *unitary transformation*.

$$\mathscr{U}\mathscr{H}\mathscr{U}^{-1} = \begin{pmatrix} \lambda_1 & 0 & \cdots & 0 \\ 0 & \lambda_2 & \cdots & 0 \\ \cdot & \cdot & & \cdot \\ \cdot & \cdot & & \cdot \\ \cdot & \cdot & & \cdot \\ 0 & 0 & \cdots & \lambda_n \end{pmatrix} \tag{15}$$

In our application the secular matrix omitting the ϵ's (or x's) is trans-formed by a unitary transformation with the matrix of the MO coefficients to a diagonal matrix containing the eigenvalues along the diagonal. This procedure is called *diagonalization of a matrix*. The unitary character arises from the quantum mechanical restriction that orthonormality be preserved.

Example. For allyl, $\mathscr{C}\mathscr{H}\mathscr{C}^{-1} =$

$$
\begin{pmatrix} \frac{1}{2} & \frac{1}{\sqrt{2}} & \frac{1}{2} \\ \frac{1}{\sqrt{2}} & 0 & \frac{-1}{\sqrt{2}} \\ \frac{1}{2} & \frac{-1}{\sqrt{2}} & \frac{1}{2} \end{pmatrix} \begin{pmatrix} 0 & 1 & 0 \\ 1 & 0 & 1 \\ 0 & 1 & 0 \end{pmatrix} \begin{pmatrix} \frac{1}{2} & \frac{1}{\sqrt{2}} & \frac{1}{2} \\ \frac{1}{\sqrt{2}} & 0 & \frac{-1}{\sqrt{2}} \\ \frac{1}{2} & \frac{-1}{\sqrt{2}} & \frac{1}{2} \end{pmatrix} = \begin{pmatrix} \sqrt{2} & 0 & 0 \\ 0 & 0 & 0 \\ 0 & 0 & -\sqrt{2} \end{pmatrix}
$$

Note: $\mathscr{C}$ is simply the table of coefficients (p. 50). $\mathscr{H}$ is the matrix corresponding to the secular determinant written in terms of x with x subtracted from each diagonal element. $\mathscr{C}^{-1}$ is $\mathscr{C}$ with rows and columns reversed and in this case happens to be equal to $\mathscr{C}$.

The trace of a matrix is stated above to be unaffected by a similarity transformation. In the HMO method, in which all α's are equal, the trace of $\mathscr{H}$ is zero. Hence the trace of the diagonalized matrix must also be zero; that is, the sum of all eigenvalues is zero. If some of the α's are given a different value, as in the case of heteroatoms (Chap. 5), the trace no longer need vanish. For such cases the sum of the eigenvalues equals Σh_X. Note that h_X is defined in Chap. 5. The general case is given as (16).

$$
\mathrm{Tr}\ \mathscr{H} = \mathrm{Tr}\ (\mathscr{C}\mathscr{H}\mathscr{C}^{-1}) = \Sigma h_X \tag{16}
$$

Incidentally, simple algebra will show that the magnitude of the trace is also equal to the magnitude of the coefficient of x^{n-1} in the characteristic equation; hence this term is missing in the secular polynomial when all α's are equal.

3.4   Use of High-Speed Computers

The matrix formulation previously discussed does not help our arithmetic operations. A satisfactory way to diagonalize a matrix by hand is still to expand the corresponding determinant and to find the roots of the resulting polynomial. A good way to find the eigenvectors by hand is still the cofactor method discussed in Sec. 2.7.

Computer centers with high-speed digital computer facilities are becoming increasingly available to chemical research groups. Solution of an arithmetic problem with such a computer is often handled differently from hand computation; in particular, iterative procedures are extremely important. It is possible to reduce the secular matrix to the characteristic equation by using an iterative procedure with a high-speed computer followed by a determination of the characteristic roots;[1] however, a more

---

[1] H. O. Pritchard and F. H. Sumner, *Proc. Roy. Soc.*, **A226**, 128 (1954).

convenient procedure is an iterative process[2] which determines all eigen-values and eigenvectors simultaneously. A method due to Jacobi[3] is especially suited for molecular orbital work.[4] In this method a pair of symmetric off-diagonal elements, usually the largest in magnitude, is used to set up a unitary matrix, which is then used in a similarity transformation of the original matrix. This process may be described as a rotation of the matrix in a plane in n-dimensional space and results in the reduction of the chosen pair of matrix elements to zero. The process is repeated with another pair of off-diagonal elements in the new matrix. By repeated operation, the original matrix converges to diagonal form and the product of the transforming matrices converges to the matrix of the eigenvectors. Unlike many other methods, this technique is not disturbed by the presence of degenerate eigenvalues. The time required for complete diagonalization depends on the speed and capacity of the computer and on the number of significant figures required in the answer. As an example, a twenty-atom system may be diagonalized to nine-decimal precision in about one minute with an IBM 704 computer. For systems of varying size the time required varies roughly as n^3 with the Jacobi method.

A more recent "rotations" method is that published by Givens.[5] The time required by this method is proportional to n^2; this method is generally faster with large systems. Note that some of these programs cannot handle degenerate eigenvectors.

Matrix diagonalization programs are becoming standard library items in many computer centers, but such a program alone is not enough. The complete solution of an eigenvalue problem requires the following mini-mum specific operations: (a) input of the matrix to be diagonalized, (b) diagonalization, and (c) printing of the eigenvalues and eigenvectors. If the complete library program includes at least these three operations, it is necessary only to punch the input cards in accordance with the directions usually associated with each library program in a well-organized computer center.

Programs are usually written in "fixed-point" or "floating point." The terms refer to the location of the decimal point. Floating-point programs are designed to handle without excessive rounding errors systems in which the magnitudes of the numerical quantities may vary by many

---

[2] For a review of iterative methods and an excellent further bibliography, see *Modern Computing Methods*, Philosophical Library, New York (1958).

[3] *Ibid.*, pp. 30–31. For a more theoretical discussion, see A. S. Householder, *Principles of Numerical Analysis*, McGraw-Hill Book Co., New York (1953), pp. 160–162.

[4] R. Pariser, *J. Chem. Phys.*, **24**, 250 (1956); G. G. Hall, *Trans. Faraday Soc.*, **53**, 573 (1957); A. Streitwieser and P. M. Nair, *Tetrahedron*, **5**, 149 (1959).

[5] W. Givens, *Numerical Computation of the Characteristic Values of a Real Symmetric Matrix*, Oak Ridge National Laboratory, ORNL-1574 (1954).

powers of ten. On some computers (for example, IBM 650 and 701) these programs are often slower than fixed-point programs in which the decimal point is located at the same convenient fixed location in all of the sets of numbers handled. Fixed-point programs are usually adequate for HMO calculations, since the input matrix consists only of zeros and ones. the eigenvalues are all in the range $-3 < m < 3$, and the eigenvectors are aggregates of fractional quantities.

A matrix diagonalization subroutine is less convenient for the casual user, since it is designed for incorporation into other programs, and its utilization requires some investment of time in learning computer programming. With such knowledge, programs may be written in which the eigenvalues and eigenvectors are manipulated to derive further quantities of interest, such as charge densities and bond orders. Writing of input. output, and diagonalization subroutines is rather difficult and requires a high order of programming facility; however, these programs are usually available as subroutines which may be incorporated simply into a master program. The further manipulations required for the calculation of other molecular-orbital quantities in general require only the most elementary of programming techniques. Furthermore, the matter is made still easier by the existence for many computers of programs that write programs (e.g. IBM's *Fortran*). Such programs convert easily learned arithmetic codes into complete operating programs.

3.5 Group Theory

If we examine the coefficients of the atomic orbitals in the butadiene MO's (p. 51), we note that the coefficients of chemically equivalent atoms are equal in magnitude and differ only in sign. A corresponding observation is general for many systems. Consider what happens if we anticipate this type of result and determine our MO's not as linear combinations of atomic orbitals but as linear combinations of intermediate orbitals, which, in turn, are linear combinations of atomic orbitals defined as follows:

$$\varphi_1' = \frac{1}{\sqrt{2}} (\varphi_1 + \varphi_4)$$

$$\varphi_2' = \frac{1}{\sqrt{2}} (\varphi_2 + \varphi_3)$$

$$\varphi_3' = \frac{1}{\sqrt{2}} (\varphi_1 - \varphi_4)$$

$$\varphi_4' = \frac{1}{\sqrt{2}} (\varphi_2 - \varphi_3)$$

(17)

There are as many intermediate orbitals as starting atomic orbitals and each is normalized. We set up our secular determinant as usual but in terms of the intermediate orbitals. Thus

$$\mathbf{H}'_{11} = \int \varphi'_1 H \varphi'_4 \, d\tau = \tfrac{1}{2} \int (\varphi_1 + \varphi_4) H (\varphi_1 + \varphi_4) \, d\tau =$$

$$\tfrac{1}{2} \left(\int \varphi_1 H \varphi_1 \, d\tau + 2 \int \varphi_1 H \varphi_4 \, d\tau + \int \varphi_4 H \varphi_4 \, d\tau \right) = \alpha$$

$$\mathbf{H}'_{12} = \mathbf{H}'_{21} = \tfrac{1}{2}(H_{12} + H_{13} + H_{24} + H_{34}) = \beta$$

$$\mathbf{H}'_{13} = \mathbf{H}'_{31} = \tfrac{1}{2}(H_{11} - H_{44}) = 0$$

$$\mathbf{H}'_{14} = \mathbf{H}'_{41} = \tfrac{1}{2}(H_{12} - H_{13} + H_{24} - H_{34}) = 0$$

$$\mathbf{H}'_{22} = \tfrac{1}{2}(H_{22} + 2H_{23} + H_{33}) = \alpha + \beta$$

$$\mathbf{H}'_{23} = \mathbf{H}'_{32} = 0$$

$$\mathbf{H}'_{24} = \mathbf{H}'_{42} = 0$$

$$\mathbf{H}_{33} = \alpha$$

$$\mathbf{H}_{34} = \mathbf{H}_{43} = \beta$$

$$\mathbf{H}_{44} = \alpha - \beta$$

After dividing through each element by β and using our definition of x (p. 39), we obtain

$$\begin{vmatrix} x & 1 & 0 & 0 \\ 1 & x+1 & 0 & 0 \\ 0 & 0 & x & 1 \\ 0 & 0 & 1 & x-1 \end{vmatrix} = 0 \tag{18}$$

The determinant has been reduced to *block form*, that is,

D_1	0
0	D_2

and may be split into its component determinants, D_1 and D_2, which may be solved separately. Hence (18) is equivalent to two second-order determinants which may be solved directly:

$$\begin{vmatrix} x & 1 \\ 1 & x+1 \end{vmatrix} = x^2 + x - 1 = 0$$

$$\begin{vmatrix} x & 1 \\ 1 & x-1 \end{vmatrix} = x^2 - x - 1 = 0$$

Instead of the labor of expanding a fourth-order determinant and finding roots for a quartic equation, we have the far easier task of expanding and finding roots for two second-order determinants. The technique of finding intermediate orbitals that will yield a secular determinant in block form can be applied generally whenever the compound contains certain elements of symmetry. The whole trick lies in the definition of appropriate intermediate orbitals. The proper definitions are obtained by group-theory procedures.

A mathematical group is simply a set of elements with a defined combining operation called a "multiplication" that satisfies the following requirements:

1. The *product* of any two elements is also an element in the set.
2. The associative law holds:

$$A(BC) = (AB)C$$

3. There is a unit element, E, such that

$$EA = AE = A$$

4. Each element A has an inverse A^{-1}, which is also an element in the set such that $AA^{-1} = A^{-1}A = E$.

The total number of elements, h, is called the *order* of the group. An example of a group is the set of positive and negative integers in which the defined group "multiplication" is algebraic addition. In this case the "unit" element is the number zero.

Consider an abstract group of four elements, A, B, C, E, which follows the following group multiplication table:

	E	A	B	C
E	E	A	B	C
A	A	E	C	B
B	B	C	E	A
C	C	B	A	E

Note that each element appears only once in each row and column.

Consider now the group of *covering operations* for the rectangular pyramid in Fig. 3.1. These operations are rotations, reflections, etc., which leave the figure unchanged. Multiplication here is defined as

successive application of operations. The identity operation is no operation at all—call this E. A rotation of 180° around the central axis we may
call A. Reflection in two perpendicular planes which contain the principal
axis we may call B and C, respectively. It is readily shown that these
operations form a group, since all our requirements are met; namely,
two successive applications of any two operations are equivalent to one

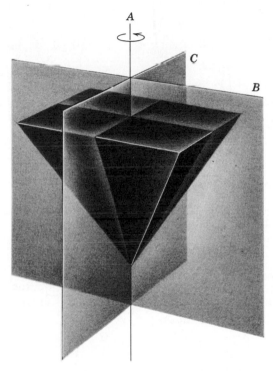

Fig. 3.1   Covering operations of a rectangular pyramid.

operation in the set, the associative law holds, the identity element is
defined, and an inverse operation may clearly be described as application
of any operation in the reverse sense. There is a one-to-one correspondence
between elements in the symmetry group and elements in the abstract
letter group; both groups obey the same multiplication table. Two such
groups are said to be *isomorphous*. For example, the product AB, which
is the operation of rotation 180° about the principal axis followed by
reflection in the B-plane, is equivalent to the single operation of reflection
in the C-plane. This equivalence may be demonstrated by labeling the
corners of the figure.

Exercise. Show that the rest of the multiplication table follows these symmetry operations.

Another group isomorphous with the foregoing is the set of four matrices labeled as follows:

$$\mathscr{E} = \begin{pmatrix} 1 & 0 & 0 & 0 \\ 0 & 1 & 0 & 0 \\ 0 & 0 & 1 & 0 \\ 0 & 0 & 0 & 1 \end{pmatrix} \qquad \mathscr{A} = \begin{pmatrix} 0 & 1 & 0 & 0 \\ 1 & 0 & 0 & 0 \\ 0 & 0 & 0 & 1 \\ 0 & 0 & 1 & 0 \end{pmatrix}$$

$$\mathscr{B} = \begin{pmatrix} 0 & 0 & 1 & 0 \\ 0 & 0 & 0 & 1 \\ 1 & 0 & 0 & 0 \\ 0 & 1 & 0 & 0 \end{pmatrix} \qquad \mathscr{C} = \begin{pmatrix} 0 & 0 & 0 & 1 \\ 0 & 0 & 1 & 0 \\ 0 & 1 & 0 & 0 \\ 1 & 0 & 0 & 0 \end{pmatrix}$$

Group multiplication here is ordinary matrix multiplication.

Exercise. Show that the four matrices obey the multiplication table given for the abstract letter group.

Any isomorphous group of matrices is called a *representation* of the group. The number of rows or columns is called the *dimensionality* of the representation. In our example the representation is of fourth dimension.

If some similarity transformation can be found which simultaneously converts each matrix, $\mathscr{X}_i$, in the group to block form, the representation is said to be *reducible*; for example (19). The individual sets of matrices,

$$\mathscr{U}\mathscr{X}_i\mathscr{U}^{-1} = \begin{pmatrix} \boxed{R_1} & & 0 \\ & \boxed{R_2} & \\ 0 & & \boxed{R_3} \end{pmatrix}_i \tag{19}$$

R_1, R_2, R_3, etc., are also representations of the group. When no further reduction is possible, the representation is said to be *irreducible*.

In our example let $\mathscr{U}$ be the unitary matrix

$$\begin{pmatrix} \tfrac{1}{2} & \tfrac{1}{2} & \tfrac{1}{2} & \tfrac{1}{2} \\ \tfrac{1}{2} & \tfrac{1}{2} & -\tfrac{1}{2} & -\tfrac{1}{2} \\ \tfrac{1}{2} & -\tfrac{1}{2} & \tfrac{1}{2} & -\tfrac{1}{2} \\ \tfrac{1}{2} & -\tfrac{1}{2} & -\tfrac{1}{2} & \tfrac{1}{2} \end{pmatrix}$$

A similarity transformation with this matrix transforms each of our four matrices as follows (note that $\mathcal{U}^{-1} = \mathcal{U}$):

$$\mathcal{U}\mathcal{E}\mathcal{U}^{-1} = \begin{pmatrix} 1 & 0 & 0 & 0 \\ 0 & 1 & 0 & 0 \\ 0 & 0 & 1 & 0 \\ 0 & 0 & 0 & 1 \end{pmatrix} \qquad \mathcal{U}\mathcal{A}\mathcal{U}^{-1} = \begin{pmatrix} 1 & 0 & 0 & 0 \\ 0 & 1 & 0 & 0 \\ 0 & 0 & -1 & 0 \\ 0 & 0 & 0 & -1 \end{pmatrix}$$

$$\mathcal{U}\mathcal{B}\mathcal{U}^{-1} = \begin{pmatrix} 1 & 0 & 0 & 0 \\ 0 & -1 & 0 & 0 \\ 0 & 0 & 1 & 0 \\ 0 & 0 & 0 & -1 \end{pmatrix} \qquad \mathcal{U}\mathcal{C}\mathcal{U}^{-1} = \begin{pmatrix} 1 & 0 & 0 & 0 \\ 0 & -1 & 0 & 0 \\ 0 & 0 & -1 & 0 \\ 0 & 0 & 0 & 1 \end{pmatrix}$$

Each of our original matrices has been transformed into block form; hence our original representation, Γ, was reducible. Indeed, the resulting matrices are diagonal, and we now have a set of four representations, Γ_1, Γ_2, Γ_3, Γ_4, which correspond to the diagonal elements of the reduced matrices. Since each of the new representations is of unit dimension, no further reduction is possible. We have four irreducible representations of our example group (Vierergruppe). These may be labeled for clarity, as in Table 3.1.

TABLE 3.1

IRREDUCIBLE REPRESENTATIONS OF THE VIERERGRUPPE*

	E	A	B	C
Γ_1	1	1	1	1
Γ_2	1	1	−1	−1
Γ_3	1	−1	1	−1
Γ_4	1	−1	−1	1

* This table should be distinguished from a character Table that looks the same.

Exercise. Show that each of these irreducible representations follows the multiplication table of the abstract letter group.

It may be shown that no further irreducible representations exist for this group which are not equivalent to one of the foregoing representations. Note that any number of equivalent sets of irreducible representations may exist.

We may indicate the structure of our representation by writing a "sum" of irreducible representations:

$$\Gamma = \Gamma_1 + \Gamma_2 + \Gamma_3 + \Gamma_4 \qquad (20)$$

If each irreducible representation of dimension l_i occurs a_i times in a reducible representation, the dimension, l of the reducible representation is given as

$$l = \sum_i a_i l_i \qquad (21)$$

The order, h, of the group is related to the dimensions of irreducible representation by

$$h = \sum_i l_i^2 \qquad (22)$$

in which the sum is taken over all of the distinct irreducible representations. Equation 22 is known as Burnside's theorem.

The importance of irreducible representations in quantum mechanics lies in the symmetry properties of the Hamiltonian. An interchange of points which enter the Hamiltonian in the same way cannot change the eigenvalues, for the resulting function "looks" like the old. The eigenfunctions must form the basis for irreducible representations of the symmetry group of the Hamiltonian. Individual functions may be said to "belong" to individual irreducible representations. An important principle is that the integrals $\int \varphi_A H \varphi_B \, d\tau$ and $\int \varphi_A \varphi_B \, d\tau$ are different from zero only when φ_A and φ_B belong to the same irreducible representation.

3.6   Characters and the Point Groups

Group-theory procedures are generally carried out not with irreducible representations but with the *characters* that are defined as the trace of the matrices of an irreducible representation. Let an element of a group be (R), and let $\Gamma_j(R)_{rs}$ be the element at the rth row and sth column in the (R)th matrix of the jth irreducible representation. Then, the corresponding character, $\chi_j(R)$, is

$$\chi_j(R) = \sum_{r=1}^{l_j} \Gamma_j(R)_{rr} = \text{Tr } \Gamma_j(R) \qquad (23)$$

The h numbers, $\chi_j(E)$, $\chi_j(A)$, $\chi_j(B)$, etc., are called the *character of the jth representation*. Since the character is a trace, it is invariant to a similarity transformation. Hence all equivalent irreducible representations have the same character. For irreducible representations of unit dimension, the character is obviously identical with the representation.

The characters form an orthogonal set of vectors; that is, it may be shown that

$$\sum_R \chi_j(R)\chi_k(R) = h\delta_{jk} \qquad (24)$$

where
$$\delta_{jk} = 1 \qquad j = k$$
$$\quad\ = 0 \qquad j \neq k$$

We recall that a reducible representation can be transformed to a set of irreducible representations by a similarity transformation, but a similarity transformation leaves the trace unchanged. Hence the character χ of a reducible representation is just the sum of characters of component irreducible representations:

$$\chi(R) = \sum_{j=1}^{k} a_j \chi_j(R) \tag{25}$$

a_j is the number of times the jth irreducible representation occurs in the reducible representation. Conversely,

$$a_j = \frac{1}{h} \sum_{R} \chi(R) \chi_j(R) \tag{26}$$

For example, the character of the matrix group on p. 74 is

$$\begin{array}{cccc} E & A & B & C \\ 4 & 0 & 0 & 0 \end{array}$$

The complete *character table* for the irreducible representations is identical with Table 3.1. For $\Gamma_1, a_1 = \frac{1}{4}(4 \cdot 1 + 0 \cdot 1 + 0 \cdot 1 + 0 \cdot 1) = 1$;

TABLE 3.2
SYMMETRY SYMBOLS

Symbol	Symmetry Element	Symmetry Operation
E	– – –	No change
C_p	p-fold axis of rotation (the principal axis of symmetry is that of largest p)	Rotation about an axis of symmetry by $360°/p$
σ_h	Plane of symmetry perpendicular to the principal axis of symmetry	Reflection in the plane of symmetry
σ_v	Plane of symmetry contains the principal axis of symmetry	Reflection in the plane of symmetry
σ_d	Plane of symmetry contains the principal axis of symmetry and bisects the angle between two a-fold axes of symmetry which are perpendicular	Reflection in the plane of symmetry
S_p	$C_p + \sigma_h$	Rotation about an axis by $360°/p$ followed by a reflection in a plane perpendicular to the axis of rotation
i	Center of symmetry	Inversion in a center of symmetry

for Γ_2, $a_2 = \frac{1}{4}(4 \cdot 1 + 0 \cdot 1 + 0 \cdot -1 + 0 \cdot -1) = 1$. Similarly, a_3 and a_4 are shown to be one each and (20) is reaffirmed.

Molecules and their corresponding Hamiltonians may be associated with the *point groups*. These groups are spacial arrays of points that possess certain *symmetry elements*. For each symmetry element there corresponds a *symmetry operation*, application of which leaves the array of points unchanged. Each set of symmetry operations constitutes a separate group in the mathematical sense and is associated with sets of irreducible representations and a character table. Character tables have been compiled for these point groups. Many of those of interest to us are collected in the Appendix.

The common symbols for operators which correspond to each symmetry operation that transforms an array of points into itself are given in Table 3.2.

For our present purposes S_p and i are not important. Point groups of interest to us are given below with the symbol for each group:

$$\mathbf{C}_p; \quad C_p \text{ only}$$

$$\mathbf{C}_{pv}; \quad C_p + p\sigma_v$$

Examples.

$$\mathbf{C}_{2v} \qquad \qquad \mathbf{C}_{2v} \qquad \qquad \mathbf{C}_{3v}$$

$$\mathbf{C}_{ph}; \quad C_p + \sigma_h$$

Examples.

$$\mathbf{C}_{2h} \qquad \qquad \mathbf{C}_{3h}$$

$$\mathbf{D}_{pd}; \quad C_p \text{ and } pC_2 \text{ perpendicular to } C_p$$

Examples.

$$\mathbf{D}_{2d} \qquad \qquad \mathbf{D}_{3d}$$

$\mathbf{D}_{ph}$; C_p, σ_h, and $p\text{-}\sigma_v$ at angles $360°/p$ to one another

Examples.

Exercise. Determine the point groups of the following compounds:

3.7 Application to LCAO Method

Any variety of intermediate linear combination orbitals can be defined, of course, but the only ones that will lead to "blocking" of the secular determinant such that splitting into smaller determinants is possible are those that form a basis for irreducible representations of the symmetry group of the molecule. Conversely, from the symmetry characteristics of the molecule and the corresponding character table we can construct the correct set of intermediate orbitals, which we may call *symmetry orbitals*. An outline of the general procedure is followed by several examples worked out in considerable detail. After some practice, some of the operations will become apparent by inspection and may be omitted.[6]

Step 1

Each orbital in the π-network of the molecule is labeled and the appropriate point group and character table are found. Note that our starting p-orbitals already have a plane of symmetry, the nodal plane. More exactly, the p-orbitals are antisymmetric with respect to this plane; that is, reflection in the plane causes an inversion of sign of the wave function. This plane is common to all systems that we treat by the simple LCAO (HMO) procedure and is present in the computation procedure described in Chap. 2. Explicit consideration of this element of symmetry does not lead to any simplification of the secular determinant; hence it may be neglected. In short, we may just as well use the character table for the corresponding point group that does not contain the molecular plane as a plane of symmetry. In some other uses we may not neglect this element of symmetry; for example, in considering allowed spectral transitions (Chap. 8).

Step 2

By applying each of the appropriate symmetry operations to each of the component atomic orbitals, we determine the character of the corresponding reducible representation. This character can be decomposed into the characters of the component irreducible representations either by inspection or by application of (26). This procedure reveals the number of symmetry orbitals belonging to each irreducible representation.

Step 3

The symmetry orbitals themselves are next obtained for each irreducible representation by applying each symmetry operation in turn to each non-equivalent atomic orbital and multiplying the resulting atomic orbital by the corresponding element in the character.

The atomic orbitals found are obviously related to the starting orbital by symmetry and are called its *partners*. The procedure, which is illustrated by the examples below, has been called the "basis function generating

---

[6] Prescriptions for application of group-theory procedures to some general systems are given by C. G. Swain and W. R. Thorson, *J. Org. Chem.*, **24**, 1989 (1959).

machine" by Van Vleck. The number of independent symmetry orbitals is that given in Step 2. Each orbital is then normalized. We shall find that the generation of basis functions is automatic only when one-dimensional representations are involved. An example with a two-dimensional representation is treated in Sec. 3.11.

Step 4

By using the symmetry orbitals in each irreducible representation, the secular determinant is set up and solved for the energies and coefficients. The resulting wave functions and coefficients are given in terms of the symmetry orbitals but may easily be converted back to the starting atomic orbitals.

In alternant systems an additional step could be considered. Potts[7] and Moffitt[8] have shown that the "symmetrical" dispositions of energy levels in such systems may sometimes be used to reduce the size of the determinants still further. The method requires several matrix multiplications, however, and especial care must be taken to avoid arithmetic errors.

3.8 C_2 Examples—Butadiene (I) and Methylenecyclopropene (II)

Butadiene—Step 1

We start by labeling the positions. It is convenient to use letter labels as in I at this stage to avoid later confusion.

Further specification of geometry is meaningless since non-neighbor interactions are neglected. The C_2 character table will give optimum use of symmetry. In the Appendix we find that the character table is:

C_2	E	C_2
A	1	1
B	1	-1

A and B are labels of irreducible representations. A refers to a *symmetric* representation; B refers to an *antisymmetric* representation. We may call these Γ_A and Γ_B, respectively.

[7] R. B. Potts, *J. Chem. Phys.*, **21**, 758 (1953); R. B. Potts and I. S. Walker, *Austral. J. Chem.*, **7**, 211 (1954). The procedure worked out in detail in the second reference, uses 3,4-benzophenanthrene as an example.

[8] W. Moffitt, *J. Chem. Phys.*, **26**, 424 (1957).

Step 2

We next apply the symmetry operations E and C_2 to each atomic orbital. We obtain the following table:

E	C_2
φ_A	φ_D
φ_B	φ_C
φ_C	φ_B
φ_D	φ_A
$\chi = 4$	0

The numbers at the bottom of each column total the number of times the symmetry operation converted a given orbital into itself. These numbers constitute the character χ of the reducible representation that represents the covering operations. The character element of E is always the total number of orbitals, n. By inspection, we see that $\chi = 2\chi_A + 2\chi_B$. Alternatively, we apply (26):

$$a_A = \tfrac{1}{2}(4 \cdot 1 + 0 \cdot 1) = 2$$
$$a_B = \tfrac{1}{2}(4 \cdot 1 + 0 \cdot -1) = 2$$

The meaning of this step may be clarified as follows: $C_2\varphi_A = \varphi_D$; $C_2\varphi_B = \varphi_C$; $C_2\varphi_C = \varphi_B$; $C_2\varphi_D = \varphi_A$. Hence

$$C_2\varphi = \begin{vmatrix} 0 & 0 & 0 & 1 \\ 0 & 0 & 1 & 0 \\ 0 & 1 & 0 & 0 \\ 1 & 0 & 0 & 0 \end{vmatrix} \begin{vmatrix} \varphi_A \\ \varphi_B \\ \varphi_C \\ \varphi_D \end{vmatrix}$$

The matrix,

$$\begin{vmatrix} 0 & 0 & 0 & 1 \\ 0 & 0 & 1 & 0 \\ 0 & 1 & 0 & 0 \\ 1 & 0 & 0 & 0 \end{vmatrix}$$

is a reducible representation of C_2 whose character is 0.

Step 3

Γ_A contains two symmetry orbitals. One is found by starting with φ_A:

$$\sum_j R_j\varphi_A\chi_A(R_j) = E\varphi_A\chi_A(E) + C_2\varphi_A\chi_A(C_2)$$

$$= \varphi_A\chi_A(E) + \varphi_D\chi_A(C_2) = \varphi_A + \varphi_D$$

Labeling and normalizing gives $\varphi_1 = (1/\sqrt{2})(\varphi_A + \varphi_D)$. Starting with φ_B,

$$\varphi_B(1) + \varphi_C(1); \qquad \therefore \ \varphi_2 = \frac{1}{\sqrt{2}}(\varphi_B + \varphi_C)$$

These symmetry orbitals are the total number of basis functions of Γ_A allowed by Step 2. If we were to proceed with φ_C,

$$\varphi_C(1) + \varphi_B(1); \qquad \text{not an independent function—identical with } \varphi_2$$

For Γ_B we proceed in a similar manner but use χ_B:

$$\varphi_A \chi_B(E) + \varphi_D \chi_B(C_2) = \varphi_A(1) + \varphi_D(-1); \qquad \therefore \varphi_3 = \frac{1}{\sqrt{2}} (\varphi_A - \varphi_D)$$

Similarly, we find, $\varphi_4 = (1/\sqrt{2})(\varphi_B - \varphi_D)$.

Step 4

From ψ_1 and ψ_2 we determine the matrix components $\mathbf{H}_{11}$, $\mathbf{H}_{12} = \mathbf{H}_{21}$, $\mathbf{H}_{22}$ and set up the secular determinant $D(\Gamma_A)$ belonging to Γ_A. A second determinant, $D(\Gamma_B)$, is found belonging to Γ_B (18).

This type of problem is an especially simple one for which we no longer need this long procedure. We can write down by inspection the symmetric and antisymmetric symmetry orbitals: a_A is equal to the number of nonequivalent positions in the molecule; a_B is equal to the number of nonequivalent positions which occur in pairs.

Methylenecyclopropene (II),

$$\begin{array}{c} \text{A} \\ | \\ \text{B} \\ \text{C}\diagdown\!\!\!\diagup\text{D} \\ \text{II} \end{array}$$

We write from inspection

$$\Gamma_A: \varphi_1 = \varphi_A \qquad\qquad \Gamma_B: \varphi_4 = \frac{1}{\sqrt{2}}(\varphi_C - \varphi_D)$$

$$\varphi_2 = \varphi_B \qquad\qquad a_B = 1$$

$$\varphi_3 = \frac{1}{\sqrt{2}}(\varphi_C + \varphi_D)$$

$$a_A = 3$$

$$D(\Gamma_A): \mathbf{H}_{11} = \mathbf{H}_{AA} = \alpha$$

$$\mathbf{H}_{12} = \mathbf{H}_{AB} = \beta$$

$$\mathbf{H}_{13} = \frac{1}{\sqrt{2}}(\mathbf{H}_{AC} + \mathbf{H}_{AD}) = 0$$

$$\mathbf{H}_{22} = \mathbf{H}_{BB} = \alpha$$

$$\mathbf{H}_{23} = \frac{1}{\sqrt{2}}(\mathbf{H}_{BC} + \mathbf{H}_{BD}) = \sqrt{2}\beta$$

$$\mathbf{H}_{33} = \tfrac{1}{2}(\mathbf{H}_{CC} + 2\mathbf{H}_{CD} + \mathbf{H}_{DD}) = \alpha + \beta$$

$$\therefore D(\Gamma_A) = \begin{vmatrix} x & 1 & 0 \\ 1 & x & \sqrt{2} \\ 0 & \sqrt{2} & x+1 \end{vmatrix} = 0$$

$$x^3 + x^2 - 3x - 1 = 0 \qquad x = -2.170, -0.311, +1.481$$

We find the coefficients in the usual way by using cofactors:

A_i	A_i for $x = -2.170$	A_i^2	c_i
$A_1: x^2 + x - 2$	0.53	0.28	0.278
$A_2: -(x + 1)$	1.17	1.36	0.612
$A_3: \sqrt{2}$	1.41	2.00	0.740
		3.64	

$$\sqrt{3.64} = 1.91$$

Hence $\psi_1 = 0.278\varphi_1 + 0.612\varphi_2 + 0.740\varphi_3$

$$= 0.278\varphi_A + 0.612\varphi_B + 0.524\varphi_C + 0.524\varphi_D$$

In a straightforward manner we can use the other two values of x and obtain two more molecular orbitals, ψ_2 and ψ_4.

$$D(\Gamma_B): \mathbf{H}_{11} = \tfrac{1}{2}(\mathbf{H}_{CC} - 2\mathbf{H}_{CD} + \mathbf{H}_{DD}) = \alpha - \beta$$
$$\therefore D(\Gamma_B) = |x - 1| = 0; \qquad x = 1$$

In this case the molecular orbital is identical to the symmetry orbital:

$$\psi_3 = \varphi_4 = \frac{1}{\sqrt{2}} (\varphi_C - \varphi_D)$$

3.9 C_{2v} Bicyclohexatriene (III)

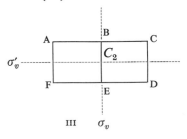

III σ_v

Step 1

The nuclei of this compound belong to the point group $\mathbf{D}_{2h}$. Eliminating the plane of this page as a plane of symmetry, the appropriate character table is that of C_{2v}, which is given as

C_{2v}	E	C_2	σ_v	σ_v'
A_1	1	1	1	1
B_2	1	−1	−1	1
A_2	1	1	−1	−1
B_1	1	−1	1	−1

Step 2

E	C_2	σ_v	σ_v'
φ_A	φ_D	φ_C	φ_F
φ_B	φ_E	φ_B	φ_E
φ_C	φ_F	φ_A	φ_D
φ_D	φ_A	φ_F	φ_C
φ_E	φ_B	φ_E	φ_B
φ_F	φ_C	φ_D	φ_A
$\chi = 6$	0	2	0

$$\chi = 2\chi(A_1) + \chi(B_2) + \chi(A_2) + 2\chi(B_1)$$

Note that we have two first-order and two second-order determinants to solve.

Steps 3 and 4

$$\Gamma(A_1): \quad \varphi_A(1) + \varphi_D(1) + \varphi_C(1) + \varphi_F(1)$$

$$\therefore \quad \varphi_1 = \tfrac{1}{2}(\varphi_A + \varphi_C + \varphi_D + \varphi_F)$$

$$\varphi_B(1) + \varphi_E(1) + \varphi_B(1) + \varphi_E(1)$$

$$\therefore \quad \varphi_2 = \frac{1}{\sqrt{2}}(\varphi_B + \varphi_E)$$

$$H_{11} = \tfrac{1}{4}(H_{AA} + 2H_{AC} + 2H_{AD} + 2H_{AF} + H_{CC} + 2H_{CD}$$
$$+ 2H_{CF} + H_{DD} + 2H_{DF} + H_{FF})$$
$$= \tfrac{1}{4}(4\alpha + 4\beta) = \alpha + \beta$$

$$H_{12} = \frac{1}{2\sqrt{2}}(H_{AB} + H_{AE} + H_{CB} + H_{CE} + H_{DB} + H_{DE}$$
$$+ H_{FB} + H_{FE})$$
$$= \frac{1}{2\sqrt{2}}(4\beta) = \sqrt{2}\beta$$

$$H_{22} = \tfrac{1}{2}(H_{BB} + 2H_{BE} + H_{EE}) = \alpha + \beta$$

$$D(\Gamma_{A_1}) = \begin{vmatrix} x+1 & \sqrt{2} \\ \sqrt{2} & x+1 \end{vmatrix} = x^2 + 2x - 1 = 0$$

$$x = -1 \pm \sqrt{2}$$

$$A_1 = x+1; \qquad A_2 = -\sqrt{2}$$

i	A_i for $x = -2.414$	A_i^2	c_i	A_i for $x = +0.414$	A_i^2	c_i
1	-1.414	2	0.707	1.414	2	0.707
2	-1.414	2	0.707	-1.414	2	-0.707
		4			4	

$$\sqrt{4} = 2 \qquad\qquad\qquad \sqrt{4} = 2$$

Hence
$$\psi_1 = 0.707(\varphi_1 + \varphi_2)$$
$$= 0.354(\varphi_A + \varphi_C + \varphi_D + \varphi_F) + 0.500(\varphi_B + \varphi_E)$$
$$\psi_4 = 0.707(\varphi_1 - \varphi_2)$$
$$= 0.354(\varphi_A + \varphi_C + \varphi_D + \varphi_F) - 0.500(\varphi_B + \varphi_E)$$

$\Gamma(B_2)$: $\varphi_A(1) + \varphi_D(-1) + \varphi_C(-1) + \varphi_F(1)$

$$\therefore \varphi_3 = \tfrac{1}{2}(\varphi_A - \varphi_C - \varphi_D + \varphi_F)$$

$$D(\Gamma_{B_2}) = |x+1| = 0$$

$$x = -1$$

$$\psi_2 = \tfrac{1}{2}(\varphi_A - \varphi_C - \varphi_D + \varphi_F)$$

$\Gamma(A_2)$: $\varphi_4 = \tfrac{1}{2}(\varphi_A - \varphi_C + \varphi_D - \varphi_F)$

$$(x - 1) = 0 \qquad x = 1$$

$$\psi_5 = \tfrac{1}{2}(\varphi_A - \varphi_C + \varphi_D - \varphi_F)$$

$\Gamma(B_1)$: $\varphi_5 = \tfrac{1}{2}(\varphi_A + \varphi_C - \varphi_D - \varphi_F)$

$$\varphi_6 = \tfrac{1}{2}(\varphi_B - \varphi_E)$$

$$\begin{vmatrix} x-1 & \sqrt{2} \\ \sqrt{2} & x-1 \end{vmatrix} = x^2 - 2x - 1 = 0$$

$$x = 1 \pm \sqrt{2}$$

As before, we find

$$\psi_3 = 0.354(\varphi_A + \varphi_C - \varphi_D - \varphi_F) + 0.500(\varphi_B - \varphi_E)$$

$$\psi_6 = 0.354(\varphi_A + \varphi_C - \varphi_D - \varphi_F) - 0.500(\varphi_B - \varphi_E)$$

Note that the subscripts of the molecular orbitals are assigned in order of increasing energy. The results are conveniently summarized in a table:

COEFFICIENTS

ψ_i	A	B	C	D	E	F	ω
1	0.354	0.500	0.354	0.354	0.500	0.354	−2.414
2	0.500	0	−0.500	−0.500	0	0.500	−1.000
3	0.354	0.500	0.354	−0.354	−0.500	−0.354	−0.414
4	0.354	−0.500	0.354	0.354	−0.500	0.354	+0.414
5	0.500	0	−0.500	0.500	0	−0.500	+1.000
6	0.354	−0.500	+0.354	−0.354	0.500	−0.354	+2.414

3.10 C_3 Trivinylmethyl (IV)

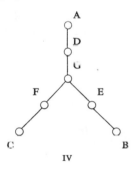

IV

The nuclei of trivinylmethyl belong to point group D_{3h}, but treatment as C_3 will give optimum use of symmetry in this case and will illustrate the special features in the use of this table. The character table is

C_3	E	C_3	C_3^2
A	1	1	1
E	1	ϵ^*	ϵ
	1	ϵ	ϵ^*

$$\epsilon = e^{2\pi i/3}$$
$$\epsilon^* = e^{-2\pi i/3}$$
$$i = \sqrt{-1}$$

We apply the symmetry operations to the π-system as usual:

E	C_3	C_3^2
φ_A	φ_B	φ_C
φ_B	φ_C	φ_A
φ_C	φ_A	φ_B
φ_D	φ_E	φ_F
φ_E	φ_F	φ_D
φ_F	φ_D	φ_E
φ_G	φ_G	φ_G
7	1	1

In determining the number of symmetry orbitals in each representation, Γ_A is straightforward: $\frac{1}{3}(7 + 1 + 1) = 3$. The Γ_E-representation involves complex characters; this representation is degenerate. The number of symmetry orbitals in one set is given by $\frac{1}{3}(7 + 1 \cdot e^{2\pi i/3} + 1 \cdot e^{-2\pi i/3})$. To evaluate this quantity we recall that

$$e^{\pm ai} = \cos a \pm i \sin a \tag{27}$$

Substituting this relation gives $\frac{1}{3}(7 + 2 \cos 2\pi/3) = 2$. The number of symmetry orbitals in the other degenerate set is the same, of course.

In representation Γ_A we find

$$\varphi_1 = \frac{1}{\sqrt{3}} (\varphi_A + \varphi_B + \varphi_C)$$

$$\varphi_2 = \frac{1}{\sqrt{3}} (\varphi_D + \varphi_E + \varphi_F)$$

$$\varphi_3 = \varphi_G$$

Hence

$$\begin{vmatrix} x & 1 & 0 \\ 1 & x & \sqrt{3} \\ 0 & \sqrt{3} & x \end{vmatrix} = x^3 - 4x = 0$$
$$x = 0, \pm 2$$

In one of the degenerate Γ_E-representations we find

$$\varphi_4 = \frac{1}{\sqrt{3}} (\varphi_A + \epsilon^* \varphi_B + \epsilon \varphi_C)$$

In normalizing this wave function, which contains complex quantities, remember that the Hermitian scalar product is taken; that is, instead of simply taking the square root of the sum of the squares of the coefficients, each coefficient is multiplied by its complex conjugate. In this connection note that

$$\epsilon \cdot \epsilon^* = 1 \tag{28}$$

and

$$\epsilon^2 = e^{4\pi i/3} = \cos\frac{4\pi}{3} + i\sin\frac{4\pi}{3} \tag{29}$$

Similarly,

$$\varphi_5 = \frac{1}{\sqrt{3}}(\varphi_D + \epsilon^*\varphi_E + \epsilon\varphi_F)$$

In setting up the secular determinant, remember that the definitions with complex conjugates must be used; for example, $\mathbf{H} = \int \varphi_r^* \mathbf{H}\varphi_s \, d\tau$. Hence

$$\begin{vmatrix} x & 1 \\ 1 & x \end{vmatrix} = 0 \qquad \begin{array}{l} x^2 - 1 = 0 \\ x = \pm 1 \end{array}$$

The last line in the character table gives the same determinant as the second line—as it must if the two representations are indeed degenerate. The seven molecular orbitals have energies that correspond to $x = 0$, ± 1, ± 1, ± 2.

The complete molecular orbitals are found in the usual way. The two degenerate Γ_E MO's, ψ_2 and ψ_3, of energy, $x = -1$, are found to be

$$\psi_2 = \frac{1}{\sqrt{6}}(\varphi_A + \epsilon^*\varphi_B + \epsilon\varphi_C + \varphi_D + \epsilon^*\varphi_E + \epsilon\varphi_F)$$

$$\psi_3 = \frac{1}{\sqrt{6}}(\varphi_A + \epsilon\varphi_B + \epsilon^*\varphi_C + \varphi_D + \epsilon\varphi_E + \epsilon^*\varphi_F)$$

Although, in practice, we may work with such orbitals in a straightforward way, the complex coefficients are difficult to visualize graphically, as in Figs. 2.7, 2.8, and 2.9. Since these orbitals are degenerate, any linear combinations are also allowable MO's. In particular, we may take the sum and difference of ψ_2 and ψ_3, whereupon the complex numbers disappear:

$$\psi_2' = \frac{1}{\sqrt{2}}(\psi_2 + \psi_3) = \frac{1}{2\sqrt{3}}(2\varphi_A - \varphi_B - \varphi_C + 2\varphi_D - \varphi_E - \varphi_F)$$

$$\psi_3' = \frac{1}{\sqrt{2}}(\psi_2 - \psi_3) = \tfrac{1}{2}(\varphi_B - \varphi_C + \varphi_E - \varphi_F)$$

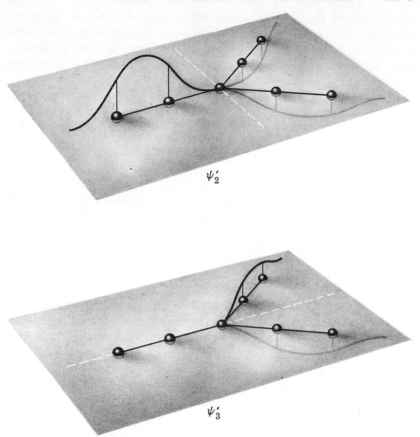

Fig. 3.2   Degenerate molecular orbitals for trivinylmethyl.

These new orbitals are still degenerate and may be diagrammed as in Fig. 3.2. Note that the nodal planes (dotted lines) are perpendicular in the two orbitals. This set of orbitals would have been obtained directly if the system were treated according to C_{3v} as in the next section.

3.11 C_{3v} Acepentylene (V)

The system of atoms is D_{3h}; we shall make most efficient use of the symmetry by treating the π-system as C_{3v}, which has the character table

C_{3v}	E	$2C_3$	$3\sigma_v$
A_1	1	1	1
A_2	1	1	-1
E	2	-1	0

This character table is actually abbreviated. There are two C_3 operations, C_3 and $C_3^2 = C_3^{-1}$. But each operation has the same character. Similarly, there are three σ_v operations: reflection in planes defined by the lines IJ, CJ and FJ, which we could call σ', σ'' and σ''', respectively. We apply all of the operations to the atomic orbitals as before:

E	C_3	C_3^2	σ'_v	σ''_v	σ'''_v
φ_A	φ_D	φ_G	φ_H	φ_E	φ_B
φ_B	φ_E	φ_H	φ_G	φ_D	φ_A
φ_C	φ_F	φ_I	φ_F	φ_C	φ_I
φ_D	φ_G	φ_A	φ_E	φ_B	φ_H
φ_E	φ_H	φ_B	φ_D	φ_A	φ_G
φ_F	φ_I	φ_C	φ_C	φ_I	φ_F
φ_G	φ_A	φ_D	φ_B	φ_H	φ_E
φ_H	φ_B	φ_E	φ_A	φ_G	φ_D
φ_I	φ_C	φ_F	φ_I	φ_F	φ_C
φ_J	φ_J	φ_J	φ_J	φ_J	φ_J
$\chi = 10$	1	1	2	2	2

$$\Gamma_{A_1}: \tfrac{1}{6}(10 + 1 + 1 + 2 + 2 + 2) = 3$$
$$\Gamma_{A_2}: \tfrac{1}{6}(10 + 1 + 1 - 2 - 2 - 2) = 1$$

Γ_E: The two in the E column of the character table indicates that this representation is doubly degenerate. The Γ_E irreducible representation is a second-order matrix. We shall see that with only the character to work with we can proceed only part way to an answer. However, a little intuition readily takes us the rest of the way. The number of times the Γ_E-representation occurs in the character for acepentylene is found in the usual way:

$$\tfrac{1}{6}(2 \cdot 10 + -1 \cdot 1 + -1 \cdot 1 + 0 \cdot 2 + 0 \cdot 2 + 0 \cdot 2) = 3$$

Therefore,

$$\chi = 3\chi_{A_1} + \chi_{A_2} + 3\chi_E$$

The A_1 and A_2 molecular orbitals and corresponding energies are determined in the usual manner:

$$\Gamma_{A_1}: \ \varphi_1 = \frac{1}{\sqrt{6}} (\varphi_A + \varphi_B + \varphi_D + \varphi_E + \varphi_G + \varphi_H)$$

$$\varphi_2 = \frac{1}{\sqrt{3}} (\varphi_C + \varphi_F + \varphi_I)$$

$$\varphi_3 = \varphi_J$$

$$\begin{vmatrix} x+1 & \sqrt{2} & 0 \\ \sqrt{2} & x & \sqrt{3} \\ 0 & \sqrt{3} & x \end{vmatrix} = 0$$

$$x^3 + x^2 - 5x - 3 = 0; \qquad x = -2.52, \ -0.57, \ +2.09$$

$$\Gamma_{A_2}: \ \varphi_4 = \frac{1}{\sqrt{6}} (\varphi_A - \varphi_B + \varphi_D - \varphi_E + \varphi_G - \varphi_H)$$

$$|x - 1| = 0 \qquad x = 1$$

Γ_E: Our function-generating machine applied to each atomic orbital gives the following nine functions:

$$f_1 = \frac{1}{\sqrt{6}} (2\varphi_A - \varphi_D - \varphi_G) \qquad f_2 = \frac{1}{\sqrt{6}} (2\varphi_B - \varphi_E - \varphi_H)$$

$$f_3 = \frac{1}{\sqrt{6}} (2\varphi_C - \varphi_F - \varphi_I) \qquad f_4 = \frac{1}{\sqrt{6}} (2\varphi_D - \varphi_G - \varphi_A)$$

$$f_5 = \frac{1}{\sqrt{6}} (2\varphi_E - \varphi_H - \varphi_B) \qquad f_6 = \frac{1}{\sqrt{6}} (2\varphi_F - \varphi_I - \varphi_C)$$

$$f_7 = \frac{1}{\sqrt{6}} (2\varphi_G - \varphi_A - \varphi_D) \qquad f_8 = \frac{1}{\sqrt{6}} (2\varphi_H - \varphi_B - \varphi_E)$$

$$f_9 = \frac{1}{\sqrt{6}} (2\varphi_I - \varphi_C - \varphi_F)$$

Our problem is to find two orthogonal sets of three independent functions that will make the same determinant. Our generating machine has constructed nine functions; however, not all are independent. We see that

$$f_1 = -(f_4 + f_7)$$
$$f_2 = -(f_5 + f_8)$$
$$f_3 = -(f_6 + f_9)$$

However, the difference function, $f_4 - f_7$, is independent of and orthogonal to f_1. Hence we construct six independent functions:

$$f_1 = \frac{1}{\sqrt{6}}(2\varphi_A - \varphi_D - \varphi_G) \qquad f_1' = \frac{1}{\sqrt{2}}(\varphi_D - \varphi_G)$$

$$f_2 = \frac{1}{\sqrt{6}}(2\varphi_B - \varphi_E - \varphi_H) \qquad f_2' = \frac{1}{\sqrt{2}}(\varphi_E - \varphi_H)$$

$$f_3 = \frac{1}{\sqrt{6}}(2\varphi_C - \varphi_F - \varphi_I) \qquad f_3' = \frac{1}{\sqrt{2}}(\varphi_F - \varphi_I)$$

These functions are still not appropriate symmetry functions. A correct set of six functions would make a six-order determinant that is factorable into two identical third-order determinants. With an actual irreducible representation, which for the E-representation would be a second-order matrix, we would have obtained these functions directly. With only the characters to work with, this is as far as we can go mechanically. At this point we must supply some intuition. The proper intuition is to take further linear combinations so as to arrive at symmetry orbitals that are symmetric and antisymmetric with respect to a common axis. Since f_1 and f_2 are functions of equivalent positions and "look" similar, we can start with them. Similar reasoning applies to f_1' and f_2'.

$$\frac{1}{\sqrt{2}}(f_1 + f_2) = \frac{1}{2\sqrt{3}}(2\varphi_A + 2\varphi_B - \varphi_D - \varphi_E - \varphi_G - \varphi_H) = g_1$$

$$\frac{1}{\sqrt{2}}(f_1 - f_2) = \frac{1}{2\sqrt{3}}(2\varphi_A - 2\varphi_B - \varphi_D + \varphi_E - \varphi_G + \varphi_H) = g_1'$$

$$\frac{1}{\sqrt{2}}(f_1' + f_2') = \tfrac{1}{2}(\varphi_D + \varphi_E - \varphi_G - \varphi_H) = g_2'$$

$$\frac{1}{\sqrt{2}}(f_1' - f_2') = \tfrac{1}{2}(\varphi_D - \varphi_E - \varphi_G + \varphi_H) = g_2$$

We note that $f_1 + f_2$ is symmetric and $f_1 - f_2$ is antisymmetric with respect to a perpendicular plane along the FJ line. Similarly, $f_1' - f_2'$ is symmetric and $f_1' + f_2'$ is antisymmetric with respect to the same plane. Calling the new functions g's, g_i is symmetric and g_i' is antisymmetric. We choose our remaining two functions with the same symmetry characteristics in respect to this plane. Hence we choose

$$g_3 = \frac{1}{\sqrt{6}} (2\varphi_F - \varphi_C - \varphi_I) \qquad g_3' = \frac{1}{\sqrt{2}} (\varphi_C - \varphi_I)$$

The symmetric functions will clearly all be orthogonal to the antisymmetric functions. We can treat each set separately and make two identical third-order determinants. However, to demonstrate the point, we can find all of the matrix elements of the complete 6×6 determinant with the columns (and rows) labeled as shown:

$$
\begin{array}{cccccc}
g_1 & g_2 & g_3 & g_1' & g_2' & g_3' \\
\end{array}
$$

$$
\begin{vmatrix}
x+1 & 0 & -\sqrt{2}/2 & 0 & 0 & 0 \\
0 & x-1 & -\sqrt{6}/2 & 0 & 0 & 0 \\
-\sqrt{2}/2 & -\sqrt{6}/2 & x & 0 & 0 & 0 \\
0 & 0 & 0 & x-1 & 0 & -\sqrt{6}/2 \\
0 & 0 & 0 & 0 & x+1 & \sqrt{2}/2 \\
0 & 0 & 0 & -\sqrt{6}/2 & \sqrt{2}/2 & x
\end{vmatrix} = 0
$$

The secular determinant is indeed in block form, and the two component third-order determinants have identical characteristic equations:

$$x^3 - 3x - 1 = 0$$

From here we can solve for x and obtain the energies and coefficients in the usual way. The technique used here is general and avoids the use of imaginary functions; the important essential is to adopt a symmetry-antisymmetry axis common to all of the symmetry orbitals.

Exercises. 1. Find the energy levels of perinaphthenyl, VI.

ns. $x = 0$, $A \pm \sqrt{6}$, $\pm\sqrt{3}$, $\pm\sqrt{3}$, ± 1, ± 1, ± 1.

2. Find energy levels and coefficients of 3-methylene-1,4-pentadiene, VII, fulvene, VIII, and benzyl, IX.

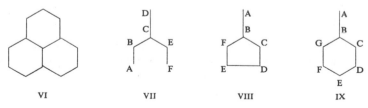

VI   VII   VIII   IX

Ans.

VII

ψ	A	B	C	D	E	F	x
1	0.230	0.444	0.628	0.325	0.444	0.230	−1.932
2	0.500	0.500	0	0	−0.500	−0.500	−1
3	0.444	0.230	−0.325	−0.628	0.230	0.444	−0.518
4	−0.444	0.230	0.325	−0.628	0.230	−0.444	0.518
5	0.500	−0.500	0	0	0.500	−0.500	1
6	0.230	−0.444	0.628	−0.325	−0.444	0.230	1.932

VIII

ψ	A	B	C	D	E	F	x
1	0.245	0.523	0.430	0.385	0.385	0.430	−2.115
2	0.500	0.500	0	−0.500	−0.500	0	−1
3	0	0	0.601	0.372	−0.372	−0.601	−0.618

IX

ψ	A	B	C	D	E	F	G	x
1	0.238	0.500	0.406	0.354	0.337	0.354	0.406	−2.101
2	0.397	0.500	0.116	−0.354	−0.582	−0.354	0.116	−1.259
3	0	0	0.500	0.500	0	−0.500	−0.500	−1
4	0.756	0	−0.378	0	0.378	0	−0.378	0

Supplemental Reading

Of the many available books on vector and matrix algebra and on group theory, most are mathematics textbooks that present the subjects as abstract concepts. The following are among the few books that treat these subjects with specific reference to quantum chemistry.

E. P. Wigner, *Group Theory*, Academic Press, New York (1959).

In the original German, this book has long been a classic. Now available in an English translation by J. J. Griffin, the book discusses vectors and matrices and the role of group theory in quantum mechanics and specifically in atomic spectra.

H. Eyring, J. Walter, and G. E. Kimball, *Quantum Chemistry*, John Wiley and Sons, New York (1944). Chaps. X and XIII.

Introduction to matrices and group theory; benzene is worked out as an example. Many character tables are given in Appendix VII.

M. Tinkham, *Advanced Quantum Mechanics of Atoms, Molecules and Solids*, A.S.U.C. Store, University of California, Berkeley, Calif. (1958).

Thorough discussion of matrices, vectors, and group theory, with application to molecular orbital and other problems. Lists character tables.

H. Margenau and G. M. Murphy, *The Mathematics of Physics and Chemistry*, D. Van Nostrand Co., Princeton, N.J. (1943).

A more "mathematics approach" to vectors, matrices, and group theory. Lists character tables.

H. Hartmann, *Theorie der chemischen Bindung*, Springer-Verlag, Berlin (1954), pp. 259–262.

Brief application of group theory to the MO's of naphthalene.

E. B. Wilson, J. C. Decius, and P. C. Cross, *Molecular Vibrations*, McGraw-Hill Book Co., New York (1955), Chaps. 5 and 6.

A discussion of group representations with specific reference to molecular vibrations. Lists character tables.

F. G. Fumi, *Nuovo cimento*, **8,** 1 (1951).

Group representations and molecular orbitals (in English).

E. Heilbronner, *Helv. Chim. Acta*, **37,** 913 (1954).

Results of the application of group theory to molecular orbitals are presented for some general systems in a pictorial manner.

R. Daudel, R. Lefebvre, and C. Moser, *Quantum Chemistry*, Interscience Publishers, New York (1959).

Application of group theory to HMO method, with character tables.

4 Variation of α and β

4.1 Significance of HMO Approximations

The calculation of electronic energies by the HMO method is relatively simple and straightforward. The question arises of the significance or worth of the calculated energies. Alternatively, we may inquire about the importance of the various approximations made. Actually, the inherent assumptions of the method are so unrealistic we marvel that the theory finds use at all! The approximations and assumptions made can be divided into two broad classes: those involved in the determination of one-electron molecular orbitals and those inherent in our subsequent use of these MO's. Some of the approximations can be reduced by further techniques within the framework of the HMO method. Others can be minimized by additional techniques that constitute various advanced MO methods. Still other approximations are inherent in any procedure that deals with a many-electron many-center system as a set of components and must be lived with as such.

One-Electron Wave Functions

A basic assumption that actually defines the LCAO method is that the wave function has the form of a linear combination of atomic orbitals. Furthermore, in the HMO method we have specifically neglected the orbitals that constitute the sigma framework of the molecule. Neglect of σ-π-interaction can be partly justified on the basis of symmetry; σ-orbitals are symmetric and π-orbitals are antisymmetric with respect to the molecular plane, and if only one electron were involved in the entire system interaction between the orbitals would vanish. Since, in practice, the sigma framework is densely occupied by electrons, interactions of two types are significant. First, Coulomb repulsions between π- and σ-electrons are important. Not only does such interaction affect the total electronic energy of the system but variations in π-electron density will affect σ-electron densities and vice-versa. We may talk of π- and σ-polarizability. Some calculations on benzene indicate that the density of σ-electrons

97

mostly exceeds the π-density above the molecular plane.[1] This type of interaction probably affects calculation of dipole moments (Sec. 6.1), but in many applications suitable account can be taken by the choice of proper empirical values for α- and β-parameters. Second, exchange of electrons between π- and σ-orbitals can occur. This type of resonance is almost always neglected but seems to have a negligible effect.[2] In at least one specific application (electron spin resonance, Sec. 6.3), however, this interaction is very important.

The assumption that the linear combinations are of carbon $2p_z$-atomic orbitals is not necessary in the HMO method because α and β are not evaluated. The exact form of the "atomic orbitals" is never required. Nevertheless, in other connections, particularly when overlap integrals are involved, the explicit assumption is made that $2p_z$-orbitals are used.[3]

In setting up the secular determinant, simplifying approximations were made for each of the integrals involved. The bond integrals, H_{rs}, were all given the same value β if atoms r and s were bonded. The bond integral is dependent on bond length; hence different bonds should have different β-values. Such variation in β can be handled within the simple MO method and is described in Sec. 4.3. H_{rs} for atoms r and s not bonded is taken as zero. This approximation is retained even in some advanced calculations. Calculations have been published in which appropriate nonzero values are used for nonbonded H_{rs} and substantial changes are evident.[4] Nevertheless, the validity of the procedure is unknown. Inclusion of such nonzero values can be accomplished within the simple MO method, but the work involved is considerably greater. All of the S_{rs} ($r \neq s$)-values were assumed to vanish, although many of these integrals have substantial values. Introduction of nonzero S_{rs}-values is also possible within the simple LCAO method (Sec. 4.2).

In an all-sp^2-carbon π-system the assumption that all H_{rr} are equal is valid for a one-electron system in an *ab initio* LCAO treatment. In polyelectron systems this assumption breaks down in effect (*vide infra*). For

---

[1] C. A. Coulson, P. W. Higgs, and N. H. March, *Nature*, **168**, 1039 (1951).

[2] I. G. Ross, *Trans. Faraday Soc.*, **48**, 973 (1952); C. M. Moser, *Trans. Faraday Soc.*, **49**, 1239 (1953). However, cf. S. L. Altmann, *Proc. Roy. Soc.*, **A210**, 327 (1952); K. Niira, *J. Chem. Phys.*, **20**, 1498 (1952); *J. Phys. Soc. Japan*, **8**, 630 (1953). See comments by C. A. Coulson, L. J. Schaad, and L. Burnelle in *Proceedings of the Third Conference on Carbon*, Pergamon Press, New York (1959), p. 27–35.

[3] One deficiency of an LCAO wave function is the failure to obey the virial theorem. See the discussion in G. W. Wheland, *Resonance in Organic Chemistry*, John Wiley and Sons, New York (1955); pp. 681–684.

[4] For examples see J. I. F. Alonso, *Compt. rend.*, **233**, 56 (1951); I. Estelles and J. I. F. Alonso, *Anales real soc. españ. fis. y quim.* (*Madrid*), **49B**, 267 (1953); J. I. F. Alonso, J. Mira, and L. Alcaniz, *Anales real soc. españ. fis. y quim.* (*Madrid*), **53B**, 101 (1957).

atoms other than carbon in the π-system $\mathbf{H}_{rr}$ will have a different value, since the energy of an electron in the isolated orbital will be different. Incorporation of such changes within the simple LCAO method is discussed in Sec. 4.4.

Polyelectron Orbitals

In the HMO method we established a series of one-electron wave functions but then proceeded to put two electrons into each. We neglected to incorporate electron-repulsion effects not only between the two electrons in one orbital but between electrons in different orbitals. This neglect is probably the single most important deficiency in the HMO method. The situation is partially relieved by changing the basis of our calculation. Instead of taking the Hamiltonian over the nuclei of the atoms in the π-lattice alone, we consider the one-electron Hamiltonian to involve as well the average field of the remaining electrons. Instead of a *core Hamiltonian*, we use an *effective self-consistent field Hamiltonian* ($\mathbf{H}_{eff}$). The usual way of stating the HMO method is in terms of $\mathbf{H}_{eff}$; however, because the electrons are now reciprocally involved, we lose precise definition of our various quantities. In practice, nothing has changed, since all of our $\mathbf{H}_{rs}$ now become $\mathbf{H}_{eff(rs)}$. The definitions of α and β are now changed in effect, but, since these quantities were not evaluated anyway, the form of the MO energies remains identical. Because of the changed and now rather more vague definitions, mathematical calculation of the α- and β-integrals is out of the question except for some particularly simple systems. They are retained as parameters to which we assign completely empirical values.

In finding a one-electron wave function for a core Hamiltonian, the electron "sees" n positive charges. With $\mathbf{H}_{eff}$, the electron is shielded from the core by other electrons and it effectively "sees" less positive charge. In this way electron repulsion is treated by decreasing the net attraction by the nuclei. However, now we must consider how α and β may depend on the electronic environment of an atom, since shielding is important. β is apparently approximately invariant to such shielding[5] (note that H_2^+ has about half the bond energy of H_2). However, α apparently depends on the attached atoms, the bond multiplicity, and net charge as well as the nature of the atomic orbital.[5,6,7] Thus $=CH_2$, $=CH\!-\!$, and $=C\diagdown^{\diagup}$

atoms may be expected to have different α-values, although each still

[5] R. S. Mulliken, *J. chim. phys.*, **46**, 497 (1949).

[6] R. S. Mulliken, *ibid.*, **46**, 675 (1949).

[7] A partial justification may be found in comparison with more precisely defined related quantities in self-consistent field MO theory; cf. Sec. 16.3 and particularly Eqs. 16.13 and 16.14.

involves an sp^2-carbon. In practice, such variation is rarely incorporated. An acetylenic carbon clearly should have α different from an ethylenic carbon. The greater effective electronegativity of a digonal carbon is reflected in the greater acidity of acetylene compared to ethylene. A variation in charge will change α considerably; for example, the α-values in ethylene and in ethylene cation will differ substantially. Consequently, we must consider with care the use made of calculations of charged species.[8] One simple method of incorporating such changes within the HMO scheme is the so-called ω-technique discussed in Sec. 4.5. In nonalternant hydrocarbons the π-charge density is not uniform; Coulson and Rushbrooke[9] warn that calculation for such systems may be less valid than for alternant hydrocarbons in which the uniform charge distribution guarantees a self-consistent field. In the more advanced SCF LCAO method[10] the field of the other electrons is considered explicitly (Chap. 16).

A further deficiency of the simple theory is apparent when we examine the total wave function for the π-system. This function, Ψ, is a product wave function made up of individual one-electron MO's (ψ). Since we put electrons 1 and 2 into ψ_1, electrons 3 and 4 into ψ_2, etc., we write

$$\Psi = \psi_1(1)\psi_1(2)\psi_2(3)\psi_2(4) \cdots \tag{1}$$

However, electrons cannot be labeled, and we must consider other product functions such as

$$\Psi'' = \psi_1(3)\psi_1(2)\psi_2(1)\psi_2(4) \cdots \tag{2}$$

The complete wave function should actually be taken as a sum of individual product wave functions in which complete permutation of electrons and orbitals is established. This procedure is followed in the antisymmetrized LCAO (ASMO) method,[11] which is also discussed further in Chap. 16.

Neither the use of a self-consistent field nor antisymmetrization provides for another feature of electronic repulsion, namely electron correlation. The use of an average field of other electrons does not provide for the tendency of electrons to keep apart. However, as we saw in Sec. 1.8, electron correlation can be provided by the inclusion of excited states. Thus, if ψ_m is the last occupied MO, we should consider adding into the total wave function the product function

$$\Psi''' = \psi_1(1)\psi_1(2) \cdots \psi_m(2m - 1)\psi_{m+1}(2m) \tag{3}$$

multiplied by a coefficient determined by a variation procedure. This procedure is the method of configuration interaction (Chap. 16). As

---

[8] Some approximations required are discussed by C. A. Coulson and M. J. S. Dewar, *Discussions Faraday Soc.*, **2**, 54 (1947).

[9] C. A. Coulson and G. S. Rushbrooke, *Proc. Cambridge Phil. Soc.*, **36**, 193 (1940).

[10] C. C. J. Roothaan, *Rev. Mod. Phys.*, **23**, 69 (1951).

[11] M. Goeppert-Mayer and A. L. Sklar, *J. Chem. Phys.*, **6**, 645 (1938).

commonly used, this method considers only correlation of electrons between component atomic orbitals and does not, except for the use of empirical quantities, consider correlation within an atomic orbital.[12,13]

The following nomenclature is convenient and is used in this book. We refer to all MO methods that neglect specific electronic repulsions as *simple LCAO* methods with or without neglect of overlap. The HMO method is the simplest of these methods, that in which all α's are equal, all β's between neighboring atoms are equal, and overlap integrals and non-neighbor β's are neglected. In other simple LCAO procedures, however, overlap integrals and variations in α and β may be included (*vide infra*). The MO procedures that specifically incorporate electronic repulsions by inclusion of e^2/r operators or their equivalent are referred to as *advanced MO* methods. Examples are the antisymmetrized MO method with configuration interaction (ASMO-CI) and the self-consistent field (SCF) methods (Chap. 16).

We see that the HMO method is left with relatively little theoretical foundation. Its success is based largely on the use of empirical quantities for α and β, which absorb many of the deficiencies arising from lack of explicit consideration of electronic repulsions. This frankly empirical basis of the theory must be considered in all of its applications. *The validity of an application cannot be assumed but must be demonstrated.*

4.2   Inclusion of Overlap

The off-diagonal matrix element for bonded atoms is $\mathbf{H}_{rs} - \mathbf{S}_{sr}E$. E is of substantial magnitude compared to $\mathbf{H}_{rs}$; $\mathbf{S}_{rs}$ for adjacent carbon Slater $2p_z$-orbitals in benzene is 0.25 (Fig. 1.11). Hence $\mathbf{S}_{rs}E$ is of substantial magnitude compared to $\mathbf{H}_{rs}$, and dropping the overlap term is a questionable procedure.[14] However, we can write

$$\beta - SE = \beta - S\alpha + S(\alpha - E) = \gamma + S(\alpha - E) \tag{4}$$

in which[15,16]

$$\gamma = \beta - S\alpha \tag{5}$$

[12] M. J. S. Dewar and C. E. Wulfman, *J. Chem. Phys.*, **29**, 158 (1958); M. J. S. Dewar and H. N. Schmeising, *Tetrahedron*, **5**, 166 (1959); M. J. S. Dewar and N. L. Hojvat, *J. Chem. Phys.*, **34**, 1232 (1961); L. C. Snyder and R. G. Parr, *J. Chem. Phys.*, **34**, 1661 (1961).

[13] Note that an electron correlation or "internal dispersion force" between adjoining but not delocalized double bonds can explain some of the properties of conjugated systems; W. T. Simpson, *J. Am. Chem. Soc.*, **73**, 5363 (1951).

[14] R. S. Mulliken, C. A. Rieke, and W. G. Brown, *J. Am. Chem. Soc.*, **63**, 41 (1941).

[15] R. S. Mulliken and C. A. Rieke, *J. Am. Chem. Soc.*, **63**, 1770 (1941).

[16] In some papers, particularly in the American literature, the symbols β and γ are reversed.

As before, we can divide through by γ and substitute, $x = (\alpha - E)/\gamma$. The diagonal elements again become x in the secular determinant, and off-diagonal bonding elements are $1 + Sx$. The approximation that $Sx \cong 0$ is less drastic than the original approximation, and the secular determinant is unchanged from the original. Hence, when we neglect overlap, we reinterpret β as γ but retain the β-symbolism.

We need not assume $Sx = 0$. If we let $1 + Sx = b$ and expand the secular determinant, the roots of the characteristic equation, now in units of b, are of the form.

$$x_j = -m_j b = -m_j(1 + Sx_j) \tag{6}$$

or
$$x_j = \frac{-m_j}{1 + m_j S} = \frac{\alpha - E_j}{\gamma} \tag{7}$$

$$E_j = \alpha + \frac{m_j}{1 + m_j S}\gamma \tag{8}$$

Making the substitution,[17]

$$n_j = \frac{m_j}{1 + m_j S} \tag{9}$$

we find
$$E_j = \alpha + n_j\gamma \tag{10}$$

The energy levels are expressed in the same form as before, but we use a new bond integral, γ, and the coefficients n_j are derived from the roots of the original characteristic equation by (9). The value $S = 0.25$ is usually assumed. The same technique can also be used when S_{rs} is not a constant but is proportional to β (Sec. 4.3). Even with this restriction removed, the general case is soluble.[18]

Inclusion of overlap requires changes in definition of charge density, bond order, etc. However, if S is given a constant value for nearest neighbors, with appropriate definitions, the numerical values for q and p remain unchanged.[19] Other definitions have been proposed.[20]

Despite the evident feasibility of incorporating overlap, nonorthogonality is still neglected in most simple LCAO work. This procedure is justified by Moffitt,[21] who argues that the definitions of α and β (and γ) are so vague that neglect of S may be and is justified empirically by the success of the simple HMO procedure. Although inclusion of overlap changes the spacing of energy levels, delocalization energies, and especially

[17] G. W. Wheland, *J. Am. Chem. Soc.*, **63**, 2025 (1941).
[18] P. O. Löwdin, *J. Chem. Phys.*, **18**, 365 (1950).
[19] B. H. Chirgwin and C. A. Coulson, *Proc. Roy. Soc.*, **A201**, 196 (1950).
[20] I. M. Bassett and R. D. Brown, *Australian J. Chem.*, **9**, 305, 315 (1956); R. McWeeny, *J. Chem. Phys.*, **19**, 1614 (1951).
[21] W. Moffitt, *J. Chem. Phys.*, **22**, 1820 (1954).

total π-electron energies, do not seem to be affected significantly.[17,22] Overlap is also neglected in the rather successful advanced MO methods of Pariser, Parr, and Pople[23] (Chap. 16), in which electron repulsion terms are explicitly incorporated.[24] Stewart[25] also concludes from his calculations that neglect of overlap is warranted in semiempirical MO methods.

Zero overlap would result from the use of orthogonal atomic-orbital basis functions.[18] Parr[26] has shown recently how the use of such functions may be expected to give results similar to and sometimes identical with those of HMO theory, especially for AH systems.

In short, overlap is usually neglected because the HMO method apparently works in many cases without it.

4.3   Variation of β

One of the assumptions made in HMO theory is the equality of bond-integral terms for nearest neighbors. Since the bond integral varies with distance, this assumption is equivalent to assuming an equality of all bond distances. Yet, in all-sp^2-carbon π-systems bond distances are known to vary from 1.33 to 1.52 A. Improvements could be made in the calculations if we knew the functional relation of β with respect to bond distance.

Lennard-Jones[27] was the first to tackle this problem; he assumed that β was directly related to bond energy and used a parabolic potential together with force constants for ethane and ethylene to derive $\beta_{1.33A}/\beta_{1.54A} = 1.22$. This ratio is undoubtedly too low, and the result has been used in only a few subsequent calculations.[28] By relating β to bond strength, effects of σ-bond compression or extension are included. An elaboration of the Lennard-Jones method with the use of Morse curve potentials has been presented by Mulliken, Rieke, and Brown.[14] It is convenient to define a standard β_0 for the benzene bond distance (1.397 A) and to scale other β's in units of the standard β_0 with the use of a dimensionless number, k:

$$\beta_{rs} = k_{rs}\beta_0 \qquad (11)$$

[22] J. DeHeer, *Phil. Mag.*, **41**, 370 (1950).

[23] R. Pariser and R. G. Parr, *J. Chem. Phys.*, **21**, 466, 767 (1953);   J. A. Pople, *Trans. Faraday Soc.*, **49**, 1375 (1953).

[24] See the comments on the success of neglect of overlap in these methods by R. S. Mulliken, *Symposium on Mol. Phys., Nikko, Japan*, **1953**, 17 and by P. O. Löwdin, **1953**, 21.

[25] E. T. Stewart, *J. Chem. Soc.*, 1850 (1959).

[26] R. G. Parr, *J. Chem. Phys.*, **33**, 1184 (1960).

[27] J. E. Lennard-Jones, *Proc. Roy. Soc.*, **A158**, 280 (1937).

[28] C. A. Coulson, *Proc. Roy. Soc.*, **A164**, 383 (1938);   J. E. Lennard-Jones and C. A. Coulson, *Trans. Faraday Soc.*, **35**, 811 (1939);   H. C. Longuet-Higgins and C. A. Coulson, *Trans. Faraday Soc.*, **42**, 756 (1946).

Although Mulliken, Rieke, and Brown gave results actually for γ, because of the effectively constant ratio of β and γ, the k's in (11) are unchanged. Their values are given in Table 4.1. Dewar and Schmeising[29] have recently

TABLE 4.1

DISTANCE VARIATION OF β PER MULLIKEN, RIEKE, AND BROWN[14]

r, A	k_r
1.20	1.71
1.33	1.23
1.397	1.00
1.45	0.83
1.54	0.57

used a Morse curve potential method to develop a new function of β with distance in which the rate of change of β is somewhat greater than in Table 4.1.

Mulliken[6,30] has proposed that β be proportional to the overlap integral S and has shown that in diatomic systems the proportionality holds rather well at bond distances of interest. For Slater overlap integrals[31,32] k as a function of distance is plotted in Fig. 4.1; values for distances of interest are listed in Table 4.2.

TABLE 4.2

k AS A FUNCTION OF DISTANCE FOR β PROPORTIONAL TO S

r, A	k
1.20	1.38
1.33	1.11
1.35	1.09
1.397	1.00
1.45	0.91
1.48	0.87
1.54	0.78

[29] M. J. S. Dewar and H. N. Schmeising, *Tetrahedron*, **5**, 166 (1959).

[30] R. S. Mulliken, *J. Phys. Chem.*, **56**, 295 (1952).

[31] The use of SCF overlap integrals gives a somewhat lesser variation of β with distance; for example, $k = 1.27$ at 1.20 A and 0.85 at 1.54 A.[32] However, SCF orbitals are known to have a rather long tail, which would tend to give overlap integrals that may be too high.

[32] R. S. Mulliken, C. A. Rieke, D. Orloff, and H. Orloff, *J. Chem. Phys.*, **17**, 1248 (1949).

Although there is no convincing evidence as to which β-distance curve to use, Mulliken's success with diatomic molecules[6,30,33] is suggestive, as is a recent successful application to calculation of ionization potentials.[34] Acceptance of the assumption further implies that β can be evalu-

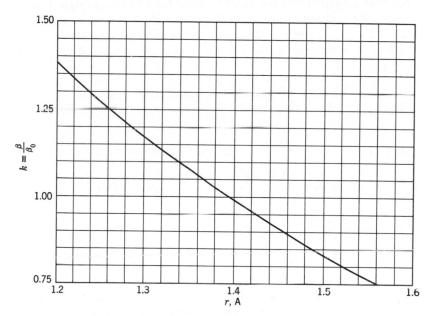

Fig. 4.1   Effect of distance on k for β proportional to S.

ated between p-orbitals twisted from coplanarity by an angle θ as (cf. Sec. 1.3)[35,36]

$$\beta = \beta_0 \cos \theta \tag{12}$$

From Table 4.2 it would appear that for most purposes the assumption that β is constant for all nearest-neighbor carbons is reasonably good.

[33] R. S. Mulliken, *J. Am. Chem. Soc.*, **72**, 4493 (1950).

[34] A. Streitwieser, Jr., and P. M. Nair, *Tetrahedron*, **5**, 149 (1959).

[35] R. G. Parr and B. L. Crawford, Jr., *J. Chem. Phys.*, **16**, 526 (1948); M. J. S. Dewar, *J. Am. Chem. Soc.*, **74**, 3345 (1952).

[36] M. Simonetta and S. Winstein [*J. Am. Chem. Soc.*, **76**, 18 (1954); cf. also M. M. Kreevoy, **80**, 5543 (1958)] have suggested that β be proportional to bond energies as obtained from Mulliken's "magic formula."[30]

$$\frac{\beta}{\beta_0} = \frac{S/1 + S}{S_0/1 + S_0} \tag{13}$$

However, Mulliken's magic formula was derived for diatomic molecules by using the assumption that β is proportional to S!

For more precise work appropriate values would seem to be $k = 1.1$ and 0.9 respectively, for double bonds and for single sp^2-sp^2 bonds, since the lengths of these bonds are relatively constant at 1.33–1.35 and 1.48–1.50 A, respectively.[37]

The HMO method with a different β-value for a bond is straightforward. Since all of the elements in the secular determinant are put in units of β, we consider each of the off-diagonal elements to be k_{rs}. We apply the technique to butadiene as an example, using $k_{C=C} = 1.1$ and $k_{C-C} = 0.9$ for structure I.[38]

$$
\begin{array}{cccc}
\text{A} & \text{B} & \text{C} & \text{D} \\
\end{array}
$$

$$
\text{H}_2\text{C}\!=\!\text{CH}\!-\!\text{CH}\!=\!\text{CH}_2
$$

$$
\begin{array}{ccc}
1.34\text{A} & & 1.34\text{A} \\
 & 1.48\text{A} & \\
\end{array}
$$

I

Symmetry orbitals may still be used, but these now yield the determinants

$$
\begin{vmatrix} x & 1.1 \\ 1.1 & x + 0.9 \end{vmatrix} = 0 = x^2 + 0.9x - 1.21
$$

$$
\begin{vmatrix} x & 1.1 \\ 1.1 & x - 0.9 \end{vmatrix} = 0 = x^2 - 0.9x - 1.21
$$

whence $x = \pm 1.64$, ± 0.74, and $E = 4\alpha + 4.76\beta$. If butadiene had two localized double bonds, to this approximation, $E = 4\alpha + 4.40\beta$; hence $DE = 0.36\beta$. Comparing this approximation to butadiene with all β's equal, $E = 4\alpha + 4.47\beta$ and $DE = 0.47\beta$, we find a substantial decrease in delocalization energy. We could go through in the usual way and calculate the coefficients, electron densities, bond orders, etc. Even with alternating bonds, butadiene is an alternant hydrocarbon, and the electron densities at each carbon are still unity.

It is possible to obtain a close approximation to the energy with varying β from calculations in which all β's are assumed equal. We expand the change in π-energy as a power series of the change in β:

$$
\delta E = \frac{\partial E}{\partial \beta_{rs}} \, \delta \beta_{rs} + \frac{1}{2} \frac{\partial^2 E}{\partial \beta_{rs}^2} \, \overline{\delta \beta}_{rs}^2 + \cdots \tag{14}
$$

In the first approximation we retain only the first term on the right; in the second approximation we retain the first two terms.

[37] L. E. Sutton, *Tetrahedron*, **5**, 118 (1959).

[38] A. Almenningen, O. Bastiansen, and M. Traettenberg, *Acta Chem. Scand.*, **12**, 1221 (1958).

First Approximation

Coulson and Longuet-Higgins[39] have shown that

$$p_{rs} = \frac{1}{2}\frac{\partial E}{\partial \beta_{rs}} \tag{15}$$

[p_{rs} is the bond order defined in (2.49).] Hence, to a first approximation,

$$\delta E \cong 2p_{rs}\delta\beta_{rs} \tag{16}$$

Example. To find the energy of butadiene as I from butadiene with all β's equal: from p. 55 we find $p_{AB} = p_{CD} = 0.894$, $p_{BC} = 0.447$. Hence $\delta E = 2(2 \times 0.894)(0.10)\beta + 2(0.447)(-0.10)\beta = 0.27\beta$ and $E_1 = E_0 + \delta E = 4\alpha + 4.47\beta + 0.27\beta = 4\alpha + 4.74\beta$. This result is very close to the energy we obtained by solving the secular equation.

Example. To estimate the π-energy of benzyl from calculations on styrene: styrene is converted to benzyl by breaking the α—β bond, that is, by converting $\beta_{\alpha-\beta}$ to 0. The bond orders for styrene, II, are

II

$\delta E/\beta = 2(0.911)(-1) = -1.82$; hence $E(\text{benzyl}) \cong 7\alpha + 8.60\beta$. This value may be compared with that obtained from solution of the secular equations: $E(\text{benzyl}) = 7\alpha + 8.72\beta$.

This approximation works rather well when the change in β is small; the error in the energy is roughly one order of magnitude less than the change made in β.

Second Approximation

The mutual bond polarizability, $\pi_{rs,tu}$, is half the second derivative of the energy with respect to β (17) and is a function of the coefficients (18).[39,40]

$$\pi_{rs,tu} = \frac{\partial p_{rs}}{\partial \beta_{tu}} = \frac{1}{2}\frac{\partial^2 E}{\partial \beta_{rs}\partial \beta_{tu}} = \frac{\partial p_{tu}}{\partial \beta_{rs}} = \pi_{tu,rs} \tag{17}$$

$$\pi_{rs,tu} = 2\sum_{j=1}^{m}\sum_{k=m+1}^{n}\frac{(c_{jr}c_{ks} + c_{js}c_{kr})(c_{jt}c_{ku} + c_{ju}c_{kt})}{\epsilon_j - \epsilon_k} \tag{18}$$

In (18) the first m-orbitals are doubly occupied. The first summation is over the occupied orbitals; the second summation is over the unoccupied

[39] C. A. Coulson and H. C. Longuet-Higgins, *Proc. Roy. Soc.*, **A191**, 39 (1947). See also the recent derivations of K. Fukui, C. Nagata, T. Yonezawa, and A. Imamura, *Bull. Chem. Soc. Japan*, **32**, 452 (1959).

[40] C. A. Coulson and H. C. Longuet-Higgins, *Proc. Roy. Soc.*, **A193**, 447 (1948).

orbitals. Because of the relationship between coefficients of bonding and antibonding orbitals, for even-AH's (18) reduces to (19).

$$\pi_{rs,tu} = 4 \sum_{j=1}^{n/2} \sum_{k=1}^{n/2} \frac{(c_{jr}c_{ks} - c_{js}c_{kr})(c_{jt}c_{ku} - c_{ju}c_{kt})}{\epsilon_j + \epsilon_k} \qquad j < k \qquad (19)$$

Values of bond polarizabilities for a number of compounds have been calculated.[40,41,42,43] Note that $\pi_{rs,tu}$ expresses the change in p_{rs} caused by a change in β_{tu}. Polarizabilities have units of reciprocal β.

Exercise. Using (19), calculate $\pi_{rs,tu}$ for butadiene.

Ans.

t, u	r, s:	1, 2	2, 3	3, 4
1, 2		0.089	−0.179	0.089
2, 3		−0.179	0.358	−0.179

To the second approximation, the power-series expansion becomes

$$\delta E = 2p_{rs}\delta\beta_{rs} + \pi_{rs,rs}\overline{\delta\beta_{rs}^2} \qquad (20)$$

Example. To find E for benzyl from styrene to the second approximation: bond polarizabilities for styrene are

t, u	r, s:	α, β	$\alpha, 1$	1, 2	2, 3	3, 4
α, β		0.073	−0.155	0.044	−0.011	0.009
$\alpha, 1$		−0.155	0.338	−0.099	0.021	−0.014
1, 2		0.044	−0.099	0.249	−0.190	0.123
2, 3		−0.011	0.021	−0.190	0.247	−0.214
3, 4		0.009	−0.014	0.123	−0.214	0.253
4, 5		0.009	−0.014	−0.081	0.127	−0.202
5, 6		−0.011	0.021	0.112	−0.091	0.127
6, 1		0.044	−0.099	−0.157	0.112	−0.081

$$\delta E = -1.82 + (0.073)(-1)^2 = -1.747$$

$$\therefore E_{\text{benzyl}} \cong 7\alpha + 8.67\beta$$

The error is now only 0.05β.

Other Approximation Methods

Several empirical formulas have been developed for estimating π-bond energies. These formulas find some use in estimating DE-values.[44]

[41] C. A. Coulson and R. Daudel, *Dictionary of Values of Molecular Constants*, Centre de Chimie Théorique de France, Paris.

[42] F. H. Sumner, *Trans. Faraday Soc.*, **51**, 315 (1955).

[43] A. Streitwieser, Jr., and J. I. Brauman, *Tables of Molecular Orbital Calculations*, Pergamon Press, New York, in press.

[44] For examples see V. M. Tatavskii, *Zhur. Fiz. Khim.*, **25**, 241 (1951); *C.A.*, **45**, 5988 (1951); C. Vroelant, *Compt. rend.*, **235**, 958 (1952); O. Chalvet, *Ann. Chim.*, **9**, 96 (1954); A. L. Green, *J. Chem. Soc.*, **1956**, 1886.

Some more ·precise approximation methods derive the π-energies of large systems from smaller ones. Brown's[45] "annelation energy," A, for example, is defined as the loss in delocalization energy which accompanies the fusion of two systems at respective bonds a and b. An example is the formation of phenanthrene from benzene and naphthalene:

DF = 2.000β 3.683β 5.448β
                   $A = 0.235\beta$

Empirically, it is found that a plot of A versus $\sqrt{p_a p_b}$ gives a good straight line (21).

$$A = (2.1533 \sqrt{p_a p_b} - 1.7325)\beta \qquad (21)$$

When p_a and p_b differ by no more than 0.1, this method appears to give results good to $\sim \pm 0.01\beta$; for example, in the case illustrated $p_a = 0.667$, $p_b = 0.725$; hence $A = 0.232$ by (21). The error in this case is only 0.003β.

Another method, due to Dewar and Pettit,[46] manufactures large polycyclic aromatic hydrocarbons by introducing crosslinks into appropriate monocyclic polyenes for which general equations have been written (2.42). The method involves a fair amount of labor and gives approximate results which lack precision.

A particularly simple and important method that has been shown by Dewar[47] to apply to the joining of two odd-AH units may be summarized as (22):

$$R_{RS} = E_{RS} - (E_R + E_S) \cong 2 a_{or} b_{os} \beta \qquad (22)$$

A linkage of bond integral β is made between atom r in R and atom s in S to manufacture the new π-system RS; a_{or} is the nonbonding MO (NBMO) coefficient of r, b_{os} is the NBMO coefficient of s. It will be recalled that these coefficients can be found virtually by inspection (p. 54). This method tends to underestimate R, but Dewar has demonstrated a good linear relationship between the approximate and the actual R's. For a series of compounds, relative magnitudes should be fairly reliable.

[45] R. D. Brown, *Trans. Faraday Soc.*, **46**, 1013 (1950).
[46] M. J. S. Dewar and R. Pettit, *J. Chem. Soc.*, **1954**, 1617.
[47] M. J. S. Dewar, *J. Am. Chem. Soc.*, **74**, 3345 (1952).

For the special case in which system S is a single atom (22) reduces to

$$R_{RS} \cong 2a_{or}\beta \tag{23}$$

Example. To find the E of styrene from benzyl. From the method on p. 54, a_o of the α carbon of benzyl is $2/\sqrt{7}$; hence $R_{styrene} \cong 2(2/\sqrt{7})\beta = 1.51\beta$. The actual value is 1.70β.

Dewar's method is important in a later discussion of reactivity (Chap. 11).

R_{RS} values for any systems R and S are found empirically to be related to p_{ab} in RS by (24):[48]

$$R_{RS} \cong \frac{p_{ab} - 0.057}{0.825}\beta \tag{24}$$

Example. Styrene formed by benzene + ethylene.

$$R_{RS} \cong \frac{0.406 - 0.057}{0.825}\beta = 0.423\beta$$

Actual $R_{RS} = 8\alpha + 10.424\beta - (6\alpha + 8.000\beta + 2\alpha + 2.000\beta)$

$$= 0.424\beta$$

4.4 Variation of α

Although α was originally defined as $\int \varphi_j H \varphi_j \, d\tau$ in which H was a core Hamiltonian, H has now been modified to include the average effects of other electrons. Hence α is expected to vary somewhat from carbon to carbon. Indeed, Mulliken[6] has derived values of α that differ by more than 20 kcal./mole between ethylene and benzene. Although these values were obtained for an SCF method and are not directly comparable to our α's, the results are suggestive of changes that should, perhaps, be considered. In practice, however, α is taken to be the same for all olefinic and aromatic carbons.[49]

The core potential is still an important part of α, however, and important variations are to be expected for atoms other than carbon in the π-lattice. In considering changes in α, rather than fractional changes, it is convenient to consider differential changes in units of β relative to a standard α_0 usually taken as the α for a benzene carbon:

$$\alpha_r = \alpha_0 + h_r\beta_0 \tag{25}$$

[48] R. D. Brown, *Australian J. Sci. Res.*, **5**, 339 (1952).
[49] As one example in which different α's have been used, H. H. Jaffé, *J. Chem. Phys.*, **20**, 778 (1952), treated naphthalene with slightly different values of α for the α, β, and central carbons.

Since α and β are negative quantities, h_r is positive for an atom effectively more electronegative than the standard carbon. With this definition, the secular equations may be solved in a straightforward way.

Example. To calculate E_π for $CH_2{=}X$, in which $\beta_{CX} = \beta_0$ and $\alpha_X = \alpha_0 + \beta_0$, the secular determinant is

$$\begin{vmatrix} \alpha_0 - E & \beta_0 \\ \beta_0 & \alpha_0 + \beta_0 - E \end{vmatrix} \equiv \begin{vmatrix} x & 1 \\ 1 & x+1 \end{vmatrix} = x^2 + x - 1$$

$$x = \frac{-1 \pm \sqrt{5}}{2} = -1.618,\ +0.618$$

Hence $E_\pi - 2\alpha_0 + 3.236\beta_0$. Compare these values with those for ethylene.

Exercise. Calculate E for C—C—C—X and C—C—X—C with all β's equal and $h_X = 1$.

Ans.   C—C—C—X: $E = 4\alpha + 5.759\beta$
                        C—C—X—C: $E = 4\alpha + 5.666\beta$

The effect of a change in α on the total π-energy may be expressed as a power series:

$$\delta E = \frac{\partial E}{\partial \alpha_r} \delta\alpha_r + \frac{1}{2} \frac{\partial^2 E}{\partial \alpha_r^2} \overline{\delta\alpha_r^2} + \cdots \tag{26}$$

in which $\delta\alpha_r = \alpha_r - \alpha_0 = h_r\beta_0$ (27)

We may again speak of first and second approximations, depending on our use of one or two terms in the expansion.

First Approximation

In the HMO theory Coulson and Longuet-Higgins[39] have shown that

$$q_r = \frac{\partial E}{\partial \alpha_r} \tag{28}$$

Hence, to the first approximation,

$$\delta E \simeq q_r \delta\alpha_r = q_r h_r \tag{29}$$

This equation is especially useful, since q_r is readily evaluated for odd- or even-AH without solving secular equations (p. 54).

Example. To calculate E for C—C—C—X and C—C—X—C from butadiene to the first approximation with all β's equal and $h_X = 1$. Butadiene is an AH; hence $q = 1$ for all positions and $\delta E \simeq (1)(1) = 1\beta$, for both compounds. This approximation may be compared with the δE's calculated from the solution of the secular equations: C—C—C—X: $\delta E = 4\alpha + 5.759\beta - (4\alpha + 4.472\beta) = 1.287\beta$; C—C—X—C: $\delta E = 4\alpha + 5.666\beta - (4\alpha + 4.472\beta) = 1.194\beta$. $\delta\alpha = 1$ is a rather large change for accommodation by only the first term in a series expansion; we see that the estimated δE's are only roughly comparable to the actual δE's.

Second Approximation

The second derivative in (26) is called the atom polarizability, $\pi_{r,s}$, and is also a function of the coefficients and energies of MO's.[39]

$$\pi_{r,s} = \frac{\partial q_r}{\partial \alpha_s} = \frac{\partial^2 E}{\partial \alpha_r \, \partial \alpha_s} = \frac{\partial q_s}{\partial \alpha_r} = \pi_{s,r} \tag{30}$$

$$\pi_{r,s} = 4 \sum_{j=1}^{m} \sum_{k=m+1}^{n} \frac{c_{jr} c_{kr} c_{js} c_{ks}}{\epsilon_j - \epsilon_k} \tag{31}$$

In (31) the first summation is over the occupied orbitals; the second summation is over the unoccupied orbitals. For use of this equation no orbital can be singly occupied. Because of the correspondence between bonding and antibonding orbitals in AH's, (32) can be derived for even-AH's; set in this equation refers to the set of starred or unstarred atoms (p. 46).

$$\pi_{r,s} = \pm 4 \sum_{j=1}^{n/2} \sum_{k=1}^{n/2} \frac{c_{jr} c_{js} c_{kr} c_{ks}}{\epsilon_j + \epsilon_k} \tag{32}$$

$+$ if r and s belong to the same set
$-$ if r and s belong to opposite sets

Exercise. Calculate $\pi_{r,s}$ for butadiene.

Ans.	s	r:	1	2	3	4
	1		0.626	−0.402	0.045	−0.268
	2		−0.402	0.402	−0.045	0.045

To the second approximation, the series expansion may be written

$$\delta E \cong q_r \delta\alpha_r + \tfrac{1}{2}\pi_{r,r}\overline{\delta\alpha_r^2} \tag{33}$$

Example. To calculate δE to the second approximation for the compounds in the previous example,

C—C—C—X: $\delta E = 1 + \tfrac{1}{2}(0.626)(1) = 1.313\beta$
C—C—X—C: $\delta E = 1 + \tfrac{1}{2}(0.402)(1) = 1.201\beta$

These results differ by only 0.01–0.02β from the true values.

$\pi_{r,s}$ values for a number of compounds have been published.[41,42,43,50,51] Of these quantities, the *self-atom* polarizabilities, $\pi_{r,r}$ are the most important because of their use in correlating reactivities (Chap. 11). Furthermore, Coulson and Longuet-Higgins[51] have established a good linear correlation between p_{rs} of a bond in RS between atoms r and s and the self-atom polarizabilities of atom r in R and atom s in S of the separated systems:

$$p_{rs} \cong 0.10 + 0.69\sqrt{\pi_{rr}\pi_{ss}} \tag{34}$$

Example. To find p of the 1—α bond in styrene.

π_{rr} of benzene $= 0.398$; π_{ss} of ethylene $= 0.5$

$$\therefore p_{rs} \cong 0.10 + 0.69\ \sqrt{(0.5)(0.398)} = 0.408$$

The true value is 0.406 (p. 107).

An empirical correlation has also been established between conjugation energy, R_{rs}, and self-atom polarizabilities:[52]

$$R_{rs} \cong 0.811\sqrt{\pi_{rr}\pi_{ss}} + 0.062 \tag{35}$$

Example. To estimate the conjugation energy in styrene.

$$R_{rs} \cong 0.811\ \sqrt{(0.5)(0.398)} + 0.062 = 0.424$$

This result is identical with the true conjugation energy (p. 110).

Atom-Bond Polarizabilities

Bond polarizabilities are related to the differential change in bond order with change in β; atom polarizabilities are the differential change in electron density with change in α. The differential change in bond order with change in α and in electron density with change in β are given by the *atom-bond polarizabilities*, $\pi_{t,rs}$, or the *bond-atom polarizabilities*, $\pi_{rs,t}$, which are related by (36):[39]

$$\pi_{rs,t} = \frac{\partial p_{rs}}{\partial \alpha_t} = \frac{1}{2}\frac{\partial^2 E}{\partial \beta_{rs}\partial \alpha_t} = \frac{1}{2}\frac{\partial q_t}{\partial \beta_{rs}} = \frac{1}{2}\pi_{t,rs} \tag{36}$$

These quantities are also functions of the coefficients and energies of MO's:[39]

$$\pi_{t,rs} = 4\sum_{j=1}^{m}\sum_{k=m+1}^{n}\frac{c_{jt}c_{kt}(c_{jr}c_{ks} + c_{js}c_{kr})}{\epsilon_j - \epsilon_k} \tag{37}$$

Note that for AH's all atom-bond polarizabilities vanish.

[50] C. A. Coulson and H. C. Longuet-Higgins, *Proc. Roy. Soc.*, **A192**, 16 (1947); T. Nakajima, *J. Chem. Phys.*, **23**, 587 (1955); G. Bessis, S. S. Sung, and O. Chalvet, *Compt. rend.*, **248**, 1523 (1959).

[51] C. A. Coulson and H. C. Longuet-Higgins, *op. cit.*, **A195**, 188 (1948).

[52] R. D. Brown, *Australian J. Sci. Research*, **5**, 339 (1952).

Example. The 9,10 bond in azulene, III, is undoubtedly a "single" bond.

III

Estimate the change in DE and in electron density distribution on changing $\beta_{9,10}$ from β_0 to $0.9\beta_0$.

$$p_{9,10} = 0.401; \qquad \pi_{9,10-9,10} = 0.334$$

t:	2	3	4	5	6	10
q_t:	1.047	1.173	0.855	0.985	0.870	1.027
$\pi_{9,10-t}$:	0.0157	0.0753	−0.0574	−0.0056	−0.0529	0.0063

$$\delta E \cong 2p\delta\beta_{rs} + \pi_{rs,rs}\overline{\delta\beta^2} = 2(0.401)(-0.1) + (0.334)(-0.1)^2$$
$$= 0.077\beta$$

$$\delta q_t \cong \pi_{t,rs}\delta\beta_{rs} = 2\pi_{rs,t}\delta\beta_{rs}$$

t	δq_t	New q_t
2	−0.0031	1.044
3	−0.0151	1.158
4	0.0115	0.867
5	0.0011	0.986
6	0.0106	0.881
10	−0.0013	1.026

The charge distribution is evened out by the change in $\beta_{9,10}$. This change is in the expected direction, since a complete change to $\beta_{9,10} = 0$ converts the system to an even-AH which has uniform charge distribution.

Alternative methods of calculating polarizabilities have been given.[53] The changes in α and β may be described as perturbations of the original system. The first and second terms of the power-series expansions are equivalent to first- and second-order perturbations. Other results of perturbation theory have been exploited[54] and other techniques for calculating the perturbations are described.[55] McWeeny's[55a] method is of particular value because its use of the method of steepest descents gives

[53] I. Samuel, *Compt. rend.*, **241**, 1464 (1955); R. D. Brown, *J. Chem. Soc.*, **1956**, 767.
[54] See particularly the series by M. J. S. Dewar, *J. Am. Chem. Soc.*, **74**, 3341, 3345, 3350, 3353, 3355, 3357 (1952). Many of these results will be described in the sequel.
[55] (a) R. McWeeny, *Proc. Royal Soc.*, **A237**, 355 (1956); (b) R. D. Brown and I. M. Bassett, *Proc. Phys. Soc.*, **71**, 724 (1958).

close prediction of the effects of change of α and β without calculation of polarizabilities.

4.5   The ω-Technique

A simple argument shows that a carbon atom in a π-lattice for which $q \neq 1$ cannot really have $\alpha = \alpha_0$. If $q < 1$, the net positive charge means that the screening "seen" by any one electron is reduced and the Coulombic attraction to the nucleus is increased. α for such a carbon should have a more negative value. Conversely, a carbon with a net negative charge should have a less negative value of α. Wheland and Mann[56] proposed that the value of α should be linearly related to the charge. Their proposal may be formulated as (38):

$$\alpha_r = \alpha_0 + (1 - q_r)\omega\beta_0 \tag{38}$$

ω is a dimensionless parameter whose value may be so chosen as to give best agreement with experiment. With one empirical parameter, this *ω-technique* introduces some electron repulsion within the framework of the simple LCAO method.[57] The use of this method in the literature is still rather limited, but it has been shown to lead to improved calculated dipole moments for hydrocarbons,[56,58] and greatly improved energy values for organic cations.[57,59]

The values proposed for the empirical parameter ω have ranged from 0.33 to 1.8;[56,57,59,60] an intermediate value, $\omega = 1.4$, seems now to be soundly established.[57,59]

In principle, one first obtains the charge distribution according to the HMO method; this charge distribution is used to calculate new α_r's according to (38), and the new charge distribution is calculated. The process is continued until the charge distribution no longer changes, that is, until the system is self-consistent. In practice, one cycle of iteration often suffices for many purposes.[57,61]

[56] G. W. Wheland and D. E. Mann, *J. Chem. Phys.*, **17**, 264 (1949).

[57] A. Streitwieser, Jr., *J. Am. Chem. Soc.*, **82**, 4123 (1960).

[58] G. Berthier and A. Pullman, *Compt. rend.*, **229**, 761 (1949); A. Streitwieser, Jr., and P. M. Nair, unpublished calculations.

[59] N. Muller, L. W. Pickett, and R. S. Mulliken, *J. Am. Chem. Soc.*, **76**, 4770 (1954); N. Muller and R. S. Mulliken, *J. Am. Chem. Soc.*, **80**, 3489 (1958); A. Streitwieser, Jr., and P. M. Nair, *Tetrahedron*, **5**, 149 (1959). A similar technique has been used in a calculation of N_2O_4; cf. O. Chalvet and R. Daudel, *J. chim. phys.*, **49**, 79 (1952).

[60] A. Laforgue, *Compt. rend.*, **228**, 1430 (1949); *J. chim. phys.*, **46**, 568 (1949); J. G. Burr, Jr., *J. Chem. Phys.*, **26**, 431 (1957); Y. Kurita and M. Kubo, *Bull. Chem. Soc., Japan*, **24**, 13 (1951); W. -C. Lin, *J. Chinese Chem. Soc.*, II, **2**, 37 (1955); E. Sparatore, *Atti accad. ligure sci. e lettere, Genoa*, **13**, 1 (1957).

[61] A. Streitwieser, Jr., and P. M. Nair, *Tetrahedron*, **5**, 149 (1959).

Exercise. Apply the ω-technique with $\omega = 1.4$ to allyl cation and determine the electron densities for each of several iterations.

Ans.

No. of Iterations	q_1	q_2
0	0.500	1.000
1	0.621	0.757
2	0.534	0.934
3	0.597	0.806
4	0.552	0.896
5	0.584	0.830
6	0.560	0.880
∞	0.571	0.858

An alternative method has been suggested by Pritchard and Sumner,[62] who propose a parabolic rather than a linear relation between α_r and q_r. This method is probably more accurate, but it is less convenient and requires in effect an additional parameter.

[62] H. O. Pritchard and F. H. Sumner, *Proc. Roy. Soc.*, **235**, 136 (1956).

5 Heteroatoms

5.1 Introduction

Formaldehyde, pyridine, and chlorobenzene are examples of compounds with π-systems which contain *heteroatoms*, X, atoms other than carbon. Heteroatoms may be treated in the HMO method by appropriate changes in the empirical α and β parameters associated with each atom and bond. As discussed in Chap. 4, these changes are incorporated in units of the standard α_0 and β_0, usually of benzene, by use of the definitions

$$\alpha_X = \alpha_0 + h_X\beta_0 \tag{1}$$

$$\beta_{CX} = k_{CX}\beta_0 \tag{2}$$

Calculations involving such changes can be made either by solving the secular equations or with the approximation methods discussed in Chap. 4. The problem arises of the specific values to be used for the h- and k-parameters of various heteroatoms. The ideal procedure to evaluate these parameters would start with a correlation between some experimental property and a calculated quantity established for hydrocarbon systems and would follow with an application of this correlation to a number of compounds containing one or more of the heteroatoms under test with a systematic variation of h_X and k_{CX}. Such a procedure would not only determine the best values of these parameters to use with the correlation but would also reveal whether unique values exist; however, cf. Sec. 5.6. Similar applications of other correlations would then determine the constancy of the parameter values in various applications. Unfortunately, this ideal procedure has never been followed *in toto* and has only rarely even been used in part. More often, parameter values have been assigned as guesses partly based on theory or on the basis of applications of assumed validity. The result has been a profusion of parameter values in the literature.

Part of the problem is that there are several variations of MO theory and each will yield somewhat different parameter values with a given application. In particular, neglect or non-neglect of overlap can make an appreciable difference. Without overlap, electron flow is often exaggerated[1] and different parameters are required to achieve the same electron density distribution, for example. The use of the ω-technique leads to parameter values that differ from those from simple LCAO theories, especially in cations. The difference is particularly marked in h_X values, presumably because these are compared in this technique with carbons that have enhanced α values due to their extra positive charge.[2] Another complicating factor is the so-called *auxiliary inductive parameter*[3] (AIP) in which positive h-values are assigned to one or more carbons close to the heteroatom to take account of inductive effects (Sec. 5.6).

Some general principles seem to be valid. In the first HMO treatment of heteroatom compounds Wheland and Pauling[4] pointed out that although the magnitude of h_X may be in doubt its sign is usually known with confidence. The effective nuclear charge of the atom is important in determining the magnitude of α. Along one row of the periodic table the increase in nuclear charge is not completely screened by the additional electrons; the increasing core potential means greater h_X. We expect $h_B < h_C < h_N < h_O < h_F$. Down one column of the periodic table the valence electron is more effectively shielded by inner shells; hence h_X should decrease in a like manner. For example, we expect $h_S < h_O$, $h_{Cl} < h_F$. Empirical electronegativities vary in the same way, and it has been suggested both that α_X be proportional to empirical electronegativities[5] and that h_X be proportional to electronegativity differences.[6,7] Some of Pauling's[8] electronegativity values are given in Table 5.1. The proportionality constant has frequently been taken as unity leading to

$$h_X = \chi_X - \chi_C \tag{3}$$

This simple prescription seems to be reasonably good in some cases but account must be taken of another important principle.

We should distinguish, for example, the nitrogen in pyridine from that in pyrrole, the oxygen in formaldehyde from that in phenol. In one case the heteroatom contributes one electron to the π-system; in the other it

[1] J. H. Goldstein, *J. Chem. Phys.*, **24**, 507 (1956).

[2] A. Streitwieser, Jr., *J. Am. Chem. Soc.*, **82**, 4123 (1960).

[3] R. D. Brown, *Quart. Revs.*, **6**, 63 (1952).

[4] G. W. Wheland and L. Pauling, *J. Am. Chem. Soc.*, **57**, 2086 (1935).

[5] C. Sandorfy, *Bull. soc. chim. France*, **1949**, 615.

[6] E. Gyoerffy, *Compt. rend.*, **232**, 515 (1951).

[7] C. A. Coulson, *Valence*, Oxford University Press, London (1952), p. 242; C. A. Coulson and J. de Heer, *J. Chem. Soc.*, **1952**, 483.

[8] L. Pauling, *The Nature of the Chemical Bond*, third edition, Cornell University Press, Ithaca, N.Y. (1960), p. 90; see also ref. 7, p. 134.

contributes two. Alternatively, in the one case the heteroatom contributes an effective nuclear charge of $+1$ to the core potential; in the other the contributed nuclear charge is $+2$. Clearly, α for a heteroatom that contributes two electrons should be considerably more negative than α for the same heteroatom in a π-system to which it contributes one electron;

TABLE 5.1

PAULING ELECTRONEGATIVITY VALUES, χ

B	C	N	O	F
2.0	2.5	3.0	3.5	4.0
	Si	P	S	Cl
	1.8	2.1	2.5	3.0
			Se	Br
			2.4	2.8
				I
				2.5

that is, $h_{\ddot{X}} > h_{\dot{X}}$. This important distinction has frequently been ignored in the literature. We might expect the difference $h_{\ddot{X}} - h_{X}$ to have about the same magnitude as ω; that is, about 1 to 1.5.[2]

k_{CX}-values are surely dependent on the C—X bond distance in various compounds as in analogous carbon cases. Because of the relative constancy of bond distances of given bond types, it seems appropriate to differentiate explicitly between C—X single bonds, double bonds, and "aromatic" bonds. For nitrogen, examples of the corresponding bonds would be those in pyrrole, an azomethine ($R_2C{=}NR$) and pyridine, respectively. The corresponding k-values could be symbolized as k_{C-N}, $k_{C=N}$, and k_{CN}, respectively.

Sandorfy[5] has suggested that k-values are a simple function of bond distance, but this method has little theoretical foundation. Wheland[9] has suggested that the assumption of the proportionality of β and S be extended to heteroatom bonds as well. Values derived with this assumption are listed in Table 5.2. The increase in β_{C-X} from fluorine to chlorine is not expected on the basis of chemical experience; it has been suggested that the assumption of proportionality to S is not appropriate with a change in principal quantum number.[2]

A common assumption has been that each k equals unity; h-values derived with this assumption will have to be changed if different k-values are used.

[9] G. W. Wheland, *J. Am. Chem. Soc.*, **64**, 900 (1942).

We shall review a number of the methods used to derive parameter values for oxygen and nitrogen from experimental quantities. Two types of application may be distinguished: those which involve the wave functions and those which use the π-energy. In any approximate quantum mechanical method the eigenvectors are always less accurate than the

TABLE 5.2

k-VALUES FROM PROPORTIONALITY TO OVERLAP INTEGRALS

Bond	Bond Distance, A	$S/S_0 = k$
C=N	1.31	0.92
C⋯N	1.36	0.84
C—N	1.42	0.75
C=O	1.21	0.89
C⋯O	1.29	0.76
C—O	1.36	0.65
C—F	1.30	0.59
C—Cl	1.70	0.65
N=N	1.23	0.87

eigenvalues. A variation in basis function generally varies the eigenvectors and the eigenvalues to different degrees. Consequently, parameter values derived from eigenvectors (charge densities, bond orders, etc.) may not apply to applications involving eigenvalues (resonance energies, activation energies, ionization potentials, etc.). In practice, however, the variations in derived values are at least as great within one type as they are between the two types of application.

5.2 Nitrogen and Oxygen

An early procedure for determining $k_{C=X}$ is based on an assumed proportionality to the difference between single- and double-bond energies; that is,[10,11]

$$k_{C=X} \cong \frac{E_{C=X} - E_{C-X}}{E_{C=C} - E_{C-C}} \qquad (4)$$

The derived values are sensitive to the way the bond energies are defined. The traditional definition has recently been criticized (see Chap. 9). Some

[10] J. E. Lennard-Jones, *Proc. Roy. Soc.*, **A158,** 280 (1937).
[11] C. A. Coulson, *Trans. Faraday Soc.*, **42,** 106 (1946).

derived k-values which probably should not be weighted highly are $k_{C-N} = 0.3-0.5$,[12] $k_{CN} = 1-1.2$,[13,14] and $k_{C-O} = 0.1-0.3$,[12b] $k_{C=O} = 1.4-2$.[11,14,15]

Resonance energies (Chap. 9) are rather insensitive functions of the parameter values and also depend on the way bond energies are defined. However, from such energies several authors have derived the following sets of values: for $k_{CN} = k_{CO} = 1$ (assumed), $h_N = 2$, $h_O = 4$;[4,16] for $h_{\ddot{N}} = h_{\ddot{N}} = 0$ (assumed), $k_{CN} = 0.4-0.55$.[17,18]

A popular technique has been to find parameter values that would reproduce experimental dipole moments (Sec. 6.1). The charge density distribution is sensitive to variations in Coulomb integrals, but, at least for systems derived from AH, it is not sensitive to the assumed β-values because the atom-bond polarizabilities for AH are zero. This method gives rather consistent values of $h_{\ddot{N}} \cong 0.5$, $h_{\ddot{O}} \cong 1$.[12a,15,19] Orgel et al.[14] deliberately sought h-values that would give π-moments too high by a factor of 1.6. Their parameter values, consequently, are somewhat larger than those above: $h_{\ddot{N}} = 1$, $h_{\ddot{O}} = 2$, $h_{\ddot{N}} = 2$, $h_{\ddot{O}} = 3.2$.

An interesting recent variation of this approach has been to reproduce the electron-density distribution given by advanced MO calculations of the SCF, ASMO, or Pariser-Parr type (Chap. 16). Values reported using this method are $h_{\ddot{N}} = 0.1-0.35$, $k_{CN} = 0.9-1.1$,[20,21] $h_{\ddot{N}} = 2$,[21,22] $h_{\ddot{O}} = 0.15$,

[12] (a) S. S. Perez, M. A. Herraez, F. J. Igea, and J. Esteve, *Anales real soc. españ. fis. y quim.* (*Madrid*), **51B**, 91 (1955); (b) F. L. Pilar and J. R. Morris, II, *J. Chem. Phys.*, **34**, 389 (1961).

[13] H. C. Longuet-Higgins and C. A. Coulson, *Trans. Faraday Soc.*, **43**, 87 (1947).

[14] L. E. Orgel, T. L. Cottrell, W. Dick, and L. F. Sutton, *Trans. Faraday Soc.*, **47**, 113 (1951).

[15] G. Battista, B. Scrocco, and E. Scrocco, *Atti accad. nazl. Lincei.*, *Rend, Classe sci. fis., mat. e nat.*, **8**, 183 (1950); *C.A.*, **44**, 7147 (1950).

[16] J. G. M. Bremner and W. C. G. Bremner, *J. Chem. Soc.*, **1950**, 2335.

[17] M. J. S. Dewar, *Trans. Faraday Soc.*, **42**, 764 (1946).

[18] S. S. Perez, M. A. Herraez, and F. J. Igea, *Anales real soc. españ. fis. y quim.* (*Madrid*), **50B**, 243 (1954).

[19] M. J. S. Dewar, *J. Chem. Soc.*, **1950**, 2329; J. I. F. Alonso, *Compt. rend.*, **233**, 403 (1951); P. Löwdin, *J. Chem. Phys.*, **19**, 1323 (1951); F. Gerson and E. Heilbronner, *Helv. Chim. Acta*, **41**, 2332 (1958); R. A. Barnes, *J. Am. Chem. Soc.*, **81**, 1935 (1959). See also ref. 18; Y. Kurita and M. Kubo, *Bull. Chem. Soc. Japan*, **24**, 13 (1951); S. Odiot and M. Roux, *J. chim. phys.*, **50**, 141 (1953). Calculations of the toluidines using an ω-technique are reported by J. G. Burr, Jr., *J. Chem. Phys.*, **26**, 431 (1957). Neglect of the distinction between σ- and π-moments (Sec. 6.1) gives different parameter values; cf. J. Ploquin, *Compt. rend.*, **226**, 245 (1948).

[20] R. McWeeny and T. E. Peacock, *Proc. Phys. Soc.*, **70A**, 41 (1957).

[21] (a) R. D. Brown and A. Penfold, *Trans. Faraday Soc.*, **53**, 397 (1957); (b) R. D. Brown and M. L. Heffernan, *Australian J. Chem.*, **10**, 211 (1957); **12**, 319 (1959).

[22] Compare with the completely different treatments of M. J. S. Dewar and L. Paoloni, *Trans. Faraday Soc.*, **53**, 261 (1957) and H. Kobayashi, *J. Chem. Phys.*, **30**, 1373 (1959).

and $k_{C=O} = 1.1.$[23] Spin density distributions of semiquinone anions as given by ESR spectra (Sec. 6.4) have recently been used to derive $k_{C-O} = 1.56$, $h_{\ddot{O}} = 1.2.$[23a]

In a rather clever approach, Jaffé[24] assumed a proportionality between charge densities and Hammett's σ-constants. The σ-constants for heteroatom substituents were then used to derive corresponding h- and k-values. This approach assumes the existence of an inductive effect of a particular form and may be criticized (Sec. 5.6). Furthermore, most of the derived values differ substantially from other derivations and apparently are not suitable for other applications.

Ultraviolet spectral correlations have been important in testing advanced MO techniques but have been used rather little with HMO methods (Chap. 8). However, recent comparisons of spectra of nitrogen heterocycles have led to $h_{\ddot{N}} \cong 1$ and 0.5 and $k_{CN} \cong 1.$[25] The energies of the lowest unoccupied and highest occupied MO's have been correlated with polarographic reduction potentials and with ionization potentials, respectively (Chap. 7). The experimental potentials of heteroatom compounds could be compared with the effect of the corresponding h- and k-parameter values in the energies of the molecular orbitals. Several such correlations have been reported but usually without a systematic study of the effect of a range of parameter values. The reduction potentials of aromatic aldehydes and ketones give good correlations with $h_{\ddot{O}} = 2$, $k_{C=O} = \sqrt{2}.$[26,27] However, the fit of the correlation is apparently not very sensitive to the particular choice of parameters used (cf. Sec. 7.1).[27] Some vinyl derivatives also give a good correlation,[28] and parameter values have been derived from stilbene derivatives by using a perturbation technique.[29] Correlations of the redox potentials of quinones-hydroquinones and of phenols are not sensitive to the parameter values chosen for oxygen.[30]

[23] R. D. Brown and M. L. Heffernan, *Trans. Faraday Soc.*, **54**, 757 (1958).

[23a] G. Vincow and G. K. Fraenkel, *J. Chem. Phys.*, **34**, 1333 (1961).

[24] H. H. Jaffé, *J. Chem. Phys.*, **20**, 279, 1554 (1952).

[25] J. N. Murrell, *Mol. Phys.*, **1**, 384 (1958); N. Mataga, *Bull. Chem. Soc. Japan*, **31**, 463 (1958). Using an approximation technique and additional assumptions, D. Peters, *J. Chem. Soc.*, **1957**, 1993, has derived h- and k-values for a number of heteroatoms. The k-values are generally smaller than those of other derivations.

[26] R. W. Schmid and E. Heilbronner, *Helv. Chim. Acta*, **37**, 1453 (1954); J. N. Chaudhuri and S. Basu, *Nature*, **182**, 179 (1958).

[27] T. Fueno, K. Morukuma, and J. Furukawa, *Bull. Inst. Chem. Research, Kyoto Univ.*, **36**, 96 (1958).

[28] T. Fueno, K. Asada, K. Morokuma, and J. Furukawa, *Bull. Chem. Soc. Japan*, **32**, 1003 (1959).

[29] T. Fueno, K. Morokuma, and J. Furukawa, *Nippon Kagaku Zasshi*, **79**, 116 (1957); T. Fueno, K. Morokuma, and J. Furukawa, *Bull. Inst. Chem. Research, Kyoto Univ.*, **36**, 87 (1958).

Ionization potentials of water and ammonia together with UV data have been used to derive $h_{\ddot{O}} = 1.5{-}1.8$, $h_{\ddot{N}} = 0.9{-}1.2$.[31,32] When the ω-technique is used, larger h-values are derived.[2]

Relative reactivities are an obvious source of derived parameter values. Nevertheless, studies in this area are surprisingly limited. The orientation effects of substituents in aromatic substitution limit the range of possible parameter values but do not define the values with any precision (Sec. 11.7).[4,5,9,33,34] From radical reactions of pyridine, Brown derived $h_{\ddot{N}} = 0.5$, $k_{CN} = 1$.[35] These values and $h_{\ddot{N}} = 2$ have been shown to be at least consistent with substitution reactions on heterocyclic rings,[36] although radical reactions of heterocycles seem to be relatively insensitive to the particular parameter values employed.[37]

The acidity of a number of heterocyclic phenols has been used to derive $h_{\ddot{N}} \simeq 0.6$.[38]

A survey of the various derived values for nitrogen shows a good clustering about the values $h_{\ddot{N}} = 0.5$, $h_{\ddot{N}} = 1.5$. For the bond integrals reasonable values appear to be $k_{C-N} = 0.8$, $k_{CN} = 1$. Less consistency is shown for the oxygen parameters. The frequently used values for carbonyl oxygen, $h_{\ddot{O}} = 2$, $k_{C=O} = \sqrt{2}$, appear to be somewhat high. Values that should be suitable for approximate work would seem better to be $h_{\ddot{O}} = 1$, $h_{\ddot{O}} = 2$, $k_{C-O} = 0.8$, $k_{C=O} = 1$.[39,40] Note that we have

[30] M. Diatkina and J. Syrkin, *Acta Physicochim. U.R.S.S.*, **21**, 921 (1946); M. G. Evans, *Trans. Faraday Soc.*, **42**, 113 (1946); M. G. Evans, J. Gergely, and J. de Heer, **45**, 312 (1949), V. Gold, **46**, 109 (1950), C. A. Coulson and J. de Heer, **47**, 681 (1951), M. G. Evans and J. de Heer, **47**, 801 (1951); N. S. Hush, *J. Chem. Soc.*, **1953**, 2375. The oxidation potentials of phenols have recently been used to derive $h_{\ddot{O}} \simeq 1$ (T. Fueno, T. Lee, and H. Eyring, *J. Phys. Chem.*, **63**, 1940 (1959); cf. Sec. 13.1.

[31] F. A. Matsen, *J. Am. Chem. Soc.*, **72**, 5243 (1950).

[32] D. P. Stevenson, personal communication.

[33] M. J. S. Dewar, *J. Chem. Soc.*, **1949**, 463; S. L. Matlow and G. W. Wheland, *J. Am. Chem. Soc.*, **77**, 3653 (1955).

[34] S. Basu and J. N. Chaudhuri, *Proc. Natl. Inst. Sci. India*, **24A**, 130 (1958).

[35] R. D. Brown, *J. Chem. Soc.*, **1956**, 272.

[36] I. M. Bassett and R. D. Brown, *J. Chem. Soc.*, **1954**, 2701; R. D. Brown and M. L. Heffernan, *J. Chem. Soc.*, **1956**, 4288; R. D. Brown, *Australian J. Chem.*, **8**, 100 (1955), R. D. Brown and M. L. Heffernan, **9**, 83 (1956), R. D. Brown and B. A. W. Coller, **12**, 152 (1959); R. D. Brown and R. D. Harcourt, *J. Chem. Soc.*, **1959**, 3451. See also ref. 4.

[37] B. Pullman and J. Effinger, *International Colloquium on Calculation of Wave Functions*, C.N.R.S., Paris (1958), p. 351.

[38] S. F. Mason, *J. Chem. Soc.*, **1958**, 674.

[39] These assignments are similar to those used by S. Nagakura and A. Kuboyama, *J. Am. Chem. Soc.*, **76**, 1003 (1954) and A. Kuboyama, *Bull. Chem. Soc. Japan*, **31**, 752 (1958).

[40] The parameter values derived by different authors were usually employed in more or less extensive calculations of heteroatom-containing systems. Various of these sets of

neglected the additional complication that the unshared electron pair on the nitrogen of aniline, for example, is not pure p; that is, the amino group is bent somewhat from the plane of the ring. We would not then expect the same parameter values to hold for aniline as for pyrrole.[8,41] For many applications, however, it is unlikely that this difference will be an important factor.

---

values were also adopted in subsequent calculations. Additional examples are the various treatments of aromatic amines [C. A. Coulson and J. Jacobs, *J. Chem. Soc.*, **1949**, 1983; J. D. Roberts and D. A. Semenow, *J. Am. Chem. Soc.*, **77**, 3152 (1955); G. Tarrago and B. Pullman, *J. chim. phys.*, **55**, 782 (1958)], cyclazines [R. J. Windgassen, Jr., W. H. Saunders, Jr., and V. Boekelheide, *J. Am. Chem. Soc.*, **81**, 1459 (1959); R. D. Brown and B. A. W. Coller, *Mol. Phys.*, **2**, 158 (1959)], nitrogen heterocycles [J. Ploquin, *Compt. rend.*, **236**, 245 (1948); G. Berthier and B. Pullman, *Compt. rend.*, **236**, 1725 (1948); H. C. Longuet-Higgins and C. A. Coulson, *J. Chem. Soc.*, **1949**, 971; O. Chalvet and C. Sandorfy, *Compt. rend.*, **228**, 566 (1949); C. Sandorfy and P. Yvan, *Compt. rend.*, **229**, 715 (1949); *Bull. soc. chim. France*, **1950**, 131; H. H. Jaffé, *J. Chem. Phys.*, **20**, 1554 (1952); P. T. Narasimhan, *J. Indian Inst. Sci.*, **35A**, 281 (1953); I. M. Bassett and R. D. Brown, *J. Chem. Soc.*, **1954**, 2701; H. Kon, *Sci. Repts.*, *Tôhoku Univ.*, *First Ser.*, **38**, 67 (1954); S. Basu, *Proc. Natl. Inst. Sci. India*, **21A**, 173 (1955); R. D. Brown, *Australian J. Chem.*, **8**, 100 (1955); R. D. Brown and M. L. Heffernan, *Australian J. Chem.*, **9**, 83 (1956); T. H. Goodwin and A. L. Porte, *J. Chem. Soc.*, **1956**, 3595; R. D. Brown and M. L. Heffernan, *J. Chem. Soc.*, **1956**, 3683, 4288; A. M. Liquori and A. Vaciago, *Gazz. chim. ital.*, **86**, 769 (1956); G. R. Seely, *J. Chem. Phys.*, **27**, 125 (1957); J. I. F. Alonso, R. Domingo, L. C. Vila, and F. Peradijordi, *Anales real soc. españ. fis. y quim.* (*Madrid*), **53B**, 109 (1957); R. McWeeny, *Proc. Phys. Soc.* (*London*), **70A**, 593 (1957); A. Pullman and B. Pullman, *Bull. soc. chim. France*, **1958**, 766; B. Pullman and A. Pullman, *ibid.*, 973; T. Nakajima and B. Pullman, *Bull. soc. chim. France*, **1958**, 1502; B. M. Lynch, R. K. Robins, and C. C. Cheng, *J. Chem. Soc.*, **1958**, 2973; J. I. F. Alonso, R. Domingo, and L. C. Vila, *Rec. trav. chim.*, **78**, 215 (1959); G. Del Re, *Tetrahedron*, **10**, 81 (1960)], azo compounds [O. Chalvet, *Compt. rend.*, **239**, 1135 (1954); J. I. F. Alonso and L. C. Vila, *Anales real soc. españ. fis. y quim.* (*Madrid*), **72B**, 617 (1956)], carbonyl compounds [E. Scrocco and P. Chiorboli, *Atti accad. nazl. Lincei.*, *Rend classe sci. fis.*, *mat. e nat.*, **8**, 248 (1950); M. G. Evans and J. de Heer, *Trans. Faraday Soc.*, **47**, 801 (1951); R. D. Brown, *J. Chem. Soc.*, **1951**, 2670; Y. Kurita and M. Kubo, *Bull. Chem. Soc. Japan*, **24**, 13 (1954); G. Berthier, B. Pullman, and J. Pontis, *J. chim. phys.*, **49**, 367 (1952); I. Estelles, J. I. F. Alonso, and J. Mira, *Anales real soc. españ. fis. y quim.* (*Madrid*), **50B**, 11 (1954); I. Estelles and J. I. F. Alonso, *Anales real soc. españ. fis. y quim.* (*Madrid*), **50B**, 151 (1954); M. L. Josien and J. Deschamps, *J. chim. phys.*, **52**, 213 (1955); T. H. Goodwin, *J. Chem. Soc.*, **1955**, 1689; I. Samuel, *Compt. rend.*, **240**, 2534 (1955); R. Blinc and E. Pirkmajer, "*J. Stefan*" *Inst. Repts.* (*Ljubljana*), **4**, 133 (1957)], and others [A. Y. Namiot and M. E. Dyatkina, *Compt. rend. acad. sci. U.R.S.S.*, **53**, 809 (1946); M. E. Dyatkina, *Doklady Akad. Nauk S.S.S.R.*, **58**, 1069 (1947); H. Kon, *Sci. Repts.*, *Tôhoku Univ.*, **36**, 203 (1952); C. M. Moser, *J. Chem. Soc.*, **1953**, 1073; O. Chalvet and M. Roux, *Compt. rend.*, **237**, 1521 (1953); T. H. Goodwin, *J. Chem. Soc.*, **1955**, 4451; S. S. Perez and F. J. Igea, *Anales real soc. españ. fis. y quim.* (*Madrid*), **52B**, 509 (1956); O. Polansky, *Monatsh. Chem.*, **88**, 670 (1957); N. L. Allinger and G. A. Youngdale, *Tetrahedron Letters*, **9**, 10 (1959); E. L. Wagner, *J. Phys. Chem.*, **63**, 1403 (1959).

[41] J. J. Elliott and S. F. Mason, *J. Chem. Soc.*, **1959**, 2352.

The difference in core potentials between $\ddot{X}$ and $\dot{X}$ compares to the corresponding difference between a carbonium ion and a neutral carbon and suggests that $h_{\ddot{X}} - h_{\dot{X}} \cong \omega$.[2] With $\omega = 1.4$, this approximation is roughly consistent with the parameter values assigned above. From the acid-base equilibria of heterocyclic amines and phenols, an approximate value of the difference in Coulomb integral between a positively charged and neutral nitrogen can be estimated to be $\Delta h \cong 1.2$–1.9.[38,41,42] Advanced SCF calculations on methylenimonium and pyridinium ions are reproduced with $h_{N^+} = 1.4$–2.4.[43] A reasonable average value for h_{N^+} would seem to be about 2.[44] This value also accounts for the orientation effects of nitration of nitrogen heterocycles in acid media.[45]

The N—O bond is involved in amine oxides and in the nitro group. In both groups the nitrogen has a formal charge of $+1$; hence h_{N^+} of about 2 would seem appropriate. Indeed, Jaffé's method based on the $\sigma\rho$-correlation yields this value for amine oxides.[46] The derived value of $h_O = 1$ also seems reasonable; $k_{N-O} = 0.6$ has been derived from the dipole moment of furazane,[14] but from a study of the substitution orientation of pyridine oxide Barnes[47] concludes that $k_{N-O} \cong 0.75$. A value of $k_{N-O} \cong 0.7$ seems reasonable by analogy with k_{C-N} and k_{C-O}. An assumed proportionality with overlap integrals yields $k_{N-O} = 0.69$.[48,49]

5.3   Halogens

Of the halogens, chlorine has been the most thoroughly studied from a MO viewpoint. A simple model for chlorine is that in which the atom contributes one p_z-like orbital and two electrons to the π-edifice. Chlorine is then treated as a heteroatom similar to nitrogen and oxygen with

[42] H. C. Longuet-Higgins, *J. Chem. Phys.*, **18**, 275 (1950).

[43] (a) R. D. Brown and A. Penfold, *Trans. Faraday Soc.*, **53**, 397 (1957); (b) S. Mataga and N. Mataga, *Z. physik. Chem. NF*, **19**, 231 (1959).

[44] The values $h_{\ddot{N}} = 0.5$ and $h_{N^+} = 2.0$ were also judged to be appropriate by R. D. Brown and M. L. Heffernan, *Australian J. Chem.*, **9**, 83 (1956) and R. D. Brown and R. D. Harcourt, *J. Chem. Soc.*, **1959**, 3451.

[45] M. J. S. Dewar and P. M. Maitlis, *J. Chem. Soc.*, **1957**, 2518.

[46] H. H. Jaffé, *J. Am. Chem. Soc.*, **76**, 3527 (1954).

[47] R. A. Barnes, *ibid.*, **81**, 1935 (1959).

[48] For N—O bond length of 1.24 A, the value for a nitro group.

[49] For other treatments of the nitro group see J. I. F. Alonso and R. Domingo, *Anales real soc. españ. fis. y quim.* (*Madrid*), **51B**, 321 (1955); M. Simonetta and A. Vaciago, *Nuovo cimento*, **11**, 596 (1954); J. I. F. Alonso, *Compt. rend.*, **233**, 403 (1951); S. L. Matlow and G. W. Wheland, *J. Am. Chem. Soc.*, **77**, 3653 (1955); G. Favini and S. Carra, *Gazz. chim. ital.*, **85**, 1029 (1955); J. Trotter, *Can. J. Chem.*, **37**, 905 (1959). Jaffé's treatment of amine oxides has been extended to other cases by T. Kubota, *Nippon Kagaku Zasshi*, **80**, 578 (1959); *Yakugaku Zasshi*, **79**, 388 (1959).

assignments of appropriate h_{Cl}- and k_{C-Cl}-values. This model, which appears to be reasonably successful, ignores any additional contribution of $3d$-orbitals to the π-system (compare sulfur, Sec. 5.4).

The difference in electronegativities between carbon and chlorine has been used to assign h_{Cl};[50,51] however, since the chlorine core potential in the π-system is $+2$, this procedure will give an h-value too low on the basis of our previous discussion. From valence state and molecular ionization potentials Howe and Goldstein derive (including overlap) $h_{Cl} = 0.8$, $k_{C-Cl} = 0.56$.[52] However, orientation effects in substitution in chlorobenzene require a h_{Cl} of about 2.[9] Chlorobenzene has given $h_{Cl} = 2.5$ from spectral data[53] and $h_{Cl} = 1.0$–1.6, $k_{C-Cl} = 0.5$–0.9 from the ionization potential.[54] The ionization potential of HCl has been used to derive $h_{Cl} = 1.6$–1.9[31,32] and $h_{Cl} = 2.8$.[2] The latter value, obtained by the ω-technique, gives $k_{C-Cl} = 0.37$ when applied to the ionization potentials of several organic compounds. This k-value is close to the value $\frac{1}{3}$, derived by Bersohn.[50] For most applications $h_{Cl} = 2$ and $k_{C-Cl} = 0.4$ seem to be reasonable approximations.

Much less has been done with the other halogens. The electronegativity suggests a value for fluorine of about $h_F = 3$. This value also results from Matsen's method[31] with the ionization potential of HF. From fluorobenzene, I'Haya[54] derives $h_F = 1.5$–2.1, $k_{C-F} = 0.5$–0.7. A value of $k_{C-F} = 0.6$–0.7 would seem reasonable on the basis of overlap integrals (Table 5.2) and by analogy with k_{C-N} and k_{C-O}.

In a similar manner, values derived for bromine are $h_{Br} = 0.9$–2.4;[53,54] k_{C-Br} is undoubtedly less than k_{C-Cl}; a value of about 0.3 is probably reasonable. Howe and Goldstein's treatment of vinyl bromide and bromobenzene included overlap; parameter values were obtained in part from quadrupole coupling data and are difficult to formulate in our terminology.[55]

5.4　Sulfur

Sulfur has been treated as a "normal" heteroatom analogous to oxygen, nitrogen, and chlorine with $h_S \cong 0$ and $k_{C-S} \cong 1$.[56] In this model sulfur

[50] R. Bersohn, *J. Chem. Phys.*, **22**, 2078 (1954).

[51] T. Anno and A. Sado, *Bull. Chem. Soc. Japan*, **28**, 350 (1955).

[52] J. H. Goldstein, *J. Chem. Phys.*, **24**, 507 (1956); J. A. Howe and J. H. Goldstein, *J. Chem. Phys.*, **26**, 7 (1957). See also M. Simonetta, G. Favini, and S. Carra, *Mol. Phys.*, **1**, 181 (1958).

[53] W. W. Robertson and F. A. Matsen, *J. Am. Chem. Soc.*, **72**, 5252 (1950). This method assumed $k_{C-X} = 1$. A smaller k_{C-X}-value would result in a lower h_X.

[54] Y. I'Haya, *J. Am. Chem. Soc.*, **81**, 6120 (1959).

[55] J. A. Howe and J. H. Goldstein, *J. Chem. Phys.*, **27**, 831 (1957).

[56] M. M. Kreevoy, *J. Am. Chem. Soc.*, **80**, 5543 (1958).

is equivalent to a carbanion. This technique ignores the core potential of $+2$ contributed by sulfur in compounds of the type of thiophene and thiophenol. Actually, the close similarity of ionization potentials of ammonia and of hydrogen sulfide (10.52 and 10.47, respectively)[57] suggests that $h_{\ddot{S}} \cong h_{\ddot{N}}$.

In one of the earliest treatments of organic sulfur compounds Longuet-Higgins[58] specifically considered the contribution of $3d$-orbitals and showed that a particular hybridization gives two pd^2-hybrids, not mutually orthogonal, having symmetry and energy appropriate for conjugation with carbon $2p_z$-orbitals. In calculations with such a model, note that a LCAO method does not require the AO's to be centered on separate atoms. If the two sulfur orbitals are symbolized by S' and S'', we now require $h_{S'}$, $h_{S''}$, $k_{C-S'}$, $k_{C-S''}$, and $k_{S'-S''}$. From arguments based on ionization potentials and resonance energies, Longuet-Higgins[58] derived $h_{S'} = h_{S''} \cong 0$, $k_{C-S'} = k_{C-S''} \cong 0.8$, $k_{S'-S''} \cong 1$. In this way, a calculation of thiophene is equivalent to that of benzene, except that $\beta_{1,2}$ and $\beta_{5,6}$ are reduced by 20%. This model has been used in calculations of several sulfur heterocycles.[59]

5.5   Other Elements

From a study of the reactivity of ethyleneboronic acid, the quantities $h_B \cong -1$, $k_{C-B} \cong 0.7$ have resulted.[60] The sign of h_B is certainly in the expected direction from the relative electronegativity of boron, and the magnitudes of the parameters are not unreasonable, judging from other first-row elements. Note that in the normal case a neutral boron contributes a $2p_z$-orbital but no electrons to the π-bond structure.

The treatment of atoms further down the periodic table involves potential complications by possible contributions of orbitals other than p

[57] Summarized by F. H. Field and J. L. Franklin, *Electron Impact Phenomena*, Academic Press, New York (1951), appendix.

[58] H. C. Longuet-Higgins, *Trans. Faraday Soc.*, **45**, 173 (1949).

[59] M. G. Evans and J. de Heer, *Acta Cryst.*, **2**, 363 (1949); J. de Heer, *J. Am. Chem. Soc.*, **76**, 4802 (1954); J. C. Patel, *J. Sci. Ind. Research*, **16B**, 370 (1957); L. Melander, *Arkiv Kemi*, **11**, 397 (1957); K. Kikuchi, *Sci. Repts.*, *Tôhoku Univ.*, **41**, 35 (1957); J. Koutecky, *Collection Czechoslov. Chem. Communs.*, **24**, 1608 (1959). For other treatments of sulfur compounds, see ref. 4; J. Metzger and A. Pullman, *Compt. rend.*, **226**, 1613 (1948); *Bull. soc. chim. France*, **1948**, 1166; G. Berthier and B. Pullman, *Compt. rend.*, **231**, 774 (1950); P. Chiorboli and P. Manaresi, *Gazz. chim. ital.*, **84**, 248 (1954); G. De Alti and G. Milazzo, *Univ. studi. Trieste, Fac. sci. Ist. chim. No.*, **24** (1958); A. Pullman, *Bull. soc. chim. France*, **1958**, 641; G. Giacometti and G. Rigatti, *J. Chem. Phys.*, **30**, 1633 (1959). Two models for sulfur have recently been compared by J. Koutecky, R. Zahradnik, and J. Paldus, *J. chim. phys.*, **56**, 455 (1959).

[60] D. S. Matteson, *J. Am. Chem. Soc.*, **82**, 4228 (1960).

(cf. sulfur, Sec. 5.4; this possibility applies, of course, to bromine as well). Little MO work has been done on organic compounds of other elements.[61]

5.6 Auxiliary Inductive Parameter (AIP)

In their original treatment of aromatic substitution Wheland and Pauling[4] found it necessary to include an "inductive effect" of a heteroatom on an attached carbon. The more electronegative atom was said to polarize the σ-bond, thereby increasing the effective electronegativity of the attached carbon.[4,9] This effect is accommodated by assigning to the attached carbon a h-value, say h_α, usually taken to be some given fraction, δ, of the h of the associated heteroatom; that is, for the group, C_α—X,

$$h_{C_\alpha} = \delta h_X \tag{5}$$

In their work Wheland and Pauling found that $\delta = \frac{1}{10}$ was satisfactory. This value was subsequently adopted by several authors,[62] although values ranging from $\frac{1}{3}$ to $\frac{1}{8}$ have also been used.[63]

The inductive effect of a substituent in aliphatic chains in affecting the acidity of carboxylic acids, for example, can be assessed by (6), which is due to Branch and Calvin.[64]

$$\Delta E = \epsilon^n D \tag{6}$$

D is the effect of a substituent on an adjacent atom, n is the number of atoms between the substituent and the functional group, and ϵ is a fraction found empirically to be $1/2.8$. Dewar[65] has suggested applying the same idea to π-systems. In this approach a heteroatom affects the Coulomb integral not only of the attached carbon but of the next carbon as well, and the next and the next, etc., with the effect dying off in an exponential fashion in accordance with the equation

$$h_{C_n} = \delta^n h_X \tag{7}$$

This approach, with $\delta = \frac{1}{3}$, has been used by several subsequent authors.[24,65,66,67,68,69,70] The approach and the specific value used for δ

---

[61] A LCAO treatment of selenophene has been reported: G. De Alti and G. Milazzo, *Univ. studi Trieste, Fac. sci. Ist. chim. No.*, **23** (1958).

[62] For example, see refs. 14, 47, 53; J. I. F. Alonso, *Compt. rend.*, **233**, 403 (1951); and C. Sandorfy, *Can. J. Chem.*, **36**, 1739 (1958).

[63] Examples: refs. 9 and 13; I. Samuel, *Compt. rend.*, **240**, 2534 (1955); S. L. Matlow and G. W. Wheland, *J. Am. Chem. Soc.*, **77**, 3653 (1955).

[64] G. E. Branch and M. Calvin, *The Theory of Organic Chemistry*, Prentice-Hall, Englewood Cliffs, N.J. (1946), Sec. 25.

[65] M. J. S. Dewar, *J. Chem. Soc.*, **1949**, 463.

[66] M. J. S. Dewar, *J. Chem. Soc.*, **1950**, 2329.

[67] V. Gold, *Trans. Faraday Soc.*, **46**, 109 (1950).

[68] H. H. Jaffé, *J. Chem. Phys.*, **21**, 415 (1953).

[69] H. H. Jaffé, *J. Am. Chem. Soc.*, **77**, 274 (1955).

[70] R. Pauncz and E. A. Halevi, *J. Chem. Soc.*, **1959**, 1967; E. A. Halevi and R. Pauncz, *J. Chem. Soc.*, **1959**, 1974.

have been justified particularly by Jaffé,[68] with extensive calculations related to Hammett's $\sigma\rho$ relation. Unfortunately, his conclusions are dependent on several assumptions of questionable validity (cf. Sec. 11.7), and the use of (7) has been criticized by Brown[71] and by Del Re.[72] The AIP concept has also been criticized as theoretically unsound by McWeeny.[73]

The various treatments of the AIP presume the operation of an inductive effect as a polarization induced along bonds by the relative electronegativity of a substituent:[74]

$$C \rightarrow C \rightarrow C \rightarrow X \quad \text{or} \quad C \leftarrow C \leftarrow C \leftarrow Y$$

Perhaps the best evidence that such an effect actually exists comes from quadrupole resonance measurements (Sec. 6.2), but the effect is rapidly damped and seems to be small two bonds away from the substituent and unmeasurable thereafter.[75]

The observations frequently associated with this type of inductive effect now seem to be due almost wholly to direct electrostatic interactions between charges or dipoles and dipoles—the so-called "field" or "direct" effect. The early suggestion of Bjerrum that such interactions are responsible for the differences between the first and second dissociation constants of carboxylic acids has been developed to give good semiquantitative agreement, despite the many mathematical and physical difficulties involved.[76] The acidities of monobasic acids have been shown to depend on the magnitude and orientation of the bond dipole associated with the substituent.[77]

Operationally, the distinction between the inductive effect and the field or direct effect is that the one operates through bonds and the other operates directly through space. Recent studies of substituted amines of known and varying spacial orientations has demonstrated the overwhelming importance of the distance between the substituent dipole and

[71] R. D. Brown, *Quart. Revs.*, **6**, 63 (1952).

[72] G. Del Re, *Tetrahedron*, **10**, 81 (1960).

[73] R. McWeeny, *Proc. Roy. Soc.*, **A237**, 355 (1956).

[74] G. N. Lewis, *Valence and the Structure of Atoms and Molecules*, The Chemical Catalog Co., New York (1923), p. 139, C. K. Ingold, *Ann. Repts. on Prog. Chem.*, **23**, 129 (1926).

[75] H. O. Hooper and P. J. Bray, *J. Chem. Phys.*, **33**, 334 (1960).

[76] N. Bjerrum, *Z. Physik. Chem.*, **106**, 219 (1923); J. G. Kirkwood and F. H. Westheimer, *J. Chem. Phys.*, **6**, 506 (1938); F. H. Westheimer and J. G. Kirkwood, *J. Chem. Phys.*, **6**, 513 (1938); F. H. Westheimer and M. W. Shookhoff, *J. Am. Chem. Soc.*, **61**, 1977 (1939).

[77] W. A. Waters, *Phil. Mag.* [7] **8**, 436 (1929); W. S. Nathan and H. B. Watson, *J. Chem. Soc.*, **1933**, 890; H. M. Smallwood, *J. Am. Chem. Soc.*, **54**, 3048 (1932); A. Eucken, *Angew. Chem.*, **45**, 203 (1932); F. H. Westheimer and M. W. Shookhoff, *J. Am. Chem. Soc.*, **61**, 555 (1939).

the functional group compared to the number of bonds separating the functions.[78]

In aromatic systems such a direct effect is complicated by interactions with charges distributed by the aromatic ring as well as by conjugation effects. It is interesting that the effects of *meta*-substituents are determined largely by the direct interaction with the substituent bond dipole, although *para*-substituents, which are more directly conjugated with the functional group, do not follow the simple electrostatic patterns.[79]

The Branch and Calvin[64] technique (6) is clearly a device to incorporate within structure an electrostatic distance effect.[80] The transplantation of this method to MO theory is an attempt to embrace within the theory a factor foreign to its nature and may well be *malentendu*. Within given applications, the use of a generalized AIP such as (7) may be successful with a suitable manipulation of parameters, but the technique probably should not be expected to be either very general or to do better than give qualitative or semiquantitative results. However, in the absence of specific considerations of the effects of direct interactions, the use of an inductive parameter is often necessary to introduce the appropriate corrections at least approximately. An example is orientation effects in aromatic substitution (Sec. 11.7).

When an AIP technique is used, charge distributions are particularly sensitive to the particular technique, (5) or (7), and the magnitude of δ used. Since charge distributions are important in calculations of dipole moments (Sec. 6.1), significant approaches are the recent comparisons of charge distributions calculated by advanced MO methods with those derived from HMO studies with appropriate parameters. Brown and Heffernan[21b] found that the SCF charge distribution in pyridine could not be adequately reproduced with (7); (5) could be used successfully, however. Recent calculations of pyrrole[22] and pyridinium cation[43b] have also demonstrated the successful use of (5), the simple equation in which only the attached carbon is altered. The value of δ used in these treatments has varied from $\frac{1}{8}$ to $\frac{1}{3}$. Brown[21,22] has suggested that an auxiliary inductive parameter is necessary not so much because of a polarization of a σ-electronic framework but rather to the effect of the core potential of the heteroatom on a π-electron at the neighboring carbon.

[78] C. A. Grob, E. Renk, and A. Kaiser, *Chem. and Ind.*, **1955**, 1222; H. K. Hall, Jr., *J. Am. Chem. Soc.*, **78**, 2570 (1956).

[79] G. Schwarzenbach and H. Egli, *Helv. Chim. Acta*, **17**, 1183 (1934); F. H. Westheimer, *J. Am. Chem. Soc.*, **61**, 1977 (1939); J. M. Sarmousakis, *J. Chem. Phys.*, **12**, 277 (1944); M. Judson and M. Kilpatrick, *J. Am. Chem. Soc.*, **71**, 3115 (1949); J. D. Roberts and W. T. Moreland, *J. Am. Chem. Soc.*, **75**, 2167 (1953).

[80] The factor of $(1/2.8)^n$ is close to the factors r_o^2/r_n^2 for the distance between the substituent bond and the functional group.

In using an AIP, particular applications should be tested individually, although for approximate work the use of (5) with $\delta = 0.1$ does not seem to be unreasonable.

5.7 Hyperconjugation

In Sec. 4.1 we mentioned that for reasons of symmetry there is no overlap between σ- and π-orbitals and the interaction that does occur is electrostatic in origin. The statement is true, provided the σ-orbital lies in the nodal plane of the π-orbital; otherwise, the contribution to the overlap integral over the positive and negative lobes will not completely cancel. The methyl group in propylene may be taken as an example.

The wave function for the methyl group may be approximated as a set of three C—H bond functions, each of which is taken as a linear combination of a hydrogen $1s$-orbital and an appropriate carbon sp^3-orbital. Alternatively we may first form *group orbitals* as linear combinations of the three hydrogen orbitals. For the structure, I, such group orbitals could take the form

$$\varphi_1 = \frac{1}{\sqrt{3}} (\varphi_{H_1} + \varphi_{H_2} + \varphi_{H_3})$$

$$\varphi_2 = \frac{1}{\sqrt{6}} (2\varphi_{H_1} - \varphi_{H_2} - \varphi_{H_3})$$

$$\varphi_3 = \frac{1}{\sqrt{2}} (\varphi_{H_2} - \varphi_{H_3})$$

$$C_1{-}C_2{-}C_3 \begin{matrix} H_1 \\ \diagup \\ \diagdown H_3 \\ H_2 \end{matrix}$$

I

These functions are mutually orthogonal if overlap integrals are neglected. We now form linear combinations with appropriate carbon orbitals: φ_1 has no node and can overlap with whatever $2s$- and $2p_x$-orbital of C_3 is "left over" from the σ-bond to C_2; φ_2 has a nodal plane roughly comparable to the nodal plane of a $2p_z$-orbital; overlap between these two orbitals can occur. Similarly, φ_3 has as its nodal plane the xz-plane; overlap can occur with the carbon p_y-orbital. This set of three bonds is equivalent to the set of three bonds to hybrid carbon orbitals. We have simply taken the methyl group as a whole and have dissected the pattern of electron probability in a different manner. However, from this point of view we perceive that the C_3—$2p_z$-orbital and the φ_2-group orbital can by their symmetry join in the π-lattice; the two electrons in this C_3—H_3 bond, which is the π-type, can be treated as part of the π-system. The overlap interaction of appropriately oriented σ-bonds with π-systems is called *hyperconjugation*.[81]

In practice, for the calculation of such systems the methyl group can be treated as a modified attached vinyl group, II, in which C_α is a "normal" π-network carbon, Y is the tetrahedral carbon of the methyl group, and Z is the hydrogen group pseudo-atom. The various α and β parameters

$$C_\alpha\text{—Y—Z}$$
$$\text{II}$$

required are handled as in heteroatom cases with appropriate assignments of h_{C_α}, h_Y, h_Z, $k_{C_\alpha\text{—Y}}$, and $k_{Y\text{—Z}}$.[82] From the location of the hydrogen nuclei in the C—H_3 pseudo-π bond, we expect the electrons to be held rather tightly; that is, $k_{Y\text{—Z}}$ should be relatively large. In their original calculations of hyperconjugation, Mulliken, Rieke, and Brown[81] derived values of $k_{C_\alpha\text{—Y}}$ and $k_{Y\text{—Z}}$ of about 0.8 and 5, respectively. They assumed all Coulomb integrals to be equal; that is, all h's equal 0. Overlap integrals were included. This method was adopted in some subsequent work,[83] although most recent calculations have used different parameter values. The k's have been assigned largely on the basis of assumed proportionality to overlap integrals. It has usually been assumed that the H_3 pseudo-orbital is relatively electropositive;[84] h_Z is usually given a small negative value. The electron-donating inductive effect of a methyl group is usually recognized by assigning a small negative value to h_{C_α}. The values used in the recent literature in calculations that include overlap are $h_{C_\alpha} = 0\text{-}-0.1$, $h_Y = 0\text{-}-0.1$, $h_Z = 0\text{-}-0.5$, $k_{C_\alpha\text{—Y}} = 0.7\text{-}1$,   and $k_{Y\text{—Z}} = 2\text{-}2.5$.[70,85]

The same parameter values have been used in several calculations with

[81] R. S. Mulliken, C. A. Rieke, and W. G. Brown, *J. Am. Chem. Soc.*, **63**, 41 (1941).

[82] For further discussion see C. A. Coulson, *Quart. Revs.*, **1**, 144 (1947); *Valence*, Oxford University Press, London (1952), pp. 310–314.

[83] J. S. Roberts and H. A. Skinner, *Trans. Faraday Soc.*, **45**, 339 (1949); H. H. Jaffé, *J. Chem. Phys.*, **20**, 778 (1952).

[84] Note the argument given by N. Muller, L. W. Pickett, and R. S. Mulliken, *J. Am. Chem. Soc.*, **76**, 4770 (1954). The opposite assignment was assumed by R. S. Mulliken and C. C. J. Roothaan, *Chem. Rev.*, **41**, 219 (1947).

[85] A. Pullman, B. Pullman, and G. Berthier, *Bull. Soc. chim. France*, **17**, 591 (1950); C. A. Coulson and V. A. Crawford, *J. Chem. Soc.*, **1953**, 2052; N. Muller, L. W. Pickett and R. S. Mulliken, *J. Am. Chem. Soc.*, **76**, 4770 (1954); Y. I'Haya, *Bull. Chem. Soc. Japan*, **28**, 376 (1955); *J. Chem. Phys.*, **23**, 1165, 1171 (1955); N. Bauman and G. J. Hoijtink, *Rec. trav. chim.*, **76**, 841 (1957); A. Lofthus, *J. Am. Chem. Soc.*, **79**, 24 (1957); N. Muller and R. S. Mulliken, *J. Am. Chem. Soc.*, **80**, 3489 (1958).

neglect of overlap,[86] although I'Haya reports that neglect of overlap requires a larger value for k_{X-Y}, close to 3.[87] Other treatments have also been published.[88] However, neglect of overlap in this model for hyperconjugation may be a serious approximation because the overlap integral between the methyl carbon and the hydrogen pseudo-orbital has the rather large value of 0.5. Indeed, recent correlations with experimental quantities demonstrate that this MO method gives very poor results with methyl-substituted carbonium ions, although the method appears reasonable for radicals.[89]

In a simple model of a methyl group, introduced by Wheland and Pauling,[90,91] conjugation is neglected and the inductive effect alone is considered by assigning to the attached carbon of the π-system a negative h-value. In several applications this simple and useful model, which may be called the *inductive model* of hyperconjugation in contrast with the *conjugation model*[92] discussed previously, apparently works reasonably well with $h = -0.3$ to -0.5,[89] even though the behavior of the methyl group as a substituent in aromatic systems seems clearly to be both conjugative and inductive.[93]

Matsen[94] and Stevenson[95] have introduced the concept of the methyl group as a pseudo-heteroatom which contributes a pair of electrons to the π-system. In this *heteroatom model* we assume that the electrons in a methyl group behave to our approximation as a single electron pair on a single atom. Justification for this model can come only from an appeal to

[86] H. Lumbroso, *Compt. rend.*, **230**, 95 (1950); B. Pullman, M. Mayot, and G. Berthier, *J. Chem. Phys.*, **18**, 257 (1950); E. Heilbronner and M. Simonetta, *Helv. Chim. Acta*, **35**, 1049 (1952); O. Chalvet, *Compt. rend.*, **244**, 1043 (1957); A. K. Chandra and S. Basu, *J. Chem. Soc.*, **1959**, 1623.

[87] Y. I'Haya, *Bull. Chem. Soc. Japan*, **28**, 369 (1955).

[88] A. Pullman and J. Metzger, *Bull. Soc. Chim.*, **15**, 1021 (1948); B. Pullman and G. Berthier, *Bull. Soc. chim. France*, **1948**, 551; H. C. Longuet-Higgins and R. G. Sowden, *J. Chem. Soc.*, **1952**, 1404; W. C. Lin, *J. Chinese Chem. Soc.*, [2], **2**, 45, 105 (1955), **4**, 14 (1957); D. Peters, *J. Chem. Soc.*, **1957**, 646, 4182; M. M. Kreevoy, *Tetrahedron*, **5**, 233 (1959); A. Streitwieser, Jr., and P. M. Nair, *Tetrahedron*, **5**, 149 (1959).

[89] E. L. Mackor, G. D. Dallinga, J. H. Kruizinga, and A. Hofstra, *Rec. trav. chim.*, **75**, 836 (1956); E. L. Mackor, A. Hofstra, and J. H. van der Waals, *Trans. Faraday Soc.*, **54**, 186 (1958); A. Streitwieser, Jr., and P. M. Nair, *Tetrahedron*, **5**, 149 (1959).

[90] G. W. Wheland and L. Pauling, *J. Am. Chem. Soc.*, **57**, 2086 (1935).

[91] This suggestion was renewed more recently by H. C. Longuet-Higgins, *J. Chem. Phys.*, **18**, 283 (1950), *Proc. Roy. Soc.*, **207**, 121 (1951).

[92] A. Streitwieser, Jr., and P. M. Nair, *Tetrahedron*, **5**, 149 (1959).

[93] R. W. Taft, Jr., *J. Am. Chem. Soc.*, **79**, 1045 (1957).

[94] F. A. Matsen, *J. Am. Chem. Soc.*, **72**, 5243 (1950).

[95] Personal communication; see F. H. Field and J. L. Franklin, *Electron Impact Phenomena*, Academic Press, New York (1957), p. 124.

experiment. Its applications to date appear very promising; in particular, associated with the ω-technique, this model gives rather good correlations of ionization potentials of a variety of compounds.[2,92] Parameter values used with this model have covered the range $h_X = 1.4$–3.3; $k_{CX} = 0.7$–0.8[2,92,94,95] For the HMO method values of $k_{CX} = 0.7$ and $h_X = 2$ seem suitable,[94,95] although with the ω-technique and cations $h_X = 3$ works better.[2]

The various methods for calculating the effects of hyperconjugation agree in yielding significant values for the conjugation energy between π-systems and methyl groups. Much controversy has developed concerning the actual extent of such conjugation.[96] Many of the older "proofs" of hyperconjugation[97] have recently been questioned.[96] Individual aspects of the questions aroused are treated in appropriate sections below; for now, it suffices to say that there seems to be little doubt that hyperconjugation or π-σ conjugation is a significant energy factor in excited states and in carbonium ions. In ground states of neutral molecules, however, it is a factor involving only rather small energy differences. That the phenomenon exists, even in such systems, seems adequately demonstrated by electron spin-resonance studies of radicals[98] (Sec. 6.4) and by nuclear magnetic resonance studies of neutral compounds.[99]

5.8   Summary of Parameter Values

For convenience, a set of "recommended" parameter values is given in Table 5.3. The suggested parameter values may be useful to suggest trends and in semiquantitative work. Because they are so imprecise, there seems little need at present to distinguish between eigenvector and eigenvalue applications. In more quantitative specific correlations the values should be adjusted for the best individual fit. The suggested values are for use with the HMO method only. Inclusion of overlap or the use of the ω-technique will normally require somewhat different values.

[96] See, for example, the various papers presented at the Conference on Hyperconjugation, Indiana University, June 1958, published in *Tetrahedron*, **5**, 105–274 (1959) and reprinted as *Conference on Hyperconjugation*, Pergamon Press, New York (1959). Properly belonging to this group also is a more recent paper by R. S. Mulliken, *Tetrahedron*, **6**, 68 (1959).

[97] See summaries by V. A. Crawford, *Quart. Revs.*, **3**, 226 (1949); J. W. Baker, *Hyperconjugation*, Oxford University Press, London (1952).

[98] H. S. Jarrett and G. J. Sloan, *J. Chem. Phys.*, **22**, 1783 (1954); B. Venkataraman and G. K. Fraenkel, *J. Chem. Phys.*, **23**, 588 (1955); *J. Am. Chem. Soc.*, **77**, 2707 (1955); J. E. Wertz and J. L. Vivo, *J. Chem. Phys.*, **24**, 479 (1956).

[99] R. W. Taft, Jr., *J. Am. Chem. Soc.*, **79**, 1045 (1957); R. A. Hoffman, *Mol. Phys.*, **1**, 326 (1958).

TABLE 3

SUGGESTED PARAMETER VALUES FOR HETEROATOMS FOR USE
WITH SIMPLE LCAO THEORY

Element	Coulomb Integral	Bond Integral
Boron	$h_B = -1$	$k_{C-B} = 0.7$
Carbon		$k_{C-C} = 0.9*$
		$k_{CC} = 1\dagger$
		$k_{C=C} = 1.1\ddagger$
Nitrogen	$h_{\dot{N}} = 0.5$	$k_{C-N} = 0.8$
	$h_{\ddot{N}} = 1.5$	$k_{CN} = 1$
	$h_N^+ = 2$	$k_{N-O} = 0.7$
Oxygen	$h_{\dot{O}} = 1$	$k_{C-O} = 0.8$
	$h_{\ddot{O}} = 2$	$k_{C=O} = 1$
	$h_O^+ = 2.5$	
Fluorine	$h_F = 3$	$k_{C-F} = 0.7$
Chlorine	$h_{Cl} = 2$	$k_{C-Cl} = 0.4$
Bromine	$h_{Br} = 1.5$	$k_{C-Br} = 0.3$
Methyl (conjugation model $\equiv$ C—Y—Z)	$h_C = -0.1$	$k_{C-Y} = 0.8$
	$h_Z = -0.5$	$k_{Y-Z} = 3$
Methyl (inductive model $\equiv$ C—Me)	$h_C = -0.5$	$k_{C-Me} = 0$
Methyl (heteroatom model $\equiv$ C—X)	$h_X = 2\S$	$k_{C-X} = 0.7$
Auxiliary inductive parameter $\simeq 0.1 h_X \|$		

* For single sp^2—sp^2 bond of about 1.47 A length.
† An "aromatic" bond of about 1.40 A length.
‡ A double bond of about 1.34 A length.
§ h_{C_α} of about -0.2 may be necessary; this point has apparently never been tested in the simple LCAO method.
‖ For use with all groups except methyl; δ may be different for the groups $\dot{X}$, $\ddot{X}$ and X^+. If $\delta = 0.1$ is suitable for $\dot{X}$, a value closer to $\delta = 0.05$ may be better for the corresponding $\ddot{X}$.

II PROPERTIES OF MOLECULES

6 Electron densities and bond orders

6.1 Dipole Moments

The *dipole moment*, μ, is a vector that is a measure of the magnitude of charge displacement and is defined as

$$\mu = e\mathbf{d} \tag{1}$$

in which $\mathbf{d}$ is the vector distance separating positive and negative charges of magnitude e. In a neutral system consisting of several charged points the total or net dipole moment is the vector sum or resultant of the component moments as in Fig. 6.1.[1]

In an atom the net dipole moment is the resultant of an infinite number of differential moments. The electronic charge within the differential volume element $d\tau$ is given as $\psi^*\psi \, d\tau$; if $\mathbf{d}$ is the vector distance from the nucleus to the volume element, the net dipole moment is given as

$$\mu = \int \mathbf{d}\psi^*\psi \, d\tau \tag{2}$$

If the nucleus is at a center of symmetry with respect to the electronic charge distribution, for every $\mathbf{d}$ there will be an identical element of charge at $-\mathbf{d}$ and the net dipole moment is zero. Because of the nodal properties of atomic wave functions, the nucleus in an isolated atom is always at a center of charge symmetry; hence atoms have no dipole moments. If the electronic distribution in a molecule were simply the sum of the distributions of the component atoms, the net dipole moment would still

---

[1] For further discussion of dipole moments and their experimental determination see G. E. K. Branch and M. Calvin, *The Theory of Organic Chemistry*, Prentice-Hall, Englewood Cliffs, N.J. (1941), Chap. 18; R. J. W. Le Fevre, *Dipole Moments*, Methuen and Co., London (1953); L. E. Sutton, Chap. 9 in E. A. Braude and F. C. Nachod, *Determination of Organic Structures by Physical Methods*, Academic Press, New York (1955). The last named gives a number of references to previous books and reviews.

be zero. The perturbation of electronic charge distribution that actually occurs in bond formation, however, results in nonzero dipole moments in many compounds. The dipole moment, as obtained experimentally, is a number that does not specify direction. Hydrogen chloride, for example, has $\mu = 1.08D$,[2,3] but the assignment of hydrogen or chlorine as the

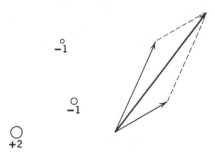

Fig. 6.1   This set of charged points has component dipole moments and a net dipole moment resultant as indicated.

positive end of the dipole must be made on theoretical grounds.[4] The use of hybrid orbitals for bond formation may produce dipole moments. If we form two hybrid sp-orbitals in a chlorine atom, two electrons can be put into one orbital and the odd electron may be placed in the other. This hypothetical atom now has an *atomic dipole*[5] in the direction $\cdot\overset{\longleftrightarrow}{Cl}$:.

[2] Electronic charge is in units of 10^{-10} esu and distance in Angstroms is 10^{-8} cm.; hence dipole moments result in units of 10^{-18} esu-cm., or D for Debye who first postulated the existence of permanent dipoles in molecules [P. Debye, *Z. Physik.*, **13**, 97 (1912)].

[3] Convenient lists of dipole moments are given in L. G. Wesson, *Tables of Electric Dipole Moments*, the Technology Press, Cambridge, Mass. (1948), and A. A. Maryott and F. Buckley, *Table of Dielectric Constants and Electric Dipole Moments of Substances in the Gaseous State*, National Bureau of Standards Circular 537, U.S. Government Printing Office, Washington, D.C. (1953).

[4] This statement is not strictly true. The direction of the dipole moment can be determined in principle by precise measurements of the effect of isotopic substitution on hyperfine Zeeman splitting effects [(C. H. Townes, G. C. Dousmanis, R. L. White, and R. F. Schwarz, *Discussions Faraday Soc.*, **19**, 56 (1955)]. The method applied to carbon monoxide, for example, showed the direction to be as expected: $C\overset{\leftarrow+}{-}O$ [B. Rosenblum, A. H. Nethercot, Jr., and C. H. Townes, *Phys. Rev.*, **109**, 400 (1958)]; similarly, deuterium iodide has been shown to be $D\overset{+\rightarrow}{-}I$ [C. A. Burns, *J. Chem. Phys.*, **30**, 976 (1959)].

[5] C. A. Coulson, *Valence*, Oxford University Press, London (1952), p. 207, introduces the atomic-dipole concept and applies it to HCl essentially as above.

If a bond is formed to a hydrogen atom with no further electron perturbation, the resulting hydrogen chloride will have a dipole moment in the same direction: $\overset{+\rightarrow}{H}Cl$. The argument does not depend on the specific hybridization used; any *s*-hybridization in the H—Cl bond will result in a dipole moment in this direction. In this hybridization argument the dipole moment is due largely to the electron density of lone-pair electrons.

Alternatively, the H—Cl bond can have ionic character by consideration of the resonance structures:

$$H—Cl \leftrightarrow H^+ Cl^- \leftrightarrow H^- Cl^+$$
$$\text{I}a \qquad\qquad \text{I}b \qquad\qquad \text{I}c$$

Since hydrogen is electropositive relative to chlorine, I*b* is more important than I*c*, and a dipole moment results in the same direction as that from hybridization. Note that both arguments are intellectual dissections of a total observable physical phenomenon. Hybridization is a device that alters the electron density pattern about individual atoms and includes distortion of unshared electron pairs; ionic character describes distortion particularly of bonding electrons. In terms of these hypothetical constructs, the dipole moment of HCl probably results both from hybridization and from ionic character.

With suitable assumptions, the total dipole moment of a molecule can be dissected into component *bond moments*[6] which are assumed to be relatively constant constitutive properties of the σ-bonds of compounds. This procedure is only of empirical utility. It neglects lone-pair electrons and the effects of *polarization*. For example, if the C—Cl bond had a constant bond moment in every compound, CH_3Cl and $CHCl_3$ would have identical dipole moments, assuming that each has a tetrahedral structure. The experimental dipole moments are 1.87 and 1.01 *D*, respectively. The electrostatic interaction between two neighboring dipoles orientated in similar directions results in a lessening of charge displacement and consequent decrease in dipole moment. In effect, the polarization induced by the neighboring dipole changes the ionic character or the hybridization of a bond or both.

For compounds containing π-bond networks it is convenient to dissect the total dipole moment into a σ-moment and a π-moment. The σ-moment is estimated from analogous systems without π-bonds, and the π-moment is given as the difference between the experimental total dipole moment and the estimated σ-moment. This procedure again neglects the polarization interaction of the σ- and π-moments; that is, if the σ-moment is

---

[6] A list of bond moments is given in G. W. Wheland, *Resonance in Organic Chemistry*, J. Wiley and Sons, New York, (1955), p. 209.

appreciable, the π-moment for the hypothetical isolated π-system is somewhat higher than the "experimental" moment so determined.

The π-moment results from a dissymmetry of the π-electron density. In Sec. 2.8 the π-electron density q_r, at atom r, was defined in terms of the molecular orbitals

$$q_r = \sum_j n_j c_{jr}^2 \tag{3}$$

in which the jth MO is occupied by n_j electrons. In an all-carbon system each atom contributes a nuclear charge of $+1$ to the π-system; hence the net charge ζ_r is

$$\zeta_r = 1 - q_r \tag{4}$$

The HMO π-moment is readily computed from the ζ_r's and the geometry of the system. All $\zeta_r = 0$ in AH; these compounds do not possess π-moments. They may have σ-moments, but such moments are normally small. Non-AH and heteroatom systems may have π-moments.

In practice, the n points of ζ_r are treated as vectors from some common origin. The direction of the vector is taken as toward the point if ζ_r is negative and toward the origin if ζ_r is positive. The resultant is the calculated π-moment.

Example. To calculate the π-moment of methylenecyclopropene, II. In Sec. 2.8 we found that the electron densities and the charge densities are

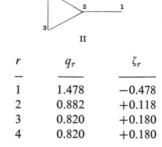

II

r	q_r	ζ_r
1	1.478	-0.478
2	0.882	$+0.118$
3	0.820	$+0.180$
4	0.820	$+0.180$

We assume that all bond distances equal 1.40 A.

Choosing atom 2 as the origin is convenient. From symmetry, there is no net moment in the y-direction. Along the x-direction

$$\mu_\pi = [2(0.180)(1.40)(\cos 30°) + (0.478)(1.40)](4.77)$$
$$= 5.26D^7$$

The factor, 4.77, is the electronic charge in appropriate units.

[7] G. Berthier and B. Pullman calculate about $5D$; *Bull. soc. chim. France*, **1949**, D457.

The answer given in this example is typical of results with the HMO procedure in that the calculated dipole moment is rather larger than the probable actual moment, which in this case is unknown.[8] Azulene, III,

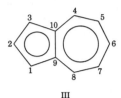

III

for example, has a calculated π-dipole moment of $6.9D$.[9] The experimental value of the molecular dipole moment is $1.0D$.[10] The σ-moment is expected to be small; hence there is no question that the simple theory gives a dipole moment here that is too large. The various elaborations within the simple theory usually improve the calculated values. Allowing for a variation of β with bond distance in azulene gives $\mu = 5.25D$.[11,12] The use of the ω-technique (Sec. 4.5) with $\omega = 1$ gives $\mu = 3.8D$.[11]

The simple hydrocarbon fulvene, IV, has an experimental dipole moment of $1.2D$.[10] HMO theory gives the charge distribution in IV

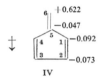

IV

which corresponds to $\mu = 4.7D$.[13] Various elaborations of the simple theory (variation of β with bond distance, the ω-technique, etc.) yield numerically lower dipole moments.[10,14,15]

[8] Calculation by the SCF MO method gives $\mu = 1.21D$ for methylenecyclopropene [A. Julg, *J. chim. phys.*, **50**, 652 (1953)].

[9] C. A. Coulson and H. C. Longuet-Higgins, *Rev. sci.*, **85**, 929 (1947).

[10] G. W. Wheland and D. E. Mann, *J. Chem. Phys.*, **17**, 264 (1949).

[11] G. Berthier and A. Pullman, *Compt. rend.*, **229**, 761 (1949).

[12] If β for the 9,10-bond in azulene is taken as 0, the system reduces to cyclodeca-pentaene, an AH which should have no dipole moment. Hence a value of $\beta_{9,10}$ can be found that will reproduce the experimental dipole moment; this value is about $\frac{1}{4}\beta_0$ [E. Sparatore, *Gazz. chim. ital.*, **88**, 671 (1958)].

[13] C. A. Coulson and H. C. Longuet-Higgins, *Rev. Sci. Instr.*, **85**, 927 (1947); A. Pullman, B. Pullman, and P. Rumpf, *Bull. soc. chim. France*, **15**, 757 (1948); C. Sandorfy, N. Q. Trinh, A. Laforgue, and R. Daudel, *J. chim. phys.*, **46**, 655 (1949).

[14] J. I. F. Alonso, *Compt. rend.*, **233**, 56 (1951); Y. Amako, *Sci. Repts., Tôhoku Univ., First Ser.*, **40**, 36 (1956).

[15] A. Julg and A. Pullman, *J. chim. phys.*, **50**, 459 (1953).

The experimental dipole moment is obtained as a scalar having no direction. The calculated dipole moments of fulvene, although varying in magnitude, depending on the particular method used, agree in predicting that the direction of the dipole moment is toward the ring (ring negative, exocyclic carbon positive). The direction of a dipole moment can frequently be determined experimentally by including phenyl groups having substituents, such as chloro and nitro, which contribute bond moments of known direction. This method has been applied to fulvene; 6,6-diphenylfulvene, V, and 6,6-bis-p-chlorophenylfulvene, VI, have $\mu = 1.34D$ and $0.68D$, respectively.[10] In chlorobenzene the bond moment of the C—Cl

V   VI

bond is directed toward the chlorine; since μ for VI is smaller than for V, μ for V must be directed in opposition to the bond moment of the C—Cl bond or toward the five-membered ring. This result agrees with the theoretical predictions of HMO theory.

Advanced MO methods of the SCF and Pariser-Parr type also give the experimental direction of the dipole moment of fulvene.[15,16] With azulene, the HMO method, its various elaborations and advanced MO procedures agree in predicting that the direction of the dipole is from the seven-membered ring toward the five-membered ring[9,10,11,12,17] (cf. Sec. 16.3). This direction has also been confirmed experimentally.[18] Various simple LCAO methods predict that the dipole moment of heptafulvene, VII, should be numerically somewhat smaller than that for fulvene and that it should be directed in the opposite sense; that is, from the ring to the exocyclic carbon.[19] Advanced MO methods give conflicting predictions about the direction of the moment but agree that it should be small.[16,20]

[16] G. Berthier, *J. Chem. Phys.*, **21**, 953 (1953); *J. chim. phys.*, **50**, 344 (1953).

[17] A. Julg, *Compt. rend.*, **239**, 1498 (1954); *J. chim. phys.*, **52**, 377 (1955); R. Pariser, *J. Chem. Phys.*, **25**, 1112 (1956).

[18] Y. Kurita and M. Kubo, *J. Am. Chem. Soc.*, **79**, 5460 (1957).

[19] G. Berthier and B. Pullman, *Trans. Faraday Soc.*, **45**, 484 (1949); E. D. Bergmann, E. Fischer, D. Ginsburg, Y. Hirshberg, D. Lavie, M. Mayot, A. Pullman, and B. Pullman, *Bull. soc. chim. France*, **1951**, 684.

[20] A. Julg, *J. chim. phys.*, **52**, 50 (1955).

An attempt to determine the direction of the dipole moment of hepta-fulvene by measurements on 8-*p*-substituted phenyl-1,2,5,6-dibenzo-heptafulvenes, VIII, failed because conflicting answers were found;[19]

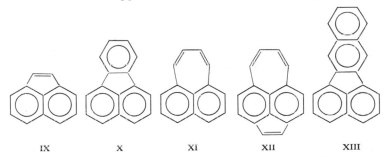

VII   VIII

however, a recent study of simple derivatives of the parent molecule, such as the 8,8-dinitrile, has led to a dipole moment of about $3D$ for VII with the ring positive.[21] The simple theory gives as good results in this case as its various elaborations!

The non-AH molecules, acenaphthylene, IX, fluoranthene, X, pleia-diene, XI, acepleiadylene, XII and 8,9-benzofluoranthene, XIII, are all pre-dicted to have substantial dipole moments according to HMO theory.[22,23] The experimental dipole moments of acepleiadylene and 8,9-benzofluor-anthene are small but appreciable, $0.5D$[24] and $1.0D$,[23] respectively.

IX   X   XI   XII   XIII

The prediction by LCAO theory in its simplest form that certain un-saturated hydrocarbons have dipole moments of appreciable magnitude in agreement with experiment is a rather encouraging achievement. Similar predictions are not possible by VB theory without taking a number

[21] M. Yamakawa, H. Watanabe, T. Mukai, T. Nozoe, and M. Kubo, *J. Am. Chem. Soc.*, **82**, 5665 (1960).

[22] C. Sandorfy, *Compt. rend.*, **227**, 198 (1948); A. Pullman, B. Pullman, E. D. Bergmann, G. Berthier, E. Fischer, Y. Hirshberg, and J. Pontis, *J. chim. phys.*, **48**, 359 (1951); B. Pullman, A. Pullman, G. Berthier, and J. Pontis, *J. chim. phys.*, **49**, 20 (1952).

[23] C. Sandorfy, N. Q. Trinh, A. Laforgue, and R. Daudel, *J. chim. phys.*, **46**, 655 (1949).

[24] D. A. Pitt, A. J. Petro, and C. P. Smyth, *J. Am. Chem. Soc.*, **79**, 5633 (1957).

of "excited" structures into account. The fact that dipole moments are predicted is of greater significance than the fact that the calculated magnitudes of the dipoles moments are usually substantially larger than the experimental values. The dipole moment depends on charge densities which are calculated from the eigenfunctions of the MO approximations. Eigenfunctions of quantum mechanical operators are generally less accurate than the corresponding eigenvalues. However, the improvement in the calculated dipole moments using the ω-technique of Wheland and Mann[10] is also encouraging. A consistent set of calculations on a variety of hydrocarbons with the currently accepted value, $\omega = 1.4$ (Secs. 4.5, 7.3) has not yet been reported but would be of interest.

HMO calculations of the μ-moments of heteroatom compounds have been reported frequently. Because of the poor quantitative agreement between calculated and experimental dipole moments with hydrocarbons, it would be surprising to find good quantitative agreement with heteroatom compounds. In practice, however, the introduction of heteroatoms involves the further empirical parameters, h_X and k_{CX} (Sec. 5.1), and it is usually possible to find parameter values that will lead to a good reproduction of experimental dipole moments. However, the use of the so-derived parameter values in other applications, particularly those which involve energy quantities (e.g., prediction of orientation in aromatic substitution) may be invalid (Chap. 5). We expect this limitation as long as the calculations of dipole moments of hydrocarbons give no better than qualitative agreement.

The most extensive application has been to nitrogen compounds[25,26,27,28] and particularly to nitrogen heterocyclics of the pyridine type.[26,27,28] This method usually produces values close to $h_{\ddot{N}} = 0.5$, $k_{CN} = 1.0$ for such a nitrogen.

An example of an explicit calculation may be helpful. For the calculation of the π-moment of pyridine we may treat pyridine as being derived from benzene by changing the Coulomb integral of the 1-carbon to $\alpha_0 + h_N\beta_0$. The effect that this change has on the π-charge densities at the other positions is given conveniently by the use of $\pi_{r,s}$-values for benzene;

[25] M. J. S. Dewar, *J. Chem. Soc.*, **1950**, 2329; J. I. F. Alonso, *Compt. rend.*, **233**, 403 (1951); J. I. F. Alonso and R. Domingo, *Anales real. soc. espan. fis. y quim.* (*Madrid*), **51B**, 321 (1955); F. Gerson and E. Heilbronner, *Helv. Chim. Acta* **41**, 2332 (1958); R. A. Barnes, *J. Am. Chem. Soc.*, **81**, 1935 (1959).

[26] J. Ploquin, *Compt. rend.*, **226**, 245 (1948); S. Odiot and M. Roux, *J. chim. phys.*, **50**, 141 (1953); S. S. Perez, M. A. Herraez, F. J. Igea, and J. Esteve, *Anales real soc. espan. fis. y quim.* (*Madrid*), **51B**, 91 (1955).

[27] L. E. Orgel, T. L. Cottrell, W. Dick, and L. E. Sutton, *Trans. Faraday Soc.*, **47**, 113 (1951).

[28] P. O. Löwdin, *J. Chem. Phys.*, **19**, 1323 (1951); **21**, 496 (1953).

that is, $-\zeta = \pi_{1,r}\delta\alpha_1 = \pi_{1,r}h_N$. Only the vector components along the symmetry axis of pyridine need be considered. Assuming a regular hexagon with sides of 1.40 A, the component distances for the α, β, and γ positions are readily shown to be 0.70, 2.10, and 2.80 A, respectively. The total π-moment is

$$\mu_\pi = -4.77 \, (2 \cdot 0.70\pi_{1,2} + 2 \cdot 2.10\pi_{1,3} + 2.80\pi_{1,4})h_N$$
$$= 2.23h_N \text{ (directed towards the N)}$$

(*Note*: for benzene, $\pi_{1,2} = -0.157$; $\pi_{1,3} = 0.009$; $\pi_{1,4} = -0.102$.)

The bond moment for a C—N bond is given as $1.0D$[6] but is derived from alkylamines. In pyridine nitrogen is bonded to an sp^2-carbon which is more electronegative than an sp^3-carbon, hence a somewhat smaller σ-bond moment is probably more appropriate for our calculation. This example illustrates some of the problems encountered in estimations of μ_σ. If $1.0D$ is used anyway as the σ-bond moment of each of the C—N bonds, the symmetry axis component of μ_σ is $1.0D$. The experimental dipole moment of pyridine is $2.22D$, which is equated to $\mu_\sigma + \mu_\pi$; hence $\mu_\pi = 1.22D$. Thus $h_N = \mu_\pi/2.23 = 1.22/2.23 = 0.55$. This calculation assumes that $k_{CN} = 1$; to a first approximation a change in k_{CN} will not affect the electron density distribution, since all $\pi_{t,rs} = 0$ for benzene, an AH.

Dipole moments of carbonyl compounds have also been treated by several authors; oxygen parameter values used for this purpose are commonly in the range $h_{\dot{O}} = 1 - 2$, $k_{C=O} = 1.4 - 2$.[27,29] In a number of calculations of dipole moments the distinction between π-moments and total moments has been ignored.

6.2   Nuclear Quadrupole Spectra

A more direct measure of the electron distribution in some molecules has recently become available through nuclear quadrupole coupling constants. Nuclei of spin of at least one have nonspherical shapes. The energy of such a system of nonspherical charge in an electric field gradient will vary, depending on the orientation of the system with respect to the field gradient. A crude analogy is a football floating in a lake; the long axis parallel to the surface of the water is a more stable orientation than that in which this axis is perpendicular to the surface. Such variation in energy does not occur if the charged system is spherical or if the electric

[29] G. Battista, B. Scrocco, and E. Scrocco, *Atti. accad. nazl. Lincei, Rend. Classe sci. fis., mat. e nat.*, **8**, 183 (1950); Y. Kurita and M. Kubo, *Bull. Chem. Soc. Japan*, **24**, 13 (1951); I. Estelles, J. I. F. Alonso, and J. Mira, *Anales real soc. españ. fis. y quim.* (*Madrid*), **50B**, 11 (1954); A. Kuboyama, *Bull. Chem. Soc. Japan*, **32**, 1226 (1959).

field is constant in the immediate vicinity of the system. The quadrupole[30] coupling constant is a measure of the energy variation. The real system of an ellipsoidal nucleus in the field at the center of an atom is quantized; only certain orientations and associated energies are allowed. Transitions from one state to another are accompanied by the absorption or emission of energy. The energy changes involved are minute. In the chlorine atom, for example, the transition energy is about 0.01 small cal./mole. The corresponding light is in the microwave and radio region.

Quadrupole coupling effects show up in the fine structure of rotation bands in microwave spectra, just as vibrational fine structure occurs in electronic transitions. The analysis of microwave spectra of linear and symmetric top[31] molecules gives a single quadrupole coupling constant that may be related to electronic structure. Because interactions at the nucleus are important, the assumption that molecular orbitals resemble atomic orbitals in this region is a good approximation; s-orbitals do not provide a field *gradient* at the nucleus and do not contribute to the coupling constants; d- and higher electrons have relatively low density at the nucleus; hence p-electrons are most important in determining the magnitude of the interactions.[32] A filled p-shell has spherical symmetry and gives no interaction with a quadrupole moment. A singly occupied p-orbital has a large field gradient at the nucleus and gives rise to relatively large coupling constants. The quadrupole coupling constant of chlorine atom, for example, is -110.4 Mc./sec., whereas that for chloride ion is zero.[32] In a number of saturated organic chlorides the constant is about -80 Mc./sec. A simple interpretation is that the decrease in the magnitude of the constant from the pure p- or atomic value is due to the s-character in the bond and suggests sp^3-hybridization in the chlorine. Actually, ionic character in the bond as well as hybridization contributes to a lowering from the atomic value.[32,33,34]

[30] Any system of electric charges may be characterized by a series of electric moments: charge, dipole, quadrupole, octopole, etc. A dipole results from the separation of two equal and opposing charges, $+-$; a quadrupole results from separation of two equal and opposing dipoles, $+--+$, etc. For further details of molecular quadrupole moments see A. D. Buckingham, *Quart. Revs.*, **13**, 183 (1959).

[31] Nonlinear molecules have three moments of inertia. In *asymmetric tops* all three moments are different; in *symmetric tops* two of the moments have the same value.

[32] C. H. Townes and B. P. Dailey, *J. Chem. Phys.*, **17**, 782 (1949).

[33] Various efforts have been made to reconcile both variables with the single experimental constant in a number of cases; cf. ref. 32; W. Gordy, *J. Chem. Phys.*, **19**, 792 (1951); **22**, 1470 (1954), *Discussions Faraday Soc.*, **19**, 14 (1955); B. P. Dailey, *Discussions Faraday Soc.*, **19**, 255 (1955); B. P. Dailey and C. H. Townes, *J. Chem. Phys.*, **23**, 118 (1955).

[34] For reviews see C. H. Townes and A. L. Schawlow, *Microwave Spectroscopy*, McGraw-Hill Book Co., New York (1955), Chap. 9; W. J. Orville-Thomas, *Quart. Revs.*,

With asymmetric top[31] molecules, analysis of the microwave spectrum determines two independent components of the quadrupole coupling constants in a coordinate system which may be taken along convenient bond axes. In this way, constants may be determined for σ- and π-bonds. The π-parameter is readily interpretable in terms of double-bond character and gives directly the π-*electron defect* of the measured atom in the molecule.[34,35,36,37] The electron defect is just the amount of electron donated to the remainder of the π-system and corresponds to our ζ_r.

Relatively few atoms of interest have the required nuclear quadrupole; most work up to now has been done with chlorine compounds. Typical results for the electron defect on chlorine are 0.04 ± 0.01 for p dichlorobenzene[35] and 0.06 ± 0.01 for vinyl chloride.[36,38,39] These values have been compared with results of LCAO theory. Goldstein's treatment of vinyl chloride and chlorobenzene included overlap.[38,40] Bersohn[35] used simple LCAO theory; assuming $h_{Cl} = 0.5$, he found $k_{C-Cl} \cong \frac{1}{3}$ reproduced the experimental electron defects for several compounds. Use of a larger h_{Cl}-value will result in a somewhat larger value for k_{C-Cl}.

Exercise. Using $h_{Cl} = 2$, $k_{C-Cl} = 0.4$, calculate ζ_{Cl} in (*a*) vinyl chloride and in (*b*) chlorobenzene. Assume all other β's = 1.
Ans. (*a*) +0.018;   (*b*) +0.015.

Microwave spectra are determined for the gas state and the derived quadrupole coupling constants pertain to this state. Direct spectral transitions between nuclear quadrupole states can be observed only in the solid state because of their low intensity. For halogens only the resonance frequency of the transition is observed; components along bond directions are not obtained. Furthermore, the interpretation of the spectra is further complicated because the results are obtained for a condensed phase. Thus fields from neighboring molecules can contribute to the field gradient resulting from p-electrons and can result in appreciable shifts in frequency. Nevertheless, a number of spectra have been reported, particularly for chlorine compounds, and have been interpreted in terms of electronic

11, 162 (1957); C. T. O'Konski in *Determination of Organic Structures by Physical Methods*, Academic Press, New York (1961), Chap. 12; R. Livingston, *Record Chem. Prog.*, **20,** 173 (1959).

[35] R. Bersohn, *J. Chem. Phys.*, **22,** 2078 (1954).

[36] J. H. Goldstein and J. K. Bragg, *Phys. Rev.*, **78,** 347 (1950).

[37] T. P. Das and E. L. Hahn, *Nuclear Quadrupole Resonance Spectroscopy*, Academic Press, New York, (1958), Chap. 7.

[38] J. H. Goldstein, *J. Chem. Phys.*, **24,** 507 (1956).

[39] SCF calculations on vinyl chloride yield 0.10. M. Simonetta, G. Favini, and S. Carra, *Mol. Phys.*, **1,** 181 (1958).

[40] J. A. Howe and J. H. Goldstein, *J. Chem. Phys.*, **26,** 7 (1957).

structure.[34,41] The direct use of HMO theory in this connection has so far been limited, but the method promises to offer unique data for the testing of valence theories generally.

6.3   Electron Spin Resonance—Introduction

The spinning electron has an associated magnetic moment which is randomly oriented in field-free space. In a magnetic field only two orientations of the electronic magnetic moment are allowed—parallel and antiparallel to the applied field. These two states correspond to different energies, and transitions between the states are accompanied by absorption or emission of radiation

$$\Delta E = h\nu \cong \frac{eh}{2\pi mc} H \tag{5}$$

in which H is the field strength. In a field of 3000 gauss, for example, typical of magnets used in this work, the transition energy is 0.80 cal./mole, which corresponds to light in the microwave region of 3.6 cm. wavelength.

This resonance peak can be split by *hyperfine interaction* with the magnetic moments of nuclei. This splitting is important for H, F, and ^{13}C but not for D or ^{12}C. Splitting by D is less than for H; it normally results in line broadening but has been resolved in a few cases; ^{12}C has no magnetic moment and causes no splitting. The proton has a spin of $\frac{1}{2}$ and an associated magnetic moment that is also aligned either parallel or antiparallel to the applied field. The magnetic field at a nearby electron, consequently, is either slightly greater or slightly less than the applied field, and four possible states result, as shown in Fig. 6.2a. The long arrow represents the electron spin, the short arrow represents the proton spin. Each state has a different energy. During transition the nuclear spin remains the same; hence transitions occur only between vertical pairs of states. There are two such transitions, as shown in Fig. 6.2b. The dotted lines correspond to the energy states of the electron in the applied magnetic field alone. We see that instead of a single spectral line we now have two, one corresponding to a higher energy, the other to a lower. Alternatively, if we keep the resonance frequency constant, one corresponds to a lower applied field strength, the other to a higher field

[41] P. J. Bray and R. G. Barnes, *J. Chem. Phys.*, **27**, 551 (1957); P. J. Bray, S. Moskowitz, H. Hooper, R. G. Barnes, and S. L. Segel, *J. Chem. Phys.*, **28**, 99 (1958); H. Negita, H. Yamamura, and H. Shiba, *Bull. Chem. Soc. Japan*, **28**, 271 (1955); H. Negita, S. Sato, T. Yonezawa, and K. Fukui, *Bull. Chem. Soc. Japan*, **30**, 721 (1957); H. Negita and S. Sato, *J. Chem. Phys.*, **27**, 602 (1957); H. Negita, *J. Sci. Hiroshima Univ.*, **21A**, 261 (1958); M. J. S. Dewar and E. A. C. Lucken, The Chemical Society Special Publication No. 12 (1958), p. 223; *J. Chem. Soc.*, **1958**, 2653; **1959**, 426.

strength. With two equivalent protons, we have eight states, as shown in Fig. 6.3a. Because the protons are equivalent, states such as $+$$+$$-$ and $+$$-$$+$ are degenerate. Again, transitions can occur only between vertical pairs in Fig. 6.3a, as shown in Fig. 6.3b. Three spectral lines of intensity, $1:2:1$, result. The generalization to greater numbers of equivalent protons is straightforward; the number and intensities of the resonance lines are

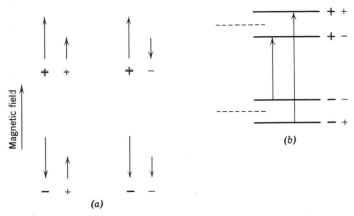

Fig. 6.2   Magnetic moments and energy levels for electron and proton in a magnetic field. The long arrows correspond to the electronic magnetic moment.

those of the coefficients of a binomial expansion; for example, $1:3:3:1$ and $1:4:6:4:1$. The spectra that result from these transitions are called *electron spin resonance* or *ESR* spectra.[12]

The magnitude of the splitting into two peaks depends on the extent of interaction of the electron with the proton. In a hydrogen atom, with the electron in a $1s$-orbital, the splitting is large, 510 gauss. For an electron in a $2p$-orbital, however, there is a node at the nucleus and the interaction vanishes. Methyl radical is planar,[43] with the odd electron undoubtedly in a $2p$-orbital of the central carbon. The ESR spectrum of this radical (Fig. 6.4) shows the four-line pattern of $1:3:3:1$ intensities expected for interaction with three equivalent protons, even though these protons are in the nodal plane of the $2p$-orbital that contains the odd electron. The explanation for this apparent anomaly lies in configuration interaction. The wave function for the ground state of the C—H σ-bond has admixed

[12] For further details and reviews see (a) D. J. E. Ingram, *Free Radicals*, Academic Press, New York (1958); (b) D. H. Whiffen, *Quart. Revs.*, **12**, 250 (1958); (c) H. C. Longuet-Higgins, Chemical Society Symposia, Bristol, 1958; Special Publication No. 12, The Chemical Society, London (1958), p. 131.

[43] T. Cole, H. O. Pritchard, N. R. Davidson, and H. M. McConnell, *Mol. Phys.*, **1**, 406 (1958).

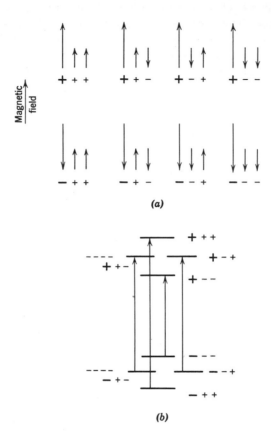

Fig. 6.3   Magnetic moments and energy levels of an electron and two equivalent protons in a magnetic field.

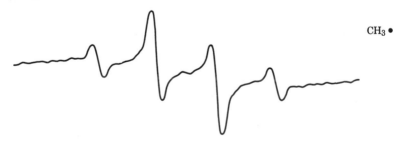

CH$_3$ •

Fig. 6.4   Derivative curve of the ESR spectrum of methyl radical (reproduced with permission from ref. 43).

orbitals of excited states (Sec. 1.5). The excited states, in turn, can interact with the spin of the $2p$-electron such that a net spin density results in a hydrogen $1s$-orbital. In short, because of electrostatic interactions, the spin of the odd electron induces a *spin polarization* in neighboring orbitals regardless of symmetry.[44,45,46,47] The splitting of about 25 gauss in methyl radical is a small fraction of that in a free hydrogen atom. The magnitude of the splitting ΔH is proportional to the net spin density at the hydrogen, which, in turn, is proportional to the unpaired spin density ρ_r at the attached carbon atom, r (6).[44,48,49]

$$\Delta H \simeq Q\rho_r \tag{6}$$

This result is important because ΔH is an experimental quantity and ρ_r can be calculated by MO theory. The calculation is not completely straightforward, however, because an unpaired spin density at one carbon can induce a net spin density of opposite sign on an adjacent carbon. Hence the sum of the absolute values of the spin densities at each carbon may be greater than unity for a free radical, although the algebraic sum, of course, is unity. Complete calculation can be accomplished with MO theory including configuration interaction.[50] Such a calculation for allyl radical gives spin densities for carbons 1, 2, and 3 of 0.622, −0.231, and 0.622, respectively.[51] Simple MO theory predicts only positive spin densities and for allyl radical would predict 0.500, 0, and 0.500, respectively. The negative spin densities are about $\frac{1}{3}$ to $\frac{1}{2}$ the adjacent densities[50,51,52] but are most likely to occur on atoms that are predicted to have zero spin in the HMO approximation.[50]

6.4   Electron Spin Resonance—Semiquinone Ions

The reduction of quinones and the oxidation of hydroquinones in alkaline media are known to involve semiquinone anions as intermediates.[53]

[44] H. M. McConnell, *J. Chem. Phys.*, **24**, 764 (1956).

[45] R. Bersohn, *J. Chem. Phys.*, **24**, 1066 (1950).

[46] S. I. Weissman, *J. Chem. Phys.*, **25**, 890 (1956).

[47] H. S. Jarrett, *J. Chem. Phys.*, **25**, 1289 (1956).

[48] H. M. McConnell, *op. cit.*, **24**, 632 (1956).

[49] H. M. McConnell and H. H. Dearman, *J. Chem. Phys.*, **28**, 51 (1958).

[50] H. M. McConnell and D. B. Chestnut, *J. Chem. Phys.*, **27**, 984 (1957); *ibid.*, **28**, 107 (1958).

[51] Ref. 42*a*, p. 113.

[52] P. Brovetto and S. Ferroni, *Nuovo cimento*, **5**, 142 (1957).

[53] L. Michaelis, M. P. Schubert, R. K. Reber, J. A. Kuck, and S. Granick, *J. Am. Chem. Soc.*, **60**, 1678 (1938); L. Michaelis and S. Granick, *J. Am. Chem. Soc.*, **70**, 624, 4275 (1948); L. Michaelis, *Ann. N.Y. Acad. Sci.*, **40**, 39 (1940).

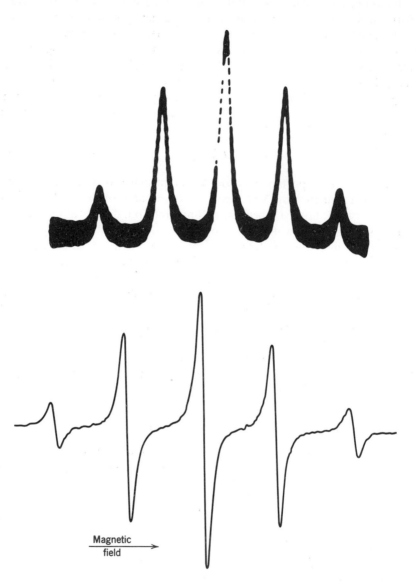

Fig. 6.5 ESR spectrum (*above*) and corresponding derivative curve (*below*) for *p*-benzo-semiquinone anion (reproduced with permission from B. Venkataramen and G. K. Fraenkel, *J. Am. Chem. Soc.*, **77**, 2707 (1955).

Semiquinone anions are radical anions. The ESR spectrum of p-benzo-semiquinone ion, XIV, shows the quintet of lines of intensity, $1:4:6:4:1$,

XIV

expected from interaction of the odd electron with four equivalent protons (Fig. 6.5). This semiquinone results from addition of one electron to the quinone π-system. The eight electrons in the p-benzoquinone π-system are paired in four bonding MO's; the new electron is placed in the lowest unoccupied MO, ψ_{m+1}. To a first approximation, the spin density ρ_r at atom r is given by the square of the coefficient of the corresponding atomic orbital in this MO. The spin densities

$$\rho_r = c_{m+1,r}^2 \tag{7}$$

may also be estimated from the experimental hyperfine splittings (hfs) by (6). For example, the splitting of 2.37 gauss found for p-benzosemi-quinone ion[54] gives $\rho = 0.07$ if $Q \simeq 34$, as assumed by McConnell.[55] McConnell showed that this value was in satisfactory agreement with a simple LCAO calculation; using $h_0 = 1$ and all β's equal, $\rho = 0.078$. Similarly, the values for o-benzosemiquinone ion derived from experiment, 0.03 and 0.11 for o- and p-hydrogens, respectively, are in qualitative agreement with the calculated values 0.049 and 0.089.

Recent data[56] allow a better test of the simple theory in this connection. Hyperfine splittings for several quinones have been determined and are summarized in Table 6.1. On the basis of HMO calculations with $h_0 = 1$, $k_{C=O} = 1$, and other β's equal, hydrogens may be assigned to corresponding splitting constants. A plot of the calculated ρ_r-values versus the experimental splitting constants (Fig. 6.6) shows a fair linear correlation; the slope of the correlation lines gives $Q = 23.8$ gauss,[57] a reasonable

[54] B. Venkataramen, B. G. Segal, and G. K. Fraenkel, J. Chem. Phys., **30**, 1006 (1959).

[55] H. M. McConnell, J. Chem. Phys., **24**, 632 (1956); some additional calculations are reported by A. Kuboyama, Bull. Chem. Soc. Japan, **32**, 1226 (1959).

[56] M. Adams, M. S. Blois, Jr., and R. H. Sands, J. Chem. Phys., **28**, 774 (1958).

[57] A. Streitwieser, Jr., and J. B. Bush, unpublished. G. Vincow and G. K. Fraenkel, J. Chem. Phys., **34**, 1333 (1961), have reported recently the results of an extensive comparison of experimental hfs constants with HMO theory including a variation of h_0 and k_{CO} parameters. Best results were obtained for $h_0 = 1.2$, $k_{CO} = 1.56$. Equation 6 was assumed and gave $Q = 22.5$ gauss.

TABLE 6.1

ESR SPECTRA OF SEMIQUINONE ANIONS

Corresponding Quinone	Hyperfine Splitting Constant (Gauss)*	Hydrogen Assigned†	Calculated ρ_r‡
o-Benzoquinone	0.95	3, 6	0.0487
	3.65	4, 5	0.0895
1,4-Naphthoquinone, XV	3.22	2, 3	0.1026
	0.57	5, 8	0.0385
		6, 7	0.0300
Acenaphthenequinone, XVI	1.24	1, 3, 4, 6	0.0227
	0.30	2, 5	0
1,4-Anthraquinone, XIX	0.46	1, 2, 3, 4, 5, 6, 7, 8	0.0357
9,10-Phenanthrenequinone, XVII	1.66	1, 8	0.0429
		3, 6	0.0399
	0.38	2, 7	0.0077
		4, 5	0.0109
Ninhydrin, XVIII	1.59	3, 6	0.0402§
	1.95	4, 5	0.0607
p-Benzoquinone	2.37	2, 3, 5, 6	0.0781
Toluquinone	1.76	3	0.0485
	2.54	5	0.0936
	2.46	6	0.0744
2,5-Dimethylbenzoquinone	1.84	3, 6	0.0441
2,6-Dimethylbenzoquinone	1.89	3, 5	0.0621
2,3-Dimethylbenzoquinone	2.60	5, 6	0.0884
2,3,5-Trimethylbenzoquinone	1.97	6	0.0556
Chlorobenzoquinone	2.21	3	0.0679
	2.45	5	0.0831
	2.21	6	0.0769
2,5-Dichlorobenzoquinone	2.03	3, 6	0.0666
2,6-Dichlorobenzoquinone	2.32	3, 5	0.0727
2,3-Dichlorobenzoquinone	2.32	5, 6	0.0817
2,3,5-Trichlorobenzoquinone	2.16	6	0.0711

* Benzoquinones: ref. 54; others: ref. 56.

† Some hydrogens were assigned by the various authors; others were assigned to correspond with the HMO calculations.

‡ Ref. 57. Parameters used: $h_O = 1$, $k_{C=O} = 1$; $h_{Cl} = 2$, $k_{C-Cl} = 0.4$; heteroatom model of methyl group was used: $h_X = 2$, $k_{CX} = 0.7$.

§ Calculated as 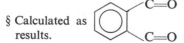 Calculation as the triketone gave poor results.

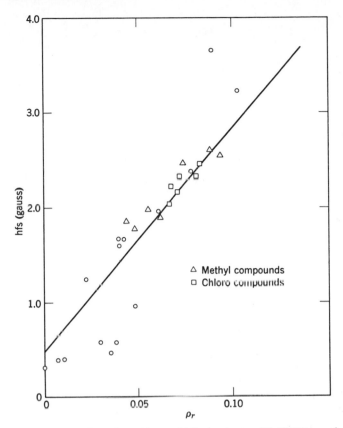

Fig. 6.6   Correlation of experimental hfs constants with HMO ρ_r values.

value based on other systems. The similar calculations of methyl- (hetero-atom model, Sec. 5.7) and chlorobenzoquinones also give satisfactory results, a finding that demonstrates the ability of MO theory to account for such hetero-substituents in a satisfactory manner.

XVIII   XIX

Additional information is derived from some other semiquinones. That from 2-phenyl-*p*-benzoquinone, XX, for example, shows an ESR spectrum

XX

expected from an inert substituent; that is, rather little unpaired spin density resides in the substituent phenyl group.[58] Since appreciable such density is indicated by HMO calculations in which β for the bond between the rings is taken as β_0 (ρ at the *para*-position of the phenyl group is calculated as 0.0227),[57] the phenyl substituent is probably far from co-planar with the quinone ring.

Methyl substituents produce additional splitting by the methyl protons. This result shows unambiguously that the bonding electrons of a methyl group interact in some way with π-electrons. In a real sense, hypercon-jugation does occur, but the ESR result does not mean that such hyper-conjugation need necessarily have the energy stabilizing effect frequently attributed thereto. Perhaps of greater significance in this connection is the finding that the effects of methyl groups are accommodated so well by the theoretical calculations that involve hyperconjugation. The magnitude of the splitting is comparable to that of ring protons in good agreement with results of MO calculations with inclusion of overlap, assuming that the effect is due to hyperconjugation.[59,60] Even protons on *t*-butyl groups can interact with the π-electron and cause splitting. In 2,5-di-*t*-butyl-*p*-benzosemiquinone, XXI, the aliphatic protons result in

[58] J. E. Wertz and J. L. Vivo, *J. Chem. Phys.*, **23**, 2441 (1955).

[59] R. Bersohn, *J. Chem. Phys.*, **24**, 1066 (1956).

[60] In valence-bond theory the splitting by methyl protons is the result of configuration interaction with excited states similar to the explanation for hyperfine splitting (hfs) by ring protons; A. D. McLachlan, *Mol. Phys.*, **1**, 233 (1958).

$$O\cdot$$

(structure XXI: a benzene ring with O· at top, C(CH$_3$)$_3$ at upper right, (CH$_3$)$_3$C at left, and O$^{\ominus}$ at bottom)

XXI

splitting by 0.06 gauss, only 3% of the magnitude of the splitting by the ring protons.[61] This small effect may be caused by a spin polarization at the proton from the spin density in the C—C bond, which in turn was induced by hyperconjugation with the ring.

6.5   Electron Spin Resonance—Hydrocarbon Radical Anions and Cations

Aromatic hydrocarbons with sufficient electron affinity react with alkali metals to form mono- or divalent salts:

$$ArH + Na \rightleftharpoons ArH^- + Na^+$$

$$ArH + Na \rightleftharpoons ArH^- + Na^!$$

The reaction is promoted by nonprotonic solvents which can effectively solvate the metal cations; examples are ethers such as dimethyl ether, tetrahydrofuran, and dimethoxyethane but not diethyl ether.[62] Liquid ammonia is also suitable, since it is a weak proton donor towards the hydrocarbon anions.

Ether solutions of the hydrocarbon salts are conducting.[62,63,64] Relative equilibrium constants for various hydrocarbons have been measured,[64,65,66] and electronic spectra for both mono- and dianions have been reported.[66,67,68,69] These ions are important intermediates in several reactions (Sec. 14.3) and are formed in polarographic reduction of the hydrocarbons (Sec. 7.1).

[61] G. K. Fraenkel, *Ann. N. Y. Acad. Sci.*, **67**, 546 (1957).

[62] N. D. Scott, J. F. Walker, and V. L. Hansley, *J. Am. Chem. Soc.*, **59**, 2442 (1936).

[63] G. J. Hoijtink, E. de Boer, P. H. Van der Meij, and W. P. Weijland, *Rec. trav. chim.*, **74**, 277 (1955).

[64] D. E. Paul, D. Lipkin, and S. I. Weissman, *J. Am. Chem. Soc.*, **78**, 116 (1955).

[65] G. J. Hoijtink, E. de Boer, P. H. Van der Meij, and W. P. Weijland, *Rec. trav. chim.*, **75**, 487 (1956).

[66] G. J. Hoijtink and P. H. Van der Meij, *Z. physik. Chem. NF*, **20**, 1 (1959).

[67] P. Balk, G. J. Hoijtink, and J. W. H. Schreurs, *Rec. trav. chim.*, **76**, 813 (1957).

[68] E. de Boer and S. I. Weissman, *Rec. trav. chim.*, **76**, 824 (1957).

[69] N. S. Hush and J. R. Rowlands, *J. Chem. Phys.*, **25**, 1076 (1956).

In dilute solution in suitable ethers the principal ionic species with most hydrocarbons are the monovalent anions. These ions result from the addition of a single electron to the lowest π-orbital of the hydrocarbon and are actually radical anions. The radical anions exhibit ESR spectra,[70] just

TABLE 6.2

ESR HYPERFINE SPLITTING CONSTANTS OF HYDROCARBON RADICAL ANIONS

Hydrocarbon	Symbol in Fig. 6.7	Position	Hyperfine Splitting Constant (Gauss)	c_{jr}^2 Simple LCAO	Reference
Benzene	Be	1	3.75	0.167	78
Naphthalene	N	1	5.01	0.181	72
		2	1.79	0.069	
Anthracene	A	1	2.74	0.096	77
		2	1.57	0.047	
		9	5.56	0.192	
Tetracene, XXIII	T	1	1.49	0.056	77
		2	1.17	0.034	
		5	4.25	0.148	
Perylene, XXII	P	1	3.09	0.083	77
		2	0.46	0.013	
		3	3.53	0.108	
Cycloöctatetraene	C	1	3.21	0.125	*
Biphenylene	Bi	1	0	0.027	†
		2	2.75	0.087	–
Pyrene, XXIV	Py	1	4.75	0.136	79
		2	1.09	0	
		4	2.08	0.087	

  * T. J. Katz and H. L. Strauss, *J. Chem. Phys.*, **32**, 1873 (1960).
  † C. A. McDowell and J. R. Rowlands, *Can. J. Chem.*, **38**, 503 (1960).

as the semiquinone ions do. The 17-line spectrum given by naphthalene radical ion[71] has been definitely interpreted to give the hfs constants for the α- and β-positions as 5.01 and 1.79 gauss, respectively.[72] The ratio

[70] First observed for sodium naphthalene by S. I. Weissman, J. Townsend, D. E. Paul, and G. E. Pake, *J. Chem. Phys.*, **21**, 2227 (1953).

[71] The two sets of four equivalent protons in naphthalene should yield a 25-line spectrum unless the ratio of the two splitting constants is near integral. Actually, all 25 lines are now easily resolved with modern equipment.

[72] T. R. Tuttle, Jr., R. L. Ward, and S. I. Weissman, *J. Chem. Phys.*, **25**, 189 (1956).

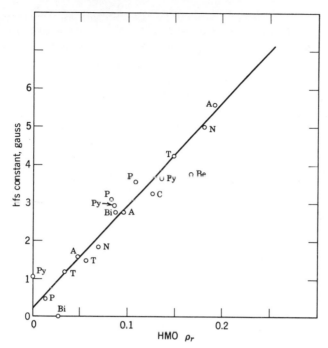

Fig. 6.7 Experimental hfs constants of hydrocarbon radical anions and HMO spin densities. Letters refer to compounds in Table 6.2.

2.80 is in excellent agreement with the ratio 2.62 of the squares of the coefficients of the α- and β-positions in the lowest vacant orbital in the HMO method.[73] Similarly, the experimental and calculated ratios for the para-/ortho-positions of biphenyl radical anion are 2.0 and 1.8, respectively.[74] HMO theory also accounts at least semiquantitatively for the total ESR spectra of the radical anions from biphenyl,[74,75,76] anthracene,[76,77] perylene, XXII,[76,77] tetracene, XXIII,[76,77] phenanthrene,[76] acenaphthylene, IX,[76] and fluoranthene, X.[76] The published hfs constants are summarized in Table 6.2. The impressive agreement between these constants and the spin densities calculated by HMO theory is shown in Fig. 6.7. The fact that the correlation line does not go through the origin is undoubtedly due to negative spin density at positions that calculate to have zero spin density.

[73] S. I. Weissman, T. R. Tuttle, Jr., and E. de Boer, J. Phys. Chem., **61**, 28 (1957).
[74] G. J. Hoijtink, Mol. Phys., **1**, 157 (1958).
[75] E. de Boer, J. Chem. Phys., **25**, 190 (1956).
[76] E. de Boer and S. I. Weissman, J. Am. Chem. Soc., **80**, 4549 (1958).
[77] A. Carrington, F. Dravnieks, and M. C. R. Symons, J. Chem. Soc., **1959**, 947.

This conclusion seems affirmed by the substantially lower than calculated hfs of benzene negative ion.[73,76,78] The symmetry of the benzene system precludes any negative spin density. The line in Fig. 6.7 should be regarded as an empirical correlation that allows for an average amount of negative spin density in more complex structures.

In practice, HMO predictions of zero spin density may correspond experimentally to hfs of substantial magnitude because of negative spin density induced by neighboring spin density. Pyrene negative ion serves as an excellent example. The hfs constants for the 1- and 4-positions[79] correspond well with the HMO correlation (Fig. 6.7). However, the lowest vacant orbital of pyrene has a node that passes through the 2-position; accordingly this position should have zero spin density. Experimentally, this position contributes a hfs constant of 1.09 gauss to the ESR spectrum. This hfs constant seems clearly to be associated with a negative spin density.[79]

XXII   XXIII   XXIV

Toluene radical anion yields a five-line ESR spectrum indicative of four equivalent protons which have been shown to be the' *ortho-* and *meta-*protons.[78] The lowest vacant orbital has a node along the line of symmetry of toluene; hence the methyl carbon and *para-*position should have approximately no spin, and the *ortho-* and *meta-*positions should predict to have the same spin density. *p*-Xylene behaves in the same way. In *m*-xylene radical anion the methyl groups cause extensive splitting, presumably because of hyperconjugation.[78] In this system the methyl groups do not lie along a node, and participation in the MO results. Unfortunately, this beautiful qualitative agreement between experiment and theory is clouded by *o*-xylene. The radical anion shows four equivalent hydrogens and little interaction with the methyl groups in complete disagreement with any simple MO picture.[78] The over-all ESR spectrum of the radical anion from acepleiadylene, XII, is also not consistent with HMO theory.[76]

[78] T. R. Tuttle, Jr., and S. I. Weissman, *J. Am. Chem. Soc.*, **80**, 5342 (1958).

[79] G. J. Hoijtink, J. Townsend, and S. I. Weissman, *J. Chem. Phys.*, **34**, 507 (1961).

Dilute solutions of aromatic hydrocarbons in concentrated sulfuric acid show ESR and electronic spectra that are strikingly similar to those of the corresponding radical anions.[80,81] This spectral similarity suggests that the radical species present are the corresponding radical cations that result from loss of one electron from the π-system. Concentrated sulfuric acid acts as an oxidizing agent in many organic reactions. In AH's the coefficients of AO's in the highest occupied and lowest unoccupied MO's differ only in sign; hence the same spin density distribution should result from the removal of an electron from the highest occupied MO as for the addition of an electron to the lowest unoccupied MO. In anthracene, for example, the ratio of hfs constants for the α- and β-positions in the cation, 2.14,[77] may be compared with the values 2.04 and 2.00 obtained for the radical anion and for HMO calculations, respectively. The main difference in ESR spectra for AH cations and anions, generally, is that the total splitting is of larger magnitude for the cation.[76,77,81] No explanation for this difference has yet been given; perhaps the change in the effective Coulomb integral of positively and negatively charged carbon causes a change in the extent of spin polarization by configuration interaction.

In non-AH's the coefficients of the highest occupied and lowest vacant MO's no longer necessarily have the same magnitude, and the ESR spectra of the radical cations and anions are expected to differ. Such a difference is observed for the ions from acepleiadylene, although neither spectrum agrees with HMO predictions.[76]

The over-all agreement between experimental ESR spectra and HMO theory is remarkable for such a crude and simple theory. We should not have been surprised if the theory totally failed in such a sensitive test of eigenfunction components. Indeed, an attempt to use a more elaborate theory gave poorer agreement,[75] although SCF theory does give better agreement.[82] The failure of the simple theory in special cases demonstrates the general rule that application of a semiempirical theory must always be made with circumspection.

[80] Y. Yokozawa and I. Miyashita, *J. Chem. Phys.*, **25**, 796 (1956); H. Kon and M. S. Blois, Jr., *J. Chem. Phys.*, **28**, 743 (1958).

[81] S. I. Weissman, E. de Boer, and J. J. Conradi, *J. Chem. Phys.*, **26**, 963 (1957); G. J. Hoijtink and W. P. Wiejland, *Rec. trav. chim.*, **76**, 836 (1957); W. I. Aalbersberg, G. J. Hoijtink, E. L. Mackor, and W. P. Weijland, *J. Chem. Soc.*, **1959**, 3049.

[82] A. D. McLachlan, *Mol. Phys.*, **3**, 233 (1960). Although McLachlan's method derives from SCF theory, it may be formulated in terms of HMO quantities as

$$\rho_r = c_{0r}^2 + \lambda \sum_s \pi_{r,s} c_{0s}^2,$$

in which λ is an adjustable parameter. This method gives good results for all of the alternant and nonalternant hydrocarbons tested with the exception of acepleiadylene.

6.6   Electron Spin Resonance—Hydrocarbon Radicals

According to HMO theory, the odd electron in odd-AH radicals is in a nonbonding MO; hence each inactive or unstarred position should have no net spin. However, substantial negative spin density is induced at these positions, and the resulting ESR spectra differ substantially from HMO predictions. In triphenylmethyl radical, the *ortho-* and *para-*positions have substantial spin density, as expected from the simple theory, but the *meta-*positions, which should be inactive according to the simple theory, actually have negative spin density of about half the magnitude of the *ortho-* and *para-*positions.[83,84] Even the ratio of spin densities at the central carbon to the *para-*position, for example, does not agree well with HMO predictions.[83]

A similar disparity is encountered with the perinaphthenyl radical, XXV. According to HMO theory, the spin density is equally distributed

XXV

among the 1, 3, 4, 6, 7, and 9 positions; hence a pattern of seven equally spaced lines is predicted for the ESR spectrum. Seven such lines are actually observed, but each is further split into a quartet corresponding to substantial spin density at the 2, 5, and 8 positions.[85] Negative spin density at these positions is predicted by VB and by advanced MO theories.[86]

HMO theory is clearly of limited application to radicals of these types. A discrepancy of zero predicted spin density and experimentally observed spin density at a position is of no significance. However, it seems reasonable to attribute significance to an experimental zero spin density at a position calculated to have appreciable odd electron density in the simple theory. An example is the pentaphenylcyclopentadienyl radical, XXVI, which shows an ESR spectrum consisting of a single line.[87] This result is

[83] F. C. Adam and S. I. Weissman, *J. Am. Chem. Soc.*, **80**, 2057 (1958).

[84] P. Brovetto and S. Ferroni, *Nuovo cimento*, **5**, 142 (1957); D. C. Reitz, *J. Chem. Phys.*, **30**, 1364 (1959); D. B. Chestnut and G. J. Sloan, *J. Chem. Phys.*, **33**, 637 (1960).

[85] P. B. Sogo, M. Nakazaki, and M. Calvin, *J. Chem. Phys.*, **26**, 1343 (1957).

[86] H. M. McConnell and H. H. Dearman, *J. Chem. Phys.*, **28**, 51 (1958); R. Lefebvre, H. H. Dearman, and H. M. McConnell, *J. Chem. Phys.*, **32**, 176 (1960).

[87] J. E. Wertz, C. F. Koelsch, and J. L. Vivo, *J. Chem. Phys.*, **23**, 2194 (1955). Recently the splitting caused by the phenyl hydrogens has been resolved; the magnitude of the splitting ($\sim$ 0.3 gauss) is consistent with twisted phenyl rings [D. C. Reitz, *J. Chem. Phys.*, **34**, 701 (1961)].

XXVI

expected only if the odd spin density is localized within the five-membered ring, which has no attached hydrogens. HMO theory predicts substantial odd electron character in the phenyl groups. The nonoccurrence of any appreciable splitting by phenyl hydrogens must mean that the phenyl groups are twisted from coplanarity with the central ring to such a degree that resonance interaction is negligible.

6.7   Bond Orders and Bond Lengths

The concept of bond order generally relates to the valency multiplicity between atoms in molecules. If the carbon-carbon bonds in ethane, ethylene, and acetylene are considered to have bond orders of one, two, and three, respectively, we note a monotonic relationship with the corresponding bond distances, 1.54, 1.33, and 1.20 A.[88] It is convenient to associate bonds of intermediate distances with fractional bond orders. Various definitions of bond order based on quantum mechanical theories have been proposed and correlations with bond distances have been suggested.[89,90,91]

As usual, we divide the observed effects into a σ- and π-part. The traditional treatment[92] has been to regard the σ-part as a constant; for carbon-carbon bonds the value of 1.54 A for a C—C bond is usually assigned. The shorter bond distances of carbon-carbon multiple bonds are attributed to the bond shortening effects of π-bond resonance. However, the σ-part of a C=C bond is approximately sp^2-sp^2 in character; the larger

[88] A summary of bond distances through 1955 is available in "Tables of Interatomic Distances and Configuration in Molecules and Ions," Special Publication No. 11, The Chemical Society, London (1958).

[89] L. Pauling, L. O. Brockway, and J. Y. Beach, *J. Am. Chem. Soc.*, **57**, 2705 (1935).

[90] W. G. Penney, *Proc. Roy. Soc.*, **A158**, 306 (1937).

[91] C. A. Coulson, *Proc. Roy. Soc.*, **A169**, 413 (1939).

[92] L. Pauling, *The Nature of the Chemical Bond*, Cornell University Press, Ithaca (1948), p. 171; second edition (1960), p. 232.

s-hybridization may be expected to produce a σ-bond shortening. Such hybridization effects were recognized and allowed for in some of Coulson's MO treatments of bond order.[93],[94] Substantial bond shortening by π-bond resonance still resulted (*vide infra*). Recently, the view has been advanced that variations in bond length are due entirely to the hybridization changes in σ-bonds[95],[96] and that π-electron effects are negligible in many hydrocarbons.[95] As of this writing an active controversy has developed about the true significance of π-bond order. Mulliken[97] and Bak and Hansen-Nygaard[98] have presented carefully reasoned arguments which recognize the effect of σ-hybridization in determining bond lengths and affirm the importance of π-electron bond order.

The controversy hinges on the value assigned to the sp^2-sp^2 C—C bond length in the absence of π-conjugation. For approximations to this value we examine the lengths of sp^2-sp^2 bonds in compounds for which π-conjugation is negligible because of geometry. An example is provided by the single bond lengths in cyclooctatetraene, 1.462 A.[99] In this hydrocarbon adjacent double bonds are almost orthogonal so that π-conjugation is negligible. However, this length may be short because of steric and electronic effects within this strained ring. In the related hydrocarbon, tetraphenylene, XXVII, the benzene rings are also isolated by the puckering of the eight-membered ring; the corresponding single bond length

XXVII

in this compound is 1.52 A.[100] In this case we could argue that the single bonds between the benzene rings are stretched by steric effects. The same objection could be applied to the 1.52 A length of the single bonds between

[93] C. A. Coulson, *Proc. Roy. Soc.*, **A207**, 91 (1951); *J. Phys. Chem.*, **56**, 311 (1952).

[94] C. A. Coulson, Chemical Society Symposia, Bristol, 1958, Special Publication No. 12, The Chemical Society (1958), p. 85.

[95] M. J. S. Dewar and H. N. Schmeising, *Tetrahedron*, **5**, 166 (1959); **11**, 96 (1960).

[96] H. J. Bernstein, *J. Phys. Chem.*, **63**, 565 (1959); M. G. Brown, *Trans. Faraday Soc.*, **55**, 694 (1959); J. Trotter, *Tetrahedron*, **8**, 13 (1960); G. R. Somayajula, *J. Chem. Phys.*, **31**, 919 (1959).

[97] R. S. Mulliken, *Tetrahedron*, **6**, 68 (1959).

[98] B. Bak and L. Hansen-Nygaard, *J. Chem. Phys.*, **33**, 418 (1960).

[99] O. Bastiansen, L. Hedberg, and K. Hedberg, *J. Chem. Phys.*, **27**, 1311 (1957).

[100] I. L. Karle and L. O. Brockway, *J. Am. Chem. Soc.*, **66**, 1974 (1944).

rings in hexaphenylbenzene. The substituent phenyls are almost perpendicular to the central ring in this hydrocarbon.[101]

Biphenyl itself is not planar in the vapor phase. The rings are twisted by about 45° and are separated by 1.48 A.[102] Resonance interaction between the rings is reduced, and we could argue that this length is close to that of the "pure" single bond. From the examples cited, we see that we can find arguments for a sp^2-sp^2 σ-length ranging from 1.46 to 1.52 A. Coulson's original proposal of 1.50 A[93] seems reasonable in this context, but he later[94] suggested that a shorter length might be better. Dewar and Schmeising,[95] relying on a variety of arguments take 1.48 A as the "natural" sp^2-sp^2-length. Mulliken,[97] arguing from theory and butadiene, adduces 1.51 ± 0.01 A, whereas Bak and Hansen-Nygaard[98] with a different theoretical position justify 1.517 A.

The natural length of the sp^2-sp^2-double bond is generally taken from ethylene, 1.335 A.[103]

Coulson[91] derived (8) which relates bond distance to bond order. In this equation s is the "natural" single-bond distance, d is the double-bond distance, and k is a parameter:

$$x = s - \frac{s - d}{1 + k\left(\dfrac{1 - p}{p}\right)} \tag{8}$$

Using the values for ethylene, benzene ($x = 1.397$, $p = 0.667$), and graphite ($x = 1.421$, $p = 0.535$), we may solve for the unknowns and find $s = 1.515$, $k = 1.05$.[104] This function is shown in Fig. 6.8. The HMO p-values are included for butadiene. These points fall far from the curve. The calculated bond length from (8), 1.437 A, is much shorter than the experimental value 1.483 A.[105]

Part of the discrepancy comes from the assumption of all β's equal in the simple theory.

Exercise. To the first approximation, what is $p_{2,3}$ for $k_{1,2} = 1.1$ and $k_{2,3} = 0.9$ in butadiene? Use the definition of $\pi_{rs,tu}$ and the values obtained previously (p. 108).

Ans. $p_{2,3} = 0.375$. The corresponding bond length (8) is 1.449 A.

[101] A. Almenningen, O. Bastiansen, and P. N. Stancke, *Acta Chem. Scand.*, **12**, 1215 (1958).

[102] O. Bastiansen, *Acta Chem. Scand.*, **3**, 408 (1949).

[103] Average of recent determinations: 1.334 A: L. S. Bartell and R. A. Bonham, *J. Chem. Phys.*, **27**, 1414 (1957); 1.337 A: H. C. Allen and E. K. Plyler, *J. Am. Chem. Soc.*, **80**, 2673 (1958); 1.334 A: B. P. Stoicheff, *Current Sci.*, **27**, 1 (1958).

[104] Dewar and Schmeising[95] use a straight line extrapolation using bond orders with overlap to extrapolate to $s = 1.488$ A.

[105] A. Almenningen, O. Bastiansen, and M. Traettenberg, *Acta Chem. Scand.*, **12**, 1221 (1958).

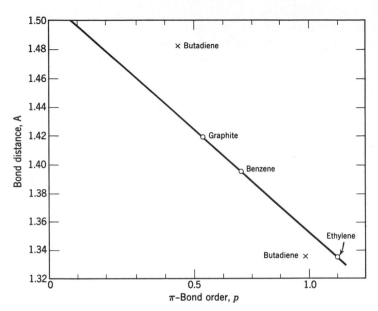

Fig. 6.8 Bond order-bond distance relation for selected points and Coulson's equation.

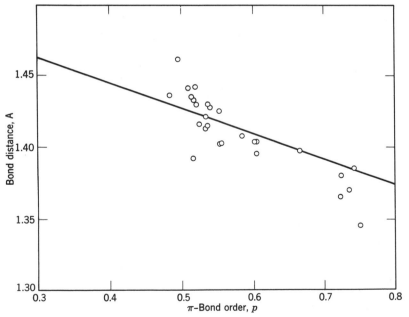

Fig. 6.9 Bond order-bond distance relations for aromatic hydrocarbons. Curve is derived from Coulson's equation (cf. Fig. 6.8).

However, even when reasonable corrections are made, a substantial π-bond order remains. Important electron correlation effects, however, appear to cause greater localization of electrons in the double bonds of classical structures of polyenes.[106] More advanced quantum mechanical calculations of butadiene yield a substantially lower π-bond order.[107]

With aromatic rings, however, carbon-carbon bonds have definite partial double-bond character and intermediate values for bond distances. Bond lengths vary from 1.36 to 1.46 A. In Table 6.3 we summarize bond distances for aromatic hydrocarbons, when known with some accuracy ($\sim \pm 0.01$ A),[108] along with corresponding p-values. The results are also plotted in Fig. 6.9. There is a definite trend towards shorter bond lengths with higher p's, but there is only a fair correlation with Coulson's equation. No reasonable change in the parameters of (8) will give a significantly better fit.

The theoretical curve may be used to predict bond lengths to ± 0.01–0.02 A. For example, the central 9–10-bond of azulene with $p = 0.401$ should be 1.445 A. The experimental value was most recently reported as 1.458 A.[109]

Similar correlations have been applied to the prediction of bond lengths of a number of polycyclic aromatic hydrocarbons.[93,110,111,112,114] Sometimes the pattern of bond-length changes follows that of the π-bond orders extremely well. A conspicuous example is that of ovalene.[112] Bond distances and bond orders change in the same manner with the

[106] M. J. S. Dewar and C. E. Wulfman, *J. Chem. Phys.*, **29**, 158 (1958).

[107] Various calculations give values of about 0.2. See the summary in ref. 97.

[108] Many of the values are summarized by J. M. Robertson, *Proc. Roy. Soc.*, **A207**, 101 (1951); *Organic Crystals and Molecules*, Cornell University Press, Ithaca (1953), Chap. 8.

[109] J. M. Robertson, H. M. M. Sherer, G. A. Sim, and D. G. Watson, *Nature*, **182**, 177 (1958).

[110] C. A. Coulson, R. Daudel, and J. M. Robertson, *Proc. Roy. Soc.*, **A207**, 306 (1951).

[111] W. E. Moffitt and C. A. Coulson, *Proc. Phys. Soc.*, **A60**, 309 (1948).

[112] A. J. Buzeman, *Proc. Phys. Soc.*, **A63**, 827 (1950).

[113] J. Barriol and J. Metzger, *J. chim. phys.*, **47**, 432 (1950).

[114] (a) C. A. Coulson, *Nature*, **154**, 794 (1944); (b) R. Pauncz and F. Berencz, *Acta Chim. Acad. Sci. Hung.*, **3**, 261 (1953); (c) *ibid.*, **4**, 333 (1954); (d) R. Pauncz and I. Wilheim, *Acta Chim. Acad. Sci. Hung.*, **11**, 63 (1956); (e) H. O. Pritchard and F. H. Sumner, *Proc. Roy. Soc.*, **A220**, 128 (1954); (f) H. O. Pritchard and F. H. Sumner, *Trans. Faraday Soc.*, **51**, 457 (1955); (g) M. A. Silva and B. Pullman, *Compt. rend.*, **242**, 1888 (1956); (h) J. Baudet and B. Pullman, *Compt. rend.*, **244**, 777 (1957); (i) G. G. Hall, *Trans. Faraday Soc.*, **53**, 573 (1957); (j) M. A. Ali, *Acta Cryst.*, **12**, 445 (1959); (k) W. N. Lipscomb, J. M. Robertson, and M. G. Rossman, *J. Chem. Soc.*, **1959**, 2601; (l) T. H. Goodwin and D. G. Watson, *J. Chem. Soc.*, **1959**, 2625; (m) J. Trotter, *Acta Cryst.*, **12**, 889 (1959).

TABLE 6.3

BOND ORDER-BOND LENGTH RELATIONS IN AROMATIC HYDROCARBONS

Hydrocarbon	Bond	Bond Distance, A	π-Bond Order, p	Reference
Benzene	1–2	1.397	0.667	88
Naphthalene	1–2	1.365	0.725	109, 110
	2–3	1.404	0.603	
	1–9	1.425	0.554	
	9–10	1.393	0.518	
Anthracene	1–2	1.370	0.738	109, 110
	2–3	1.408	0.586	
	1–11	1.423	0.535	
	9–11	1.396	0.606	
	11–12	1.436	0.485	
Coronene, XXVIII	1–2	1.385	0.745	111
	1–13	1.415	0.538	
	13–19	1.430	0.538	
	19–20	1.430	0.522	
Ovalene, XXIX	1–2	1.345	0.753	109, 112
	1–15	1.441	0.511	
	2–16	1.433	0.519	
	3–4	1.38	0.726	
	3–16	1.403	0.557	
	14–15	1.404	0.604	
	15–23	1.426	0.526	
	16–24	1.435	0.508	
	23–24	1.428	0.541	
	23–28	1.403	0.557	
	24–25	1.442	0.521	
	27–28	1.461	0.497	
Graphite		1.421	0.535	88, 113

XXVIII   XXIX

bond (Fig. 6.10). More often, however, one or more bonds in a molecule exhibits a substantial change in distance from that anticipated. An example is apparent even in naphthalene (Table 6.3). The 9–10-bond should be the longest bond on the basis of its p-value, but it is actually of intermediate length. These discrepancies are usually not improved in more elaborate MO calculations.[114ef] Moreover, the average deviation

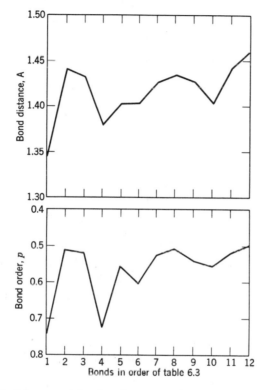

Fig. 6.10   Bond distance and bond order pattern in ovalene (after Buzeman [112]).

between experimental and calculated lengths is usually 0.01–0.02 A, about that of the usual experimental error but not much better than the average deviation from the mean distance, which is usually 0.02–0.03 A! At present it would seem that the MO method has only limited application to the prediction of bond lengths. Results just about as good are derived from examinations of the Kekulé structures of the hydrocarbons.[96,115] Because of the approximate nature of the MO bond-order–bond-length

[115] O. Chalvet and R. Daudel, *J. Phys. Chem.*, **56**, 365 (1952); R. Daudel in *Advances in Chemical Physics*, Interscience Publishers, New York (1958), Vol. 1, p. 165.

relationship, a simple nomenclature may be suggested. In π-systems
single bonds have $p < \sim 0.4$ and lengths of 1.49 ± 0.03 A. *Aromatic
bonds* have $p = 0.5$–0.7 and lengths of 1.42 ± 0.02 A. *Double bonds*
have $p > 0.8$ and lengths of 1.34 ± 0.01 A. According to this prescription,
the bonds in butadiene are *double* and almost *single*; the indicated bond
in perylene, XXX, with $p = 0.414$ and a distance of 1.50 A, is effectively
single, the 9,10-bond in azulene is *single*, etc. Clar et al.[116] have recently
called attention to a general class of aromatic hydrocarbons, which, like
triphenylene, XXXI, may be regarded as being composed only of benzene

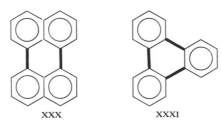

XXX   XXXI

rings joined together. Recent HMO calculations[117] of an extensive group of
such compounds shows that the bonds connecting the benzene rings have
p-values of 0.42 to 0.49 and would be classed as *single* bonds by the foregoing definition. Goodwin[118] has pointed out that the relatively long
lengths reported for such bonds (~ 1.53 A) add further fuel to the controversy over the natural length of C_{sp^2}—C_{sp^2} bonds.

MO bond length-bond order correlations have also been applied to
bonds with heteroatoms.[115,119] Much the same principles and limitations
apply, but fewer data are available for critical test.

[116] E. Clar and M. Zander, *J. Chem. Soc.*, **1958**, 1861; E. Clar, C. T. Ironside, and
M. Zander, *J. Chem. Soc.*, **1959**, 142; E. Clar, *Tetrahedron* **5**, 98 (1959); **6**, 355 (1959).

[117] R. Pauncz and A. Cohen, *J. Chem. Soc.*, **1960**, 3288.

[118] T. H. Goodwin, *J. Chem. Soc.*, **1960**, 485.

[119] M. G. Evans and J. de Heer, *Acta Cryst.*, **2**, 363 (1949); E. G. Cox and G. A.
Jeffrey, *Proc. Roy. Soc.*, **A207**, 110 (1951); T. Anno and A. Sado, *Bull. Chem. Soc.
Japan*, **28**, 350 (1955); F. H. Herbstein and G. M. J. Schmidt, *Acta Cryst.*, **8**, 406 (1955);
T. H. Goodwin, *J. Chem. Soc.*, **1955**, 4451; T. H. Goodwin and A. L. Porte, *J. Chem.
Soc.*, **1956**, 3595; J. C. Patel, *J. Sci. Ind. Research*, **16B**, 370 (1957); T. Anno, M. Ito,
R. Shimada, A. Sado, and W. Mizushima, *Bull. Chem. Soc. Japan*, **30**, 638 (1957);
E. L. Wagner, *J. Phys. Chem.*, **63**, 1403 (1959).

7 Electron affinity and ionization potential

7.1 Polarographic Reduction Potentials

Polarographic reductions are usually carried out with a dropping mercury electrode. The cathode is a mercury reservoir which opens through a capillary into the electrolyte solution. The anode is a pool of mercury or an associated standard half cell such as the saturated calomel electrode (SCE). The solution is unstirred and usually contains an electrolyte and frequently a buffer. Mercury droplets of constant size issue from the capillary at a constant rate; hence the electrode surface is constantly renewed. Organic molecules diffusing to the cathode will become reduced if the potential at the mercury surface is sufficient for reaction. The solution contains an electrolyte, so that the potential drop between electrodes is small and the applied potential essentially becomes the potential at the mercury surface. If this potential is gradually increased, a point is reached at which the organic compound is reduced at a rate that depends on the rate of diffusion of molecules to the mercury droplet. The increased current then flowing is called the *diffusion current*. A plot of current versus potential gives an *S*-shaped curve, shown in Fig 7.1. The potential at the mid-point of the diffusion current is known as the half-wave potential, $\epsilon_{1/2}$, and is relatively constant to operating parameters. The diffusion current depends on the drop size and rate, the diffusion constant and concentration of the substance being reduced, and the number of electrons involved in the reduction. Because of the last dependency, polarography is an important analytical tool. The half-wave potential is characteristic of the substance being reduced. In a *reversible reduction* the reduction product is sufficiently stable to diffuse into the solution; the half-wave potential for such a reduction depends on the relative diffusion constants of reactant and reduced product and on the standard potential for the reaction. An example is the polarographic reduction of quinones (*vide infra*). Most reductions of organic

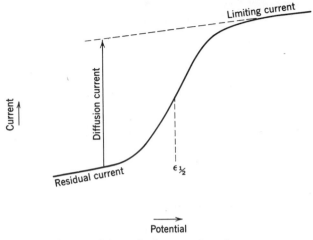

Fig. 7.1   Schematic diagram of a polarogram.

compounds are *irreversible*, usually because of appreciable overvoltage due to some slow kinetic step in the reaction.[1,2]

Laitinen and Wawzonek found that phenyl-substituted olefins and acetylenes[3] and aromatic polynuclear hydrocarbons[4] are reduced at the dropping mercury electrode and give reproducible half-wave potentials. The first reduction wave of these compounds corresponds to the approximately reversible addition of one or two electrons to the compound.[3,4,5,6,7,8,9,10]

[1] For further details of polarographic techniques and applications to organic systems see J. E. Page, *Quart. Revs.*, **6**, 262 (1952); I. M. Kolthoff and J. J. Lingane, *Polarography*, second edition, Interscience Publishers, New York (1952), especially part 4 in vol. 2; L. Meites, *Polarographic Techniques*, Interscience Publishers, New York (1955); G. W. C. Milner, *The Principles and Applications of Polarography and Other Electroanalytical Processes*, Longmans, Green, New York (1957), especially part III; H. W. Nürnberg, *Angew. Chem.*, **72**, 433 (1960).

[2] A summary of half-wave potentials of organic compounds is available in K. Schwabe, *Polarographie und chemische Konstitution organischer Verbindungen*, Akademic Verlag, Berlin (1957).

[3] H. A. Laitinen and S. Wawzonek, *J. Am. Chem. Soc.*, **64**, 1765 (1942).

[4] S. Wawzonek and H. A. Laitinen, *J. Am. Chem. Soc.*, **64**, 2365 (1942).

[5] S. Wawzonek and J. W. Fan, *J. Am. Chem. Soc.*, **68**, 2541 (1946).

[6] G. J. Hoijtink and J. van Schooten, *Rec. trav. chim.*, **71**, 1089 (1952).

[7] *Ibid.*, **72**, 691 (1953).

[8] G. J. Hoijtink, J. van Schooten, E. de Boer and W. I. Aalbersberg, *Rec. trav. chim.*, **73**, 355 (1954).

[9] G. J. Hoijtink, *Rec. trav. chim.*, **73**, 895 (1954).

[10] A. C. Aten, C. Büthker, and G. J. Hoijtink, *Trans. Faraday Soc.*, **55**, 324 (1959); K. Schwabe and E. Schmidt, *Z. physik. Chem.* (Leipzig), Sonderheft, **1958**, 278; K. Schwabe, *Z. physik. Chem.* (Leipzig), Sonderheft, **1958**, 289; H. J. Gardner, *Nature*, **183**, 320 (1959).

This process is clearly related to the reaction of the hydrocarbons with alkali metals (Sec. 6.5). This relationship has been affirmed not only by product studies (Sec. 14.3) but also by comparison of the ESR and optical spectra of the resulting ion with those of the alkali salts.[11,12] The polarographic reduction potentials parallel closely the qualitative relative stabilities of the alkali salts.[12,13]

Hückel[14] showed rather early how the chemistry of the alkali addition compounds of aromatic hydrocarbons agrees with the simple molecular-orbital picture in which additional electrons have been added to anti-bonding MO's of the π-systems. Following this lead, Maccoll[15] showed that the polarographic half-wave reduction potentials of several unsaturated hydrocarbons are linearly correlated with the HMO energy of the lowest vacant MO's of the hydrocarbons. This correlation was rapidly confirmed and interpreted.[6,16,17]

The *electron affinity*, A, of a molecule is the energy liberated when an electron adds to the molecule in the gas phase (1). A is usually a positive quantity.

$$R + e \rightarrow R^- \qquad \Delta H^\circ = -A \tag{1}$$

The free-energy change represented by the half-wave potential of a reversible one-electron addition is

$$\Delta F^\circ = (F_R^\circ)_{aq} - (F_{R^-}^\circ)_{aq} + (F_{electron}^\circ)_{Hg} \tag{2}$$

or
$$\Delta F^\circ = (F_R^\circ)_{gas} - (F_{R^-}^\circ)_{gas} + (F_{electron}^\circ)_{Hg} + \Delta\Delta F_{solv}^\circ \tag{3}$$

$\Delta\Delta F_{solv}^\circ$ is the difference in free energies of solvation of the hydrocarbon and its anion. In the absence of entropy effects, we may write (3) as

$$\Delta F^\circ = A + (F_{electron}^\circ)_{Hg} + \Delta\Delta F_{solv}^\circ \tag{4}$$

The half-wave potential, $\epsilon_{1/2}$, is given by

$$\epsilon_{1/2} = \frac{\Delta F^\circ}{\mathscr{F}} - \frac{RT}{\mathscr{F}} \ln \frac{D_R}{D_{R^-}} \tag{5}$$

[11] D. E. G. Austen, P. H. Given, D. J. E. Ingram, and M. E. Peover, *Nature*, **182,** 1784 (1958).

[12] G. J. Hoijtink and P. H. van der Meij, *Z. physik. Chem.*, NF., **20,** 1 (1959).

[13] G. J. Hoijtink, E. de Boer, P. H. van der Meij, and W. P. Weijland, *Rec. trav. chim.*, **75,** 487 (1956).

[14] E. Hückel, International Conference on Physics, London, 1934, Vol. II, The Physical Society, London (1935).

[15] A. Maccoll, *Nature*, **163,** 178 (1949).

[16] A. Pullman, B. Pullman, and G. Berthier, *Bull. soc. chim. France*, **17,** 591 (1950).

[17] L. E. Lyons, *Nature*, **166,** 193 (1950).

D_R and D_{R^-} are the diffusion constants of R and R^-, respectively, and $\mathscr{F}$ is the Faraday. In comparing a series of hydrocarbons, the term involving the diffusion constants will be small because D_R and D_{R^-} are nearly the same. The free energy of an electron in the mercury droplet is constant for the series. It would be nice if $\Delta\Delta F^\circ_{solv}$ were also a constant for the series, for then we could write

$$\epsilon_{1/2} = -(\alpha_0 + m_{m+1}\beta_0) + c' \tag{6}$$

in which we equate the electron affinity with the negative of the energy of the lowest vacant MO, $\epsilon_{m+1} = \alpha_0 + m_{m+1}\beta_0$; c' is then a constant for the series.[18] Actually, however, $\Delta\Delta F^\circ_{solv}$ is probably not independent of the hydrocarbon. It will become smaller as the hydrocarbon becomes larger;[19] however, as π-systems become larger, ϵ_{m+1} usually becomes smaller. Hence we may expect a rough proportionality between $\Delta\Delta F^\circ_{solv}$ and ϵ_{m+1}. Equation 6 then becomes

$$\epsilon_{1/2} = -bm_{m+1} + c \tag{7}$$

The slope of the correlation line is now also a parameter which may be regarded as the effective value of β for this system.

Table 7.1 summarizes polarographic half-wave reduction potentials for a number of hydrocarbons in aqueous dioxane and in 2-methoxyethanol. The m_{m+1} coefficient of the energy of the lowest vacant MO is also listed for each compound. Data for 75% aqueous dioxane include a variety of types of hydrocarbons—polycyclic aromatics, polyenes, and even triphenylmethyl radical. A number of measurements made in 96% aqueous dioxane gave only slight differences in half-wave potentials from 75% aqueous dioxane.[7,8,9,20] This result implies that differences in solvation energies must be small. A plot of the aqueous-dioxane data against m_{m+1} gives a good linear correlation (Fig. 7.2).[20] The least squares correlation line is

$$\epsilon_{1/2} \text{ (aq. diox.)} = (2.368 \pm 0.099)m_{m+1} - 0.924 \pm 0.109 \tag{8}$$

The slope corresponds to an effective value of β for this correlation of -2.37 ev or -54.6 kcal./mole. At least part of the scatter in Fig. 7.2

[18] The above derivation is essentially that of Hoijtink and Van Schooten, ref. 6.

[19] To the approximation that the difference in solvation energies is the result of a Born charging process, $\Delta\Delta F^\circ_{solv}$ is proportional to the reciprocal of the effective radius of the molecule. For estimates of this energy see L. E. Lyons, *Nature*, **166**, 193 (1950) and R. M. Hedges and F. A. Matsen, *J. Chem. Phys.*, **28**, 950 (1958).

[20] G. J. Hoijtink, *Rec. trav. chim.*, **74**, 1525 (1955).

[21] C. A. Coulson and R. Daudel, *Dictionary of Values of Molecular Constants*, Centre de Chimie Théorique de France, Paris.

[22] A. Streitwieser, Jr., and J. B. Bush, unpublished calculations.

[23] M. V. Stackelburg and W. Strakke, *Z. Elektrochem.*, **53**, 118 (1949).

[24] I. Bergman, *Trans. Faraday Soc.*, **50**, 829 (1954).

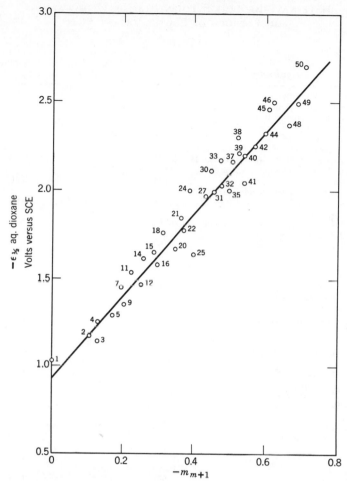

Fig. 7.2   Half-wave potentials in aqueous dioxane versus energies of lowest vacant orbitals. Numbers correspond to compounds in Table 7.1.

results from combining aromatic compounds and polyenes in the same correlation. The alternating bonds in the latter compounds require different values of β which will then change m_{m+1}. To the first approximation, a correction in m_{m+1} can be made by the first term of the power series discussed earlier (Sec. 4.3) but applied to the energy of a single MO:

$$\delta m_{m+1} = 2 \sum_{\substack{\text{all} \\ \text{bonds}}} c_{m+1,r} c_{m+1,s}\, \delta\beta_{rs} \tag{9}$$

Applied to styrene, for example, changing β of the single and double bond of the side chain to $0.9\beta_0$ and $1.1\beta_0$, respectively, changes $-m_{m+1}$ from

TABLE 7.1

HALF-WAVE REDUCTION POTENTIALS FOR HYDROCARBONS
AND ENERGIES OF LOWEST VACANT MOs

Nos. in Figs. 2 and 3	Hydrocarbon	$-m^*_{m+1}$	$-\epsilon_{1/2}$ versus SCE 75% aq. diox.†	$-\epsilon_{1/2}$ versus Hg pool 2-Methoxyethanol‡
1	Triphenylmethyl	0	1.05	
2	1,6-Dibiphenylenehexatriene, I	0.105	1.17§	
3	1,4-Dibiphenylenebutadiene	0.128	1.14§	
4	1-Phenyl-12-biphenylenedodecahexaene	0.129	1.25§	
5	1-Phenyl-8-biphenyleneoctatetraene	0.170	1.29§	
6	3,4-(peri-Phenylene)fluoranthene, II	0.186		0.90
7	1,12-Diphenyldodecahexaene	0.196	1.45§	
8	Zethrene, III	0.199		0.46
9	1-Phenyl-6-biphenylenehexatriene	0.202	1.35	
10	Pentacene, IV	0.220		0.86
11	1,10-Diphenyldecapentaene	0.224	1.54§	
12	1-Phenyl-4-biphenylenebutadiene	0.251	1.46	
13	1,2-Benzfluoranthene, V	0.252		0.975
14	1,8-Diphenyloctatetrene	0.260	1.62§	
15	Acenaphthylene	0.285	1.65	
16	Anthanthrene, VI	0.291		1.19
17	Tetracene	0.295	1.58	1.135
18	1,6-Diphenylhexatriene	0.312	1.76§	
19	7,8-Benzfluoranthene	0.312		1.165
20	Perylene, VII	0.347	1.67	1.25
21	1,2-Benzpyrene, VIII	0.365	1.85	1.36
22	Fluoranthene	0.371	1.77	1.345
23	2,3-Benzfluoranthene	0.377		1.375
24	1,4-Diphenylbutadiene	0.386	2.00	
25	Azulene	0.400	1.64	
26	8,9-Benzfluoranthene	0.401		1.39
27	Anthracene	0.414	1.96	1.46
28	2,3,6,7-Dibenzophenanthrene	0.437		1.525
29	1,12-Benzperylene, IX	0.439		1.485
30	Pyrene	0.445	2.11	1.61
31	1,2-Benzanthracene	0.452	2.00	1.53
32	1,2,5,6-Dibenzanthracene	0.474	2.03	1.545
33	1-Phenylbutadiene	0.474	2.17§	
34	1,2,7,8-Dibenzanthracene	0.492		1.57
35	4,5-Benzpyrene	0.497	2.00	1.67
36	Picene, X	0.502		1.79
37	Stilbene	0.504	2.16	
38	Chrysene, XI	0.520	2.30	1.805
39	2,2'-Binaphthyl	0.521	2.21	
40	p-Quaterphenyl	0.536	2.20	1.765
41	Coronene, XII	0.539	2.04	1.64
42	1,1-Diphenylethylene	0.565	2.25	
43	3,4-Benzphenanthrene	0.568		1.745
44	p-Terphenyl	0.593	2.33	1.905
45	Phenanthrene	0.605	2.46	1.935
46	Naphthalene	0.618	2.50	1.98
47	Butadiene	0.618	2.63	
48	Styrene	0.662	2.37	
49	Triphenylene, XIII	0.684	2.49	1.97
50	Biphenyl	0.705	2.70	2.075

* $\alpha_0 + m_{m+1}\beta_0$ is the energy of the lowest vacant orbital in units of the standard β in the HMO approximation with all α's and β's equal. Data taken from refs. 20, 21, and 22.
† Data of refs. 3, 4, 5, 6, 7, 8, 9, 20, and 23.
‡ Data of ref. 24.
§ In 96% aqueous dioxane.

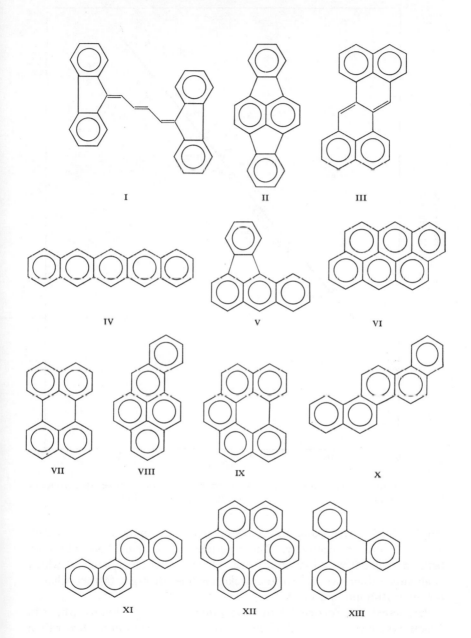

I   II   III

IV   V   VI

VII   VIII   IX   X

XI   XII   XIII

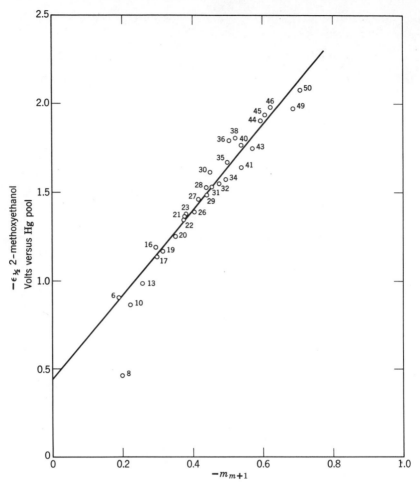

Fig. 7.3   Half-wave reduction potentials for aromatic hydrocarbons versus energies of lowest vacant orbitals. Numbers correspond to compounds in Table 7.1.

0.662 to 0.589. According to (8), these values correspond to $-\epsilon_{1/2} = 2.49$ and 2.32 volts, respectively. The experimental value of 2.36 volts is in better agreement with the latter value. On the other hand, azulene, which is already rather far from the correlation line, diverges farther when β for the 9,10-bond is decreased.

Bergman[24] has reported half-wave potentials for 78 hydrocarbons in 2-methoxyethanol. MO calculations have been carried out for fewer than half of these compounds. The values are also summarized in Table 7.1 and are plotted in Fig. 7.3. The thirty compounds in this correlation are all polycyclic aromatic hydrocarbons and give an excellent correlation, which is represented by the least squares line,[25]

$$\epsilon_{1/2}(\text{2-methoxyethanol}) = (2.414 \pm 0.092)m_{m+1} - 0.435 \pm 0.065 \quad (10)$$

The correlation is rather good; the standard deviation of a point from the line is only 5% of the total range covered. All of these benzenoid hydrocarbons may be considered to represent a family of related compounds. Even the non-AHs are benzenoid in type (e.g., fluoranthene). Correlations are clearly improved by restricting compounds to a particular class. The slope of the line which represents the effective value of β for this correlation is -2.41 ev, or -55.6 kcal./mole, in rather good agreement with the aqueous-dioxane results. The difference in the absolute values of $\epsilon_{1/2}$, compared with the aqueous-dioxane results, is due to a difference in the reference electrode. The aqueous-dioxane results are referred to the SCE, those in 2-methoxyethanol are referred to the mercury-pool anode. The correction amounts to about 0.5 volt. Indeed, there seems to be surprisingly little difference in half-wave potentials of hydrocarbons from one solvent to another. Reduction in dimethylformamide gives values varying by only hundredths of a volt from similarly referenced values in the other solvents.[26] This constancy in absolute energy values as well as slopes of correlation lines in widely varying solvents must mean that solvation energies, if not small, are changing in the same way from system to system. The contribution of solvation energies to the "effective" β's may well be small; these values may be close to the "true" value for β_0.

Compounds that are seriously nonplanar do not fit the correlation. 9,9'-Bianthryl has $\epsilon_{1/2}$ identical with anthracene; the two anthracene rings are expected to be at nearly right angles to one another in this hydrocarbon. Hoijtink[20] has demonstrated that in several other partially twisted hydrocarbons, such as dibiphenyleneethylene, XIV, and 1,1'-binaphthyl,

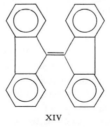

XIV

the experimental half-wave potentials are intermediate between those of the planar and the perpendicular systems.

[25] Zethrene was omitted from this correlation because of its extreme deviation.
[26] P. H. Given, *J. Chem. Soc.*, **1958**, 2684.

Half-wave potentials may also be compared with orbital energies with overlap included. We then obtain correlations as good as, but no better than, those obtained with neglect of overlap.[16,20,27,28]

The observed correlations with hydrocarbons can, in principle, be used with substituted compounds to derive h- and k-parameter values for heteroatoms. This application is complicated by the reduction of many functional groups at the dropping mercury electrode. Halides, nitro groups, etc., give reduction waves of their own. Fueno et al.[29] have demonstrated reasonable correlations with substituted styrenes and ethylenes by using a perturbation technique with assumed parameter values.

Methyl groups offer no complications. By using the inductive model (Sec. 5.7), the α_r of the attached carbon is altered by an amount $h\beta$. A reasonable value of h for the present purpose seems to be about -0.2.[28] The effect of this change on the energy of the lowest vacant orbital to the first approximation is represented by[30]

$$\delta m_j = \sum_r c_{jr}^2 h_r \tag{11}$$

From δm_{m+1} the change in $\epsilon_{1/2}$ is predicted from (8) and (10). In Table 7.2 we compare experimental and calculated $\Delta\epsilon_{1/2}$ for compounds for which suitable good data are available.[31] The prediction that all methyl compounds should be more difficult to reduce than the parent hydrocarbons is apparently borne out. The magnitudes of the shifts are small but are calculated fairly well for the benzenoid derivatives; however, the calculated changes for the methylazulenes are in poor agreement with the

[27] G. J. Hoijtink and J. van Schooten, *Rec. trav. chim.*, **72**, 903 (1953); P. Balk, S. de Bruijn, and G. J. Hoijtink, *Rec. trav. chim.*, **76**, 860 (1957).

[28] Correlation has also been proposed between half-wave potentials and *para*-localization energies (Sec. 15.1) [S. Basu and R. Bhattacharya, *J. Chem. Phys.*, **25**, 596 (1956), *Naturwissenschaften*, **45**, 208 (1958)]. This proposal has been vigorously criticized [A. Pullman and B. Pullman, *Compt. rend.*, **243**, 1632 (1956); J. I. F. Alonso and R. Domingo, *Nature*, **179**, 829 (1957)]. The empirically observed correlation has little physical significance and has been explained by P. H. Given, *Nature*, **181**, 1001 (1958).

[29] T. Fueno, K. Morokuma, and J. Furukawa, *Bull. Inst. Chem., Research, Kyoto Univ.*, **36**, 87 (1958), *Nippon Kagaku Zasshi*, **79**, 116 (1957); T. Fueno, K. Asada, K. Morokuma, and J. Furukawa, *Bull. Chem. Soc. Japan*, **32**, 1003 (1959), *J. Polymer Sci.*, **40**, 511 (1959). Parameter values used are $h_{O}' = 2, h_{O}'' = 1, k_{CO} = \sqrt{2}, h_{C1} = 1.8$, $h_N = 1$. An inductive parameter of 0.1 was used. On the basis of the discussion in Sec. 5.1, we should judge that the values used for h_{O}' and h_{O}'' should be reversed.

[30] $c_{m+1,r}^2$ also determines the hfs constant of the ESR spectrum of the corresponding radical anion (Sec. 6.5). Hence we expect a correlation between the hfs constant of a particular hydrogen and the change in $\epsilon_{1/2}$ produced by methyl substitution at that position. Insufficient data are available at present to test this prediction.

[31] This method was used by L. E. Lyons, *Research*, **2**, 587 (1949), and by Hoijtink and van Schooten, ref. 7.

extensive experimental results of Chopard-dit-Jean and Heilbronner.[32] It will be recalled that azulene itself does not fit the correlation of Fig. 7.2.

When the potential-determining step includes the addition of a proton, the half-wave potential is pH-dependent. This situation obtains, for example, with heterocyclic compounds such as pyridine and quinoline.

TABLE 7.2

EFFECT OF METHYL SUBSTITUENTS ON HALF-WAVE
REDUCTION POTENTIALS

Hydrocarbon	$-\Delta\epsilon_{1/2}$ Experimental*	$-\Delta\epsilon_{1/2}$ Calculated
Isoprene	0.04–0.10	0.07
α-Methylstyrene	0.04	0.07
β-Methylstyrene	0.19	0.17
Indene	<0.19	0.22
3-Methylcholanthrene	0.08	0.16
3-Phenylindene	0.10–0.16	0.21
1,2-Dihydronaphthalene	0.22	0.22
Acenaphthene	0.11	0.17
1-Methylazulene	0.06	0.00
2-Methylazulene	0.12	0.05
4-Methylazulene	0.08	0.11
5-Methylazulene	0.01	0.05
6-Methylazulene	0.08	0.13
1,2-Dimethylazulene	0.14	0.07
1,3-Dimethylazulene	0.10	0.00
1,4-Dimethylazulene	0.12	0.13
1,8-Dimethylazulene	0.10	0.13
4,7-Dimethylazulene	0.08	0.11
4,8-Dimethylazulene	0.12	0.21

* Refs. 3, 4, 5, 29, 33.

However proton donation does not occur in the potential-determining step in dimethylformamide,[26] and the observed difference in $\epsilon_{1/2}$ between naphthalene and quinoline, $+0.46$ volt,[26] may be combined with (11) to derive $h_{\ddot{N}} = 1.1$. This value is somewhat higher than the usual value, $h_{\ddot{N}} = 0.5$ (Sec. 5.2), and further tests with additional data would be useful.

Carbonyl compounds also give a pH-dependent wave indicative of the potential-determining reaction:

$$RR'C{=}O + H^+ + e \rightarrow RR'COH$$

[32] L. H. Chopard-dit-Jean and E. Heilbronner, *Helv. Chim. Acta*, **36**, 144 (1953).

At high pH, however, $\epsilon_{1/2}$ becomes independent of further pH change and apparently results from the reaction[33]

$$RR'C{=}O + e \rightarrow RR'\dot{C}{-}O^-$$

Schmid and Heilbronner[33] found that the $\epsilon_{1/2}$ for the latter reaction correlates well with an MO energy difference calculated between the aldehyde and the radical.[34] Their energy values also produce reasonable correlations with the energies of the lowest vacant orbitals. With the same parameter values and other $\epsilon_{1/2}$ data for ketones, Fueno et al.[35] also observed similar correlations. Although the slopes of the correlation lines are reasonably close to those obtained above, the positions of the lines are greatly shifted from those of the hydrocarbon correlations; that is, points for $\epsilon_{1/2}$ and m_{m+1} (for $h_0 = 2$, $k_{CO} = \sqrt{2}$) fall far from the previous correlation lines.

TABLE 7.3
POLAROGRAPHY OF AROMATIC ALDEHYDES

Aldehyde	$-\epsilon_{1/2}$ versus SCE*	$-m_{m+1}$†
Benzaldehyde	1.592	0.386
Cinnamaldehyde	1.335	0.260
1-Naphthalenealdehyde	1.476	0.303
2-Naphthalenealdehyde	1.505	0.350
9-Anthracenealdehyde	1.207	0.199
9-Phenanthrenealdehyde	1.449	0.300

* Data of ref. 33. The $\epsilon_{1/2}$ given in this reference have been corrected for the difference between the normal calomel electrode used and the SCE.

† Ref. 36.

Benzophenone may be regarded as a perturbed diphenylethylene. The observed $\epsilon_{1/2}$, -1.42 volts, together with (8) and (11), yields $h_0 = 0.9$. This value is close to the values, $k_{C=O} = 1$, $h_0 = 1$, considered in our earlier discussion (Sec 5.2). Use of the latter values in calculations with Schmid and Heilbronner's aromatic aldehydes produces the results summarized in Table 7.3 and plotted in Fig. 7.4.[36] The divergence of the correlation line from that in Fig. 7.2 probably is caused by a difference

[33] R. W. Schmid and E. Heilbronner, *Helv. Chim. Acta*, **37**, 1453 (1954).

[34] Aldehydes were calculated with $h_0 = 2$, $k_{CO} = \sqrt{2}$. The radical anions were treated as the corresponding hydrocarbon radical; i.e., the π-system with the —O— dropped. Overlap was included.

[35] T. Fueno, K. Morokuma, and J. Furukawa, *Bull. Inst. Chem. Research, Kyoto Univ.*, **36**, 96 (1958).

[36] A. Streitwieser, Jr., and J. B. Bush, unpublished calculations.

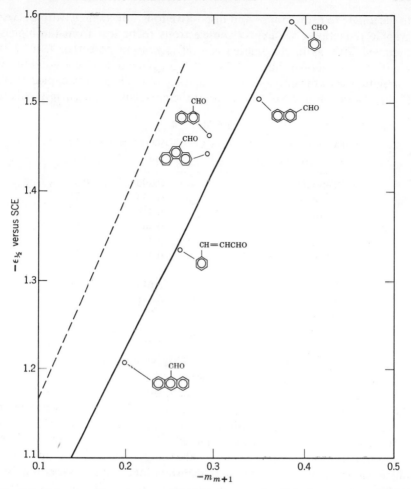

Fig. 7.4 Polarographic reduction potentials of aromatic aldehydes versus energies of lowest vacant MO's calculated with $h_O = 1$, $k_{C=O} = 1$. The dotted line is the correlation line in Fig. 7.2.

in solvent and an inadequate $h_{\dot{O}}$. Use of $h_{\dot{O}} \simeq 1.6$ would give better agreement.

7.2 Polarographic Oxidation Potentials

Polarographic oxidations of aromatic hydrocarbons in acetonitrile solution have been carried out at a rotating platinum electrode as an anode.[37] Electrochemical oxidation involves loss of one or two electrons

[37] H. Lund, *Acta Chem. Scand.*, **11**, 1323 (1957).

to the anode to form mono- and di-positive ions. By analogy with polaro-graphic reductions, we expect the electrons to be lost from the highest occupied MO, as in the related case of ionization potentials (Sec. 7.3). Hoijtink[38] has shown that Lund's half-wave potentials give a good linear correlation with the energy coefficients, m_m, of the highest occupied MO's (Table 7.4 and Fig. 7.5). The slope of the correlation line corresponds to

TABLE 7.4

POLAROGRAPHIC OXIDATION OF AROMATIC HYDROCARBONS

Hydrocarbon	m_m	$\epsilon_{1/2}$
Tetracene	0.295	0.54
1,2-Benzopyrene	0.365	0.76
Anthracene	0.414	0.84
Pyrene	0.445	0.86
1,2-Benzanthracene	0.452	0.92
1,2,5,6-Dibenzanthracene	0.477	1.00
Phenanthrene	0.605	1.23
Fluoranthene	0.618	1.18
Naphthalene	0.618	1.31
Biphenyl	0.704	1.48
Benzene	1.000	2.00

$\beta = -2.05$ ev, or -47.4 kcal./mole. This value is not far from those obtained from polarographic reductions.

For AH the m-values of highest occupied and lowest vacant MO's differ only in sign; hence a plot of polarographic reduction potentials versus oxidation potentials should give a straight line. Hoijtink[38] showed that such a linear correlation is indeed obtained and, moreover, that fluoranthene, a non-AH for which the magnitudes of m_m and m_{m+1} differ widely, deviates far from this correlation. This observation is a striking success of the simple Hückel theory and justifies the validity of the AH-non-AH distinction.

Fieser[39] has developed a reproducible electrolytic potential, called the "critical potential," which seems to be a measure of the reaction

$$ArO^- \rightarrow ArO \cdot + e$$

We would expect this potential to correlate with the energy of the highest occupied MO in the phenolate anion. ArO^- can be derived from $ArCH_2^-$ by altering the Coulomb value of the exocyclic carbon; the corresponding change in m_j is then given by (11). However, the highest occupied orbital of the isoconjugate odd-AH anion is the NBMO; $c_{m,r}^2$ is the charge

[38] G. J. Hoijtink, *Rec. trav. chim.*, **77**, 555 (1958).
[39] L. F. Fieser, *J. Am. Chem. Soc.*, **52**, 5204 (1930).

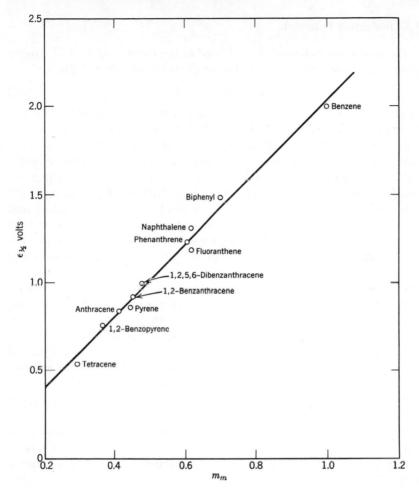

Fig. 7.5   Polarographic oxidation half-wave potential versus energies of highest occupied MO's.

density, ζ_r, of the exocyclic carbon. A plot of Fieser's critical potentials for phenols versus ζ_r for the isoconjugate carbanions does indeed give a good straight line with a slope reported to be -3.0[40] and -2.807.[41] To a first approximation, this slope is identified with the product of the effective β and $h_{\ddot{0}-}$. If β is taken as -2.0 to -2.4 ev, $h_{\ddot{0}-} = 1.2$–1.5, a value that does not seem unreasonable for a negatively charged oxygen donating two electrons (Sec. 5.2).

[40] N. S. Hush, *J. Chem. Soc.*, **1953**, 2375.
[41] T. Fueno, T. Lee, and H. Eyring, *J. Phys. Chem.*, **63**, 1940 (1959).

7.3   Ionization Potentials

The ionization potential, I, is defined as the energy required to remove an electron from a molecule or atom in the dilute gas phase (12).

$$R \rightarrow R^+ + e; \qquad \Delta H = I \qquad\qquad (12)$$

In principle, any number of ionization potentials correspond to various energies of the liberated electron and to different states of the resulting cation. Unless otherwise qualified, the ionization potential is assumed to refer to the most weakly bound electron in the molecule. Ionization potentials are important theoretically because they correspond at least approximately to the energy of the highest occupied MO or AO of the system.[42]

Ionization potentials of organic molecules have been determined principally by three methods: spectroscopy (UV), electron impact (EI), and photoionization (PI). The ultraviolet spectra of organic compounds depend on electronic transitions from the ground state to various excited states (Chap. 8). Transitions of higher energy (usually in the vacuum ultraviolet) concern excited states in which an electron is in a MO far from the remainder of the molecule. In such situations the MO resembles a high-energy AO around a "united atom" (in which the remaining electrons and nuclei are collapsed into a single nucleus). Hence such transitions resemble the familiar Rydberg series of atomic spectra.[43] Extrapolation of the effective quantum number for the series gives the transition that corresponds to the ionization potential of the molecule.[44] If the excited states can be properly identified, this method can give accurate measures of I (± 0.01 ev or better).

In the electron-impact method a low pressure of the compound in the ionization chamber of a mass spectrograph is bombarded with electrons from a heated filament. The energy of the electrons is varied by a potential drop. Collision of a molecule with an electron having an energy greater than I can result in ionization:

$$R + e \rightarrow R^+ + 2e$$

The current produced by the cations in the collection chamber is measured. Naïvely, we would expect a relation between electron energy and cation current of the form of Fig. 7.6A. The energy at which cations are just formed is the *appearance potential* and corresponds to I. In practice, the electrons generally have an energy spread with a Boltzmann distribution

[42] R. S. Mulliken, *Phys. Rev.*, **74**, 736 (1948).

[43] G. Herzberg, *Atomic Spectra and Atomic Structure*, second edition, Dover Publications, New York (1944).

[44] G. Herzberg, *Spectra of Diatomic Molecules*, second edition, D. Van Nostrand Co., Princeton, N. J. (1950), p. 327.

and a curve, shown in Fig. 7.6*B*, is usually obtained. Determination of the appearance potential is rather difficult, but procedures have been developed that yield *I* values reproducible to ±0.1–0.2 ev or better.[45]

Unfortunately, EI values are generally higher than UV-values, sometimes by 0.4 ev or more. The usual explanation for this difference is that EI-ionization is a Franck-Condon transition and yields a vertical ionization potential, whereas the spectroscopic value is an adiabatic ionization

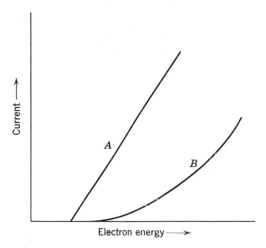

Fig. 7.6 Ionization efficiency curves in mass spectrometry.

potential.[46] The distinction is exemplified in Fig. 7.7. The transition is considered to be so fast that the nuclei do not effectively change their positions during the transition. The most probable transition, therefore, is to a vibrational level for which this nuclear configuration is relatively probable. This vibrational level is frequently above that of the zeroth level, the energy of which is determined by a proper analysis of the UV-spectra; hence vertical transitions are equal or greater in magnitude than adiabatic transitions. Recent doubt has been raised concerning the identification of EI-appearance potentials with vertical ionization potentials.[47] Thus there is considerable question of the exact significance of EI *I*-values, although they are reproducible and internally self-consistent.[48]

[45] For more detailed summaries see A. J. B. Robertson, *Mass Spectrometry*, Methuen and Co., London (1954); W. J. Dunning, *Quart. Revs.*, **9**, 23 (1955); and ref. 46.

[46] F. H. Field and J. L. Franklin, *Electron Impact Phenomena*, Academic Press, New York (1957).

[47] J. D. Morrison, *Rev. Pure and Appl. Chem.*, **5**, 22 (1955); *J. Chem. Phys.*, **21**, 1767 (1953); **29**, 1312 (1958).

[48] All *I*-values published through 1955 are summarized in an appendix of ref. 46.

In the photoionization method the vapor is irradiated with monochromatic light and the current generated by ionization is measured. Ideally, curve A in Fig. 7.6 illustrates the functional dependency of the current on the energy of the incident light. In practice, vibrational states produce irregularities in the curve, but proper analysis gives the lowest vibrational

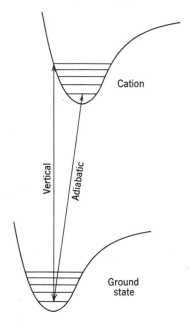

Fig. 7.7   Vertical (Franck-Condon) and adiabatic transitions in mass spectroscopy.

level of the cation and, therefore, an adiabatic ionization potential.[49] Monochromatic light is easier to obtain than monochromatic electrons; hence PI I-values can be obtained with precision approaching that of the UV-method. PI I-values are available for several hundred organic compounds, especially through the work of Watanabe.[49,50,51] These values generally agree closely with UV-values, when available. The principal deficiency of this method is that the cationic state is not identified. As usually analyzed, the procedure actually gives I-values intermediate between adiabatic and vertical transitions, although the differences from

[49] K. Watanabe, *J. Chem. Phys.*, **26,** 542 (1957).

[50] K. Watanabe and J. R. Mottl, *J. Chem. Phys.*, **26,** 1773 (1957);   K. Watanabe and T. Nakayama, ASTIA report AD-152934.

[51] Additional recent values have been measured by W. C. Price, R. Bralsford, P. V. Harris, and R. G. Ridley, *Spectrochim. Acta*, **14,** 45 (1959).

TABLE 7.5

I-VALUES FOR AROMATIC HYDROCARBONS

I, ev

Hydrocarbon	Wacks and Dibeler Ref. 53	Stevenson Ref. 54	Others	UV or PI	m_m
Benzene	9.38	9.57	9.43*	9.245**	1.000
			9.46†		
			9.52‡§‖		
			9.21¶		
Naphthalene	8.26	8.68		8.12**	0.618
Anthracene	7.55	8.20			0.414
Phenanthrene	8.03	8.62			0.605
Tetracene		7.71			0.295
3,4-Benzphenanthrene		8.40			0.568
Styrene			8.86‡	8.46††	0.662
Biphenyl				8.27††	0.705
Butadiene			9.24¶	9.07**	0.618
Ethylene			10.56¶	10.516**	1.000

* R. E. Honig, *J. Chem. Phys.*, **16**, 105 (1948).

† I. Omura, H. Baba, and K. Higasi, *J. Phys. Soc. Japan*, **10**, 317 (1955).

‡ J. D. Morrison and A. J. C. Nicholson, *J. Chem. Phys.*, **20**, 1021 (1952).

§ F. H. Field and J. L. Franklin, *J. Chem. Phys.*, **22**, 1895 (1954).

‖ K. Higasi, I. Omura, and H. Baba, *J. Chem. Phys.*, **24**, 623 (1956); I. Omura, K. Higasi, and H. Baba, *Bull. Chem. Soc. Japan*, **29**, 501 (1956); H. Baba, I. Omura, and K. Higasi, *Bull. Chem. Soc. Japan*, **29**, 521 (1956).

¶ Selected as "best value" by F. H. Field and J. L. Franklin.[48]

** Ref. 49.

†† Ref. 50.

true adiabatic values are generally small.[52] This deficiency is removed in what is apparently the most elegant technique yet used for the measurement of ionization potentials—a combination of the EI- and PI-methods in which vapor is irradiated with monochromatic light within a mass spectrometer, the product cations being collected in the usual manner.[52] Unfortunately, only a few *I*-values are available from this recent technique

I-values for a series of polycyclic aromatic hydrocarbons would be highly desirable to test a correlation similar to that for polarographic oxidation (Sec. 7.2). EI-values only are available for any extensive series of such compounds, and, unfortunately, the results from two sets of investigators are in almost total disagreement. In Table 7.5 we summarize

[52] H. Hurzeler, M. G. Inghram, and J. D. Morrison, *J. Chem. Phys.*, **28**, 76 (1958).

the I-values reported by Wacks and Dibeler[53] and by Stevenson.[54] Other values are included for comparison. Equating the ionization potential with the energy of the highest occupied MO gives

$$I = -(\alpha_0 + m_m\beta_0) \tag{13}$$

or, for a family of compounds,

$$I = bm_m + c \tag{14}$$

in which b is equated with $-\beta_0$ for this correlation. Both Stevenson's[54] and Wacks and Dibeler's data give good linear correlations with the energy of the highest occupied MO (Fig. 7.8). The correlations are represented respectively by

$$I = (2.48 \pm 0.17)m_m + 7.07 \pm 0.09^{56} \tag{15}$$

$$I = (3.14 \pm 0.24)m_m + 6.24 \pm 0.10 \tag{16}$$

The effective values of β_0 from the two correlations, -2.48 ev and -3.14 ev, respectively, differ substantially, and it is difficult at present to choose between them. Styrene lies closer to the Stevenson correlation, but the Wacks and Dibeler values are closer to the PI-values. Both values of β_0, however, are of the same order of magnitude as the values derived from polarographic reduction, indicating again that differences in energies of solvation cannot be important in the latter case.

Butadiene and ethylene fall far from either correlation line and demonstrate the limitations of application of HMO theory in this area. The UV I-values for ethylene, butadiene, and hexatriene, 10.51,[57] 9.07,[58] and 8.2[59] ev respectively, give a good linear correlation within themselves with m_m having a slope of about 4 ev. The points are shifted, but the slope is little changed if k-values for alternating double and single bonds are introduced.[60] Styrene does not fit this correlation.

The inadequacy of HMO theory is further demonstrated by the behavior of free radicals. All radicals of odd-AH's have one electron in a NBMO; hence, to the HMO approximation, all should have the same I. EI I-values are known for several such radicals and vary over a 2.6 ev range:

[53] M. E. Wacks and V. H. Dibeler, *J. Chem. Phys.*, **31**, 1557 (1959).
[54] D. P. Stevenson, private communication; cf. ref. 55.
[55] A. Streitwieser, Jr., and P. M. Nair, *Tetrahedron*, **5**, 149 (1959).
[56] A. Streitwieser, Jr., *J. Am. Chem. Soc.*, **82**, 4123 (1960).
[57] W. C. Price and W. T. Tutte, *Proc. Roy. Soc.*, **A174**, 207 (1940).
[58] W. C. Price and A. D. Walsh, *Proc. Roy. Soc.*, **A174**, 220 (1940).
[59] *Ibid.*, **A185**, 182 (1946).
[60] A. Streitwieser, Jr., unpublished calculations.

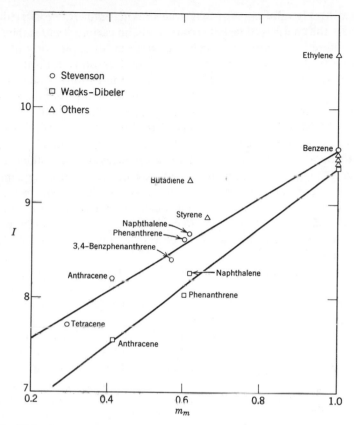

Fig. 7.8   EI I-values versus energy of highest occupied MO's for unsaturated hydrocarbons.

methyl, 9.95 ev;[61a] allyl, 8.16 ev;[61a] pentadienyl, 7.73 ev;[54] benzyl, 7.76 ev;[61b] and benzhydryl, 7.32 ev.[62] This situation undoubtedly results from neglect of electron repulsion effects in the simple HMO theory.[55,56] Explicit introduction of electron repulsion terms in more advanced MO methods produces an excellent correlation of I-values for various hydrocarbons and radicals.[63] Identification of $-I$ with the energy of the highest

[61] (a) F. P. Lossing, K. U. Ingold, and I. H. S. Henderson, *J. Chem. Phys.*, **22**, 621 (1954). (b) J. B. Farmer, I. H. S. Henderson, C. A. McDowell, and F. P. Lossing, *J. Chem. Phys.*, **22**, 1948 (1954).

[62] A. G. Harrison and F. P. Lossing, *J. Am. Chem. Soc.*, **82**, 1052 (1960).

[63] For a notable example see N. S. Hush and J. A. Pople, *Trans. Faraday Soc.*, **51**, 600 (1955). This method also gives a good account of the experimental I or *work function* of graphite, 4.39 ev,[64] which none of the simple LCAO methods can approach.

[64] A. Braun and G. Busch, *Helv. Phys. Acta*, **20**, 33 (1947).

occupied MO is valid only for SCF MO's. Alternatively, we can calculate separately the energies of neutral compound and cation; the I is then given as the difference. This is the technique used in an application of the ω-technique to I-values of these radicals as well as other radicals, olefins, and aromatic hydrocarbons. The energies of the cations were calculated after alteration of Coulomb integrals and, for $\omega = 1.4$, a good correlation was found with a slope corresponding to $\beta_0 = -2.11 \pm 0.05$ ev.[55,56]

Clearly, HMO theory can be applied to I only for a family of related compounds for which changes in electron repulsion will either remain constant or will vary roughly as does I; the sizes of the systems are important. The greatest change for the all-AH radical occurs in going from methyl to the others. The I-values for the comparably sized pentadienyl and benzyl radicals are almost identical. Furthermore, the observed ionization potentials of cyclopentadienyl (8.69 ev) and tropylium (6.60 ev) radicals agree qualitatively with the corresponding m_m-values, 0.618 and -0.445, respectively.[65]

Little has been published concerning LCAO theory and I of heteroatom systems. Matsen[66] has used I of HX systems to derive h_X-values, but these values have not been used in further calculations of I for other compounds. Using an *equivalent orbital* method which utilizes a secular determinant similar to that in simple LCAO practice, Hall[67] achieved a satisfactory correlation with methyl-substituted ethylenes, although a similar treatment of the chloroethylenes was unsuccessful. Hall used what amounts to the heteroatom model for a methyl group. With this model and the parameter values, $h_{Me} = 1.65$, $k_{C-Me} = 0.64$, Stevenson[68] got good agreement in calculations of I-values for substituted ethylenes and alkyl radicals. Franklin[69] has extended the equivalent orbital method to a variety of compounds by assigning empirical "group parameters" to structural units. Although rather similar in form to a LCAO technique, this method differs in concept and in the significance attached to the parameters used. The ω-technique of the simple LCAO method has been applied recently to the calculation of ionization potentials of a variety of Cl-, O-, and N-substituted hydrocarbons and radicals.[56]

[65] A. G. Harrison, L. R. Honnen, H. J. Dauben, Jr., and F. P. Lossing, *J. Am. Chem. Soc.*, **82**, 5593 (1960). It is interesting to note that by interpolation of these values alone a radical having $m_m = 0$ should have $I = 7.5$ ev, a value close to those observed for benzyl and benzhydryl radicals.

[66] F. A. Matsen, *J. Am. Chem. Soc.*, **72**, 5243 (1950).

[67] G. G. Hall, *Trans. Faraday Soc.*, **49**, 113 (1953).

[68] D. P. Stevenson, private communication; cf. Symposium on Mechanisms of Homogeneous and Heterogeneous Hydrocarbon Reactions, American Chemical Society, Kansas City, March, 1954, Division of Petroleum Chemistry, Abstracts, No. 29, p. 19.

[69] J. L. Franklin, *J. Chem. Phys.*, **22**, 1304 (1954).

The simple theory might find some use in following the way I changes for substitution of a given system. For example, using the heteroatom model of a methyl group with $h = 2$, $k = 0.8$, the change from benzene to toluene, m-xylene, and p-xylene is accompanied by a change in m_m from 1.00 to 0.83, 0.81, and 0.74, respectively. For $\beta_0 \cong -3$ ev, the corresponding calculated changes in I, -0.51, -0.57, and -0.78 ev may be compared with the experimental PI-changes, -0.42, -0.68, and -0.80 ev,[49] respectively.[60] Similarly, the calculated and experimental[49,50] PI I-changes from ethylene to the series, propylene, 2-butene, 2,3-dimethyl-2-butene, are, respectively, -0.69, -0.78; -1.32, -1.38; -2.22, -2.21 ev.[60]

Another possible application results from an examination of the nature of the highest occupied MO. For example, for compounds of the furan type, XV, in which both h_X and k_{CX} are greater than 0.5, the highest occupied MO has a node through X because of the symmetry of the molecule (Sec. 3.8).

XV

Hence, to the simple LCAO approximation, I should be independent of X and, indeed, should be the same as that for butadiene. EI I-values for butadiene, furan, pyrrole and cyclopentadiene are 9.24,[48] 9.03,[70,71] 8.97,[48,72] and 8.97[73] ev, respectively, in excellent agreement with the prediction. Agreement is almost as good with the UV-values: 9.07,[58] 9.06,[74] 8.95,[74] and 8.58[74] ev, respectively.[75] The same method can be applied to the simple alkyl radicals, but $\beta_0 = -3.5$ ev is required for a satisfactory correlation. Starting with I for methyl radical as 9.95 ev,[61] calculated and EI experimental values are ethyl, 8.97, 8.67–8.78[76] ev; isopropyl, 7.99, 7.90[76a] ev; t-butyl, 7.47, 7.42[77] ev. The pronounced effect of methyl substitution is explicable in terms of hyperconjugation stabilization of the cation; part of the positive charge is borne by the methyl group. It may be mentioned that there is no support in ionization potentials for the usual

[70] J. D. Morrison and A. J. C. Nicholson, *J. Chem. Phys.*, **20**, 1021 (1952).

[71] I. Omura, K. Higasi, and H. Baba, *Bull. Chem. Soc. Japan*, **29**, 501, 521 (1956).

[72] I. Omura, H. Baba, and K. Higasi, *J. Phys. Soc. Japan*, **10**, 317 (1955).

[73] J. Hissel, *Bull. soc. roy. sci., Liége*, **21**, 457 (1952).

[74] W. C. Price and A. D. Walsh, *Proc. Roy. Soc.*, **A179**, 201 (1941).

[75] The butadiene and furan data are in good accord with PI-measurements[50] but pyrrole has the much lower PI I of 8.20 ev.

[76] (*a*) J. B. Farmer and F. P. Lossing, *Canad. J. Chem.*, **33**, 861 (1955). (*b*) J. A. Hipple and D. P. Stevenson, *Phys. Rev.*, **63**, 121 (1943).

[77] F. P. Lossing and J. B. de Sousa, *J. Am. Chem. Soc.*, **81**, 281 (1959).

Baker-Nathan concept that C—H hyperconjugation is more effective than C—C hyperconjugation.[78] Recent PI-measurements[51] show that n-butylbenzene, isobutylbenzene, sec-butylbenzene, and t-butylbenzene have the same value of I of 8.68–8.69 ev.

Additional complications appear with other heteroatom functions. We have assumed in the foregoing that the electron being ionized is lost from a π-MO. When the alternative is a bonding σ-MO, this assumption is not unreasonable. If unshared electron pairs are available, however, ionization from these might require lower energy. Compared to the benzene series, methyl substitution in pyridine has little effect on I; hence the first ionization potential must pertain to loss of an electron from the nitrogen lone pair.[71,79] Moreover, I for a number of nitrogen heterocycles parallel the basicities.[80] The conclusion is supported by SCF MO calculations,[80] although other studies indicate that π-ionization is not much higher in energy, at least for pyridine and pyrazine.[81] Similarly, the first ionization potential of carbonyl compounds apparently involves the lone-pair electrons.[82]

There is no doubt that I-values of compounds are important and useful quantities; hence further studies with simple LCAO calculations should be encouraged. Nevertheless, it is also apparent that the simple treatment of electron affinities is more successful than that of ionization potentials. Addition of an electron to a neutral system may well cause less of a disruption of molecular energetics than the removal of an electron.[83] Such a difference is reflected in the usually far smaller magnitude of A than I. For example, A for naphthalene is only 0.65 ev[84] compared with I of >8 ev.

7.4   Saturated Hydrocarbons and Ionization Potentials

A quantum mechanical treatment of ionization potentials of alkanes was first developed by Hall,[85] who used the equivalent orbital representation in which a linear combination of C—H and C—C σ-orbitals leads to a

[78] For the most recent summary see E. Berliner, *Tetrahedron*, **5**, 202 (1959).

[79] K. Higasi, I. Omura, and H. Baba, *J. Chem. Phys.*, **24**, 623 (1956).

[80] T. Nakajima and B. Pullman, *Bull. soc. chim. France*, **1958**, 1502; T. Nakajima and A. Pullman, *Compt. rend.*, **246**, 1047 (1958); *J. chim. phys.*, **55**, 793 (1958).

[81] K. Maeda, *Bull. Chem. Soc., Japan*, **31**, 890 (1958).

[82] A. D. Walsh, *Trans. Faraday Soc.*, **42**, 56 (1946); K. Higasi, I. Omura, and H. Baba, *Nature*, **178**, 652 (1956); *Bull. Chem. Soc. Japan*, **28**, 504 (1955); K. Higasi, T. Nozoe, and I. Omura, *Bull. Chem. Soc., Japan*, **30**, 408 (1957).

[83] However, see the argument of G. J. Hoijtink, *Mol. Phys.*, **1**, 157 (1958).

[84] Quoted by J. A. Pople, *J. Phys. Chem.*, **61**, 6 (1957), who also gives some additional values.

[85] G. G. Hall, *Proc. Roy. Soc.*, A205, 541 (1951); J. Lennard-Jones and G. G. Hall, *Discussions Faraday Soc.*, **10**, 18 (1951).

usual secular equation. Parameters are assigned for the "Coulomb integrals" and interaction integrals between bonds. For alkanes as many as four such parameters are required. The values assigned gave an excellent reproduction of experimental EI I-values. Franklin[69] has applied his related method involving group parameters not only to alkanes but to numerous functional derivatives. A linear combination of σ-orbitals treatment has also been published by Brown.[86]

It should be emphasized that in these treatments ionization involves loss of an electron from a MO encompassing the entire molecule. For most purposes alkanes may be treated satisfactorily from a localized bond viewpoint. The effect of structure on ionization potential demonstrates convincingly, however, that the positive charge in an alkane cation is distributed over the entire molecule. For example, EI I-values for alkanes are methane, 13.12 ev, ethane, 11.65 ev, propane, 11.21 ev, butane, 10.80 ev, pentane, 10.55 ev, etc., down to decane, 10.19 ev.[87] In the usual approximations a molecular-orbital concept is required to interpret adequately alkane cations and energy changes involving such cations. Incidentally, the same effect of extension of an alkyl chain occurs to a much lesser degree when other functions are involved; for example, the PI I-values for propylene, 1-butene, 1-pentene, and 1-hexene are 9.73, 9.58, 9.50, and 9.46 ev. respectively.[50] Similarly, EI I-values for ethyl, n-propyl, and n-butyl radicals are 8.67–8.78,[76] 8.69, and 8.64 ev,[77] respectively. In these cases the molecular orbital from which the electron is lost is apparently largely localized around the unsaturated group and does not spread to a significant extent down a saturated chain.

The heteroatom model of a methyl group is a united atom model in which it is assumed that two electrons are contained in a pseudo-π-orbital. This model was used originally by Matsen[66] and by Hall,[67] and the concept was extended by Stevenson[68] to alkanes, which are treated as though a pseudo-π-system exists to which each carbon group donates two pseudo-π-electrons. With β_0 determined from spectra, $h_C = 1.65$ derived from methane, and $k_{C-C} = 0.35$, good agreement was obtained with other alkanes. It is possible to test this type of model rather simply. Since the α- and β-parameters involved are those only of the saturated carbons, the form of the MO's is identical with that of the corresponding olefins in the HMO approximation. For the alkanes all of the MO's are occupied; hence the total energy is simply the sum of the Coulomb terms; that is, bond additivity holds as in a localized bond viewpoint. The highest occupied orbital is that of the most positive root of the characteristic equation or of most negative m. Values of m and PI I for a number of

[86] R. D. Brown, *J. Chem. Soc.*, **1953**, 2615.
[87] Ref. 46, p. 108.

TABLE 7.6
IONIZATION POTENTIALS OF ALKANES PI I

Alkane	$-m_m$	(Refs. 49 and 50) ev
Methane	0	12.99
Ethane	1.000	11.63
Propane	1.414	11.08
Butane	1.618	10.63
Pentane	1.732	10.33
Hexane	1.802	10.17
Heptane	1.848	10.06
Cyclopentane	1.618	10.51
Cyclohexane	2.000	9.88
Methylcyclohexane	2.101	9.86
Isobutane	1.732	10.55
Isopentane	1.848	10.30
2-Methylhexane	1.902	10.09
3-Methylhexane	1.932	10.06
2,3-Dimethylbutane	2.000	10.00

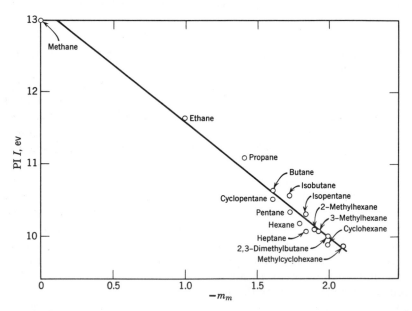

Fig. 7.9 Ionization potentials of alkanes and pseudo-π MO model. m-Values are those of the corresponding unsaturated hydrocarbons in the HMO method.

alkanes are summarized in Table 7.6 and plotted in Fig. 7.9. A good linear correlation is obtained, the slope of which corresponds to $\beta_{C-C} = -1.6$ ev, a surprisingly large value. This type of model of alkanes has also been used with the ω-technique.[56] In this case β_{C-C} turned out to be close to zero.

7.5   Charge-Transfer Complexes and Ionization Potentials

In 1948 Benesi and Hildebrand[88] reported that a solution of iodine in benzene produced a new absorption band which they showed was the result of a complex formed from one molecule of iodine and one molecule of benzene. Shortly thereafter, Mulliken worked out a detailed theory for complexes of this type in which one molecule acts as an electron donor, D, and the other acts as an electron acceptor, A.[89] Such a complex may be described as a resonance hybrid of two structures, XVI and XVII.

$$D \ A \qquad\qquad D^+ {-} A^-$$
$$\text{XVI} \qquad\qquad\quad \text{XVII}$$

XVI is just a close association of the two molecules, D and A; in XVII an electron from D has been transferred to A. Hence such complexes are known as *charge transfer complexes*. Alternatively, the wave function of the complex is taken as a linear combination of the wave functions for XVI and XVII, ψ_{XVI} and ψ_{XVII}, respectively.

$$\psi_{\text{complex}} = \psi_{XVI} + b\psi_{XVII} \tag{17}$$

b^2 is a measure of the amount of charge transfer. These complexes are often associated with an intense absorption band which corresponds approximately to a transition from the ground state of the complex in which b is rather small to an excited state in which b is larger; that is, the excited state has more "charge transfer character" than the ground state. Both the bonding energy of the complex and the energy required for transition to the electronically excited state depend in part on the energy required to transfer electron density from D to A; this energy in turn depends on the ionization potential of D and the electron affinity of A. Relations that demonstrate this dependence have been derived theoretically.[89,90,91] Furthermore, excellent correlations have been found between

[88] H. A. Benesi and J. H. Hildebrand, *J. Am. Chem. Soc.*, **70**, 2832 (1948); **71**, 2703 (1949).

[89] R. S. Mulliken, *J. Am. Chem. Soc.*, **72**, 600 (1950); **74**, 811 (1952). See also *J. Phys. Chem.*, **56**, 801 (1952); *Rec. trav. Chim.*, **75**, 845 (1956).

[90] S. H. Hastings, J. L. Franklin, J. C. Schiller, and F. A. Matsen, *J. Am. Chem. Soc.*, **75**, 2900 (1953).

[91] H. McConnell, J. S. Ham, and J. R. Platt, *J. Chem. Phys.*, **21**, 66 (1953).

TABLE 7.7
IODINE-HYDROCARBON COMPLEXES

Hydrocarbon	λ_{max} $m\mu$	$h\nu_{max}$ (ev)	m_m
Benzene	297	4.184	1.000
Naphthalene	360	3.452	0.618
Anthracene	430	2.890	0.414
Phenanthrene	378	3.288	0.605
Chrysene	394	3.154	0.520
Pyrene	420	2.959	0.445
Triphenylene	394	3.154	0.684
Biphenyl	340	3.654	0.705
Stilbene	373	3.331	0.504

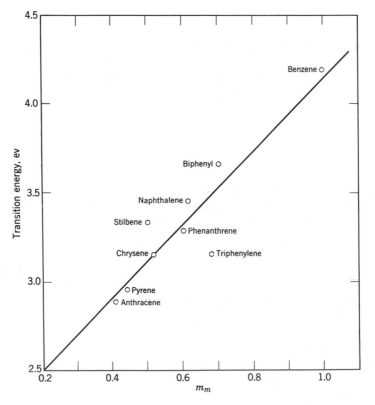

Fig. 7.10 Energy of charge-transfer transition of iodine-hydrocarbon complexes and energy of highest occupied MO.

I_D and both the stabilities of various charge transfer complexes[92] and the frequencies of the charge transfer bands.[90,91,92,93,94] Bhattacharya and Basu[95] have shown that the transition energies of complexes of iodine with aromatic hydrocarbons give a good linear correlation with the energies of the highest occupied MO's. The data are summarized in Table 7.7 and plotted in Fig. 7.10.[96] The slope corresponds to $\beta_0 = -2.0 \pm 0.3$ ev. This approach, although relatively new, should prove to be a fruitful area for MO applications.[97]

[92] R. E. Merrifield and W. D. Phillips, *J. Am. Chem. Soc.*, **80**, 2778 (1958).

[93] G. Briegleb and J. Czekalla, *Z. Electrochem.*, **63**, 6 (1959); *Angew. Chem.*, **72**, 401 (1960).

[94] J. Walkley, D. N. Glew, and J. H. Hildebrand, unpublished results.

[95] R. Bhattacharya and S. Basu, *Trans. Faraday Soc.*, **54**, 1286 (1958).

[96] Bhattacharya and Basu's correlation was established for MO's with $S = 0.25$. The HMO approximation used in Table 7.7 and Fig. 7.10 apparently gives as good a correlation.

[97] For an additional application with the ω-technique, see ref. 56.

8 Spectra

8.1 Electronic Spectra. Introduction

Thus far we have considered the properties of molecules in their ground electronic states, that is, that in which MO's are occupied in order of their energies. Systems in which one or more MO's of higher energy are occupied while MO's of lower energy are left partly vacant are known as *electronically excited states*. In principle, transitions between such states can take place with emission or absorption of radiation. Such *electronic spectra* for transitions involving the ground state are generally in the visible and ultraviolet region.

Several types of transition involving π-systems are important. Transitions in which the principal quantum number of the electron is unchanged involve the "valence shell" and the ground or "normal state" and are termed $N \rightarrow V$ transitions.[1] V_1, V_2, V_3, etc., are excited states of successively higher energy. In addition to $\pi \rightarrow \pi^*$ transitions, transitions that involve σ-electrons may take place. $\sigma \rightarrow \sigma^*$ transitions are normally in the far ultraviolet. However, transitions from lone-pair electrons to π-MO's are important in the near UV and visible region. Such transitions have been termed $N \rightarrow Q$ transitions.[1] A change in the principal quantum number of the electron can involve transitions to states of such high energy that the molecule acts toward the electron as a united atom. Successive states take on the energies of a Rydberg series, which converge to an ionization potential (Sec. 7.3). Transitions involving such states have been symbolized as $N \rightarrow R$.[2]

Mulliken[1] has suggested the symbols N for ground state and V, L, Q, R, etc., for excited states among a series of recommendations for a systematic spectroscopic nomenclature. Several of these recommendations

[1] R. S. Mulliken, *J. Chem. Phys.*, **8**, 234 (1940); **23**, 1997 (1955).

[2] cf. W. West in "Chemical Applications of Spectroscopy," Vol. IX of *Technique of Organic Chemistry*, edited by A. Weissberger, Interscience Publishers, New York (1956), p. 61. See also p. 641.

are at variance with current spectroscopic literature and Mulliken's counsel does not seem to have been embraced by spectroscopists despite a growing plea for a systematic and universally acceptable nomenclature.[3] The Mulliken symbols are convenient for describing several types of transitions but they appear to be inadequate to detail various properties of observed spectra. Of the several other systems that have been proposed in the literature, two seem to be of particular importance. One is an essentially empirical nomenclature developed by Clar and used particularly in England; this simple system is particularly useful for aromatic hydrocarbons and is described in Sec. 8.3. Platt's nomenclature[4] is a more completely developed symbolism. Although founded in theory, it can describe experimental aspects of a transition. This nomenclature appears to be gaining acceptance, particularly among American spectroscopists. Although a complete description of Platt's system is beyond the scope of this book, his state symbols are included wherever appropriate in this chapter.

Finally, the student should note that the use of a symbol for an excited state preceding that for a ground state with $\rightarrow$ to denote emission and $\leftarrow$ to denote absorption is a common notation in the spectroscopic literature. Because of the overwhelming importance of absorption spectroscopy in organic chemistry, the use of this convention would fill this chapter with sequences in the "unnatural" order (e.g., $V \leftarrow N$). The use of such unnatural symbolism would appear to be unnecessary, particularly among organic chemists; accordingly the "natural" order is used in this book (e.g., $N \rightarrow V$).

Low-lying excited states are frequently labeled by the symbol for the corresponding group representation. Such labels have been applied both to individual MO's and to the total π-wave function, which is a product function of constituent MO's. We used such notation earlier in describing MO's during the application of group theory for simplifying secular equations (Sec. 3.7). An abbreviated form sufficed for the problem then at hand. For spectroscopic purposes the complete symmetry classification must be used for the molecular geometry as it actually exists; for example, cis- and trans-butadiene must be described separately. The label of a particular MO is that of the representation which describes the behavior of the MO to all of the covering operations inherent in the symmetry of the system. To find the corresponding symbol for the total wave function, we need only know that for one-dimensional representations the character

[3] For example, see J. R. Platt, *Ann. Rev. Phys. Chem.*, **10**, 354 (1959).

[4] J. R. Platt, *J. Chem. Phys.*, **17**, 484 (1949); *J. Opt. Soc. Am.*, **43**, 252 (1953); *J. Chem. Phys.*, **25**, 80 (1956); Chap. 2 in *Radiation Biology*, Vol. III, ed. by A. Hollander, McGraw-Hill Book Co., New York (1956).

of the product wave function is the product of the characters of the component MO's. Since in the ground state of an even system each MO is doubly occupied and the character is generally either 1 or -1, the product representation must be all ones; hence the ground state is always *totally symmetric*, or *A*, in the MO approximation.

I

Naphthalene, I, serves as a suitable example. This molecule has D_{2h} symmetry; the symmetry axes are taken as in I.[5] The D_{2h} character table is given in Table 8.1.[6]

TABLE 8.1

D_{2h} CHARACTER TABLE

	E	$C_2(z)$	$C_2(y)$	$C_2(x)$	i	$\sigma(xy)$	$\sigma(xz)$	$\sigma(yz)$
A_g	1	1	1	1	1	1	1	1
A_u	1	1	1	1	-1	-1	-1	-1
B_{1g}	1	1	-1	-1	1	1	-1	-1
z B_{1u}	1	1	-1	-1	-1	-1	1	1
B_{2g}	1	-1	1	-1	1	-1	1	-1
y B_{2u}	1	-1	1	-1	-1	1	-1	1
B_{3g}	1	-1	-1	1	1	-1	-1	1
x B_{3u}	1	-1	-1	1	-1	1	1	-1

The two highest occupied MO's of naphthalene, ψ_4 and ψ_5, and the two lowest vacant orbitals, ψ_6 and ψ_7, have these forms:

$$\psi_4 = 0.408(\varphi_2 + \varphi_3 + \varphi_6 + \varphi_7) - 0.408(\varphi_9 + \varphi_{10})$$
$$\psi_5 = 0.425(\varphi_1 - \varphi_4 + \varphi_5 - \varphi_8) + 0.263(\varphi_2 - \varphi_3 + \varphi_6 - \varphi_7)$$
$$\psi_6 = 0.425(\varphi_1 + \varphi_4 - \varphi_5 - \varphi_8) + 0.263(-\varphi_2 - \varphi_3 + \varphi_6 + \varphi_7)$$
$$\psi_7 = 0.408(-\varphi_2 + \varphi_3 + \varphi_6 - \varphi_7) + 0.408(\varphi_9 - \varphi_{10})$$

[5] Note that the symbolic description of a state depends on the labeling of the axes. Mulliken[1] has recommended a general procedure for choosing such axes for any system. When applied to the linear acenes, however, his system produces an assignment that differs from that used in virtually all papers on the spectra of aromatic hydrocarbons. For the convenience of the reader who wants to refer to some of this literature the conventional axes have been retained.

[6] This table differs from the D_2 table in the inclusion of the operation of inversion at a center of symmetry (i). The character table for i alone is the same as that for C_2. The D_{2h} table is the supermatrix product, $D_2 \times i$. σ_{xy} is reflection in the xy-plane; $C_2(z)$ is rotation about the z-axis.

If we start with φ_1 and apply each of the covering operations in Table 8.1, we obtain

E	$C_2(z)$	$C_2(y)$	$C_2(x)$	i	$\sigma(xy)$	$\sigma(xz)$	$\sigma(yz)$
φ_1	$-\varphi_4$	$-\varphi_8$	φ_5	$-\varphi_5$	φ_8	φ_4	$-\varphi_1$

To convert this sequence into that of ψ_5, we must multiply the sequence by the numbers: $1\ 1\ 1\ 1\ -1\ -1\ -1\ -1$, which are the characters of the A_u representation. Hence ψ_5 belongs to A_u. Similarly ψ_4, ψ_6, and ψ_7 belong to B_{3u}, B_{2g}, and B_{1g}, respectively.

It will be convenient to refer to transitions in two ways which are equivalent when dealing with transitions from the ground state. We can speak of the transition of an electron from one MO to another, for example from ψ_5 to ψ_6 in naphthalene. Or we can speak of a transition from the ground *state* as represented by a total wave function to an excited *state*, in which the states are represented by *total* wave functions. The same example could then be symbolized as $\Psi_0 \rightarrow \Psi_{5,6}$, in which $\Psi_{5,6}$ indicates a wave function in which ψ_5 and ψ_6 are singly occupied and the other normal MO's of the ground state are doubly occupied. Ψ_0 of naphthalene belongs to the A_g representation of the D_{2h} symmetry group. To find the corresponding representation of $\Psi_{5,6}$ we may neglect the contribution of the doubly occupied MO's, since each is A_g. The representation of $\Psi_{5,6}$, for example, is given as the product of the representations of ψ_5 and ψ_6; that is, $\Gamma(\Psi_{5,6}) = \Gamma(\psi_5)\Gamma(\psi_6)$. The product of the characters of A_u and B_{2g}, $(1\ 1\ 1\ 1\ -1\ -1\ -1\ -1)(1\ -1\ 1\ -1\ 1\ -1\ 1\ -1) = (1\ -1\ 1\ -1\ -1\ 1\ -1\ 1)$, which are the characters of the B_{2u} representation; that is, $\Gamma(\Psi_{5,6}) = B_{2u}$. Similarly $\Gamma(\Psi_{5,7}) = B_{1u}$.

Transitions to these two states may be represented as $A_g \rightarrow B_{2u}$ and $A_g \rightarrow B_{1u}$, respectively. All of the excited states we have mentioned refer to the transition of a single electron and are known as singly excited states.

Three characteristics of the radiation corresponding to a transition are its energy or frequency, its intensity, and its direction of polarization. Each is associated with a corresponding molecular property. The frequency or wavelength of the radiation depends on the difference in energy between initial and final states,

$$h\nu = E^* - E \tag{1}$$

The intensity of the radiation or the probability that the transition can occur depends on the transition moment, $e\mathbf{Q}$, from state a to state b as given by

$$e\mathbf{Q} = e \int \Psi_a \sum_i \mathbf{r}_i \Psi_b \, d\tau \tag{2}$$

$\mathbf{r}_i$ is the vector distance of the ith electron to an origin. $\mathbf{Q}$ is a vector that has components along the x-, y-, and z-coordinates. Transition occurs only if Q differs from zero. If the components along the x- and y-directions vanish, we say that the transition is z-polarized. The physical significance is that a classical oscillator will undulate in resonance with the component of the electric vector of radiation oriented in the same manner.

Q differs from zero only when the integrand is totally symmetric; that is, when it belongs to an A_g representation. In our cases Ψ_a is the ground state and is always totally symmetric; hence transition occurs only if $\Gamma(\mathbf{r}) \cdot \Gamma(\Psi'_b) = A_g$.

Consider the transition $\Psi_0 \to \Psi_{5,6}$ for naphthalene as an example. If we apply the D_{2h} covering operations to a vector oriented along the z-axis, we find it transforms as follows:

$$E \quad C_2(z) \quad C_2(y) \quad C_2(x) \quad i \quad \sigma(xy) \quad \sigma(xz) \quad \sigma(yz)$$
$$1 \quad\quad 1 \quad\quad -1 \quad\quad -1 \;\; -1 \quad -1 \quad\quad 1 \quad\quad\quad 1 \;\; \equiv B_{1u}$$

The characters of the z-component of Q are found as the product of those of $\Psi_{5,6}$ and z; that is, $B_{2u} \cdot B_{1u} = B_{3g}$.[7] The transition is not allowed with z-polarization. The symmetry properties of the x-, y-, and z-vectors for this system are recorded in Table 8.1. The representation of Q_y is $B_{2u} \cdot B_{2u} = A_g$. Hence this transition is allowed with y-polarization. The foregoing constitutes the principal *selection rule* for electronic transitions. It follows that for molecules with a center of symmetry allowed transitions occur only between g- and u-states, that is, states that are symmetric and anti-symmetric, respectively, with respect to inversion in a center of symmetry.

Exercise. What is the predicted polarization of the $\Psi_0 \to \Psi_{5,7}$ transition of naphthalene?
Ans. z.

Exercise. Determine the symmetries of the following three excited states of anthracene: $\Psi_{7,8}$, $\Psi_{7,9}$, and $\Psi_{6,8}$.
Ans. B_{2u}, B_{1u}, B_{1u}.

Exercise. Determine the symmetries of all four singly excited states of *trans*-butadiene. Which are the allowed transitions? The C_{2h} character table is

	E	C_2	σ_h	i
A_g	1	1	1	1
$x\ A_u$	1	1	-1	-1
B_g	1	-1	-1	1
$y, z\ B_u$	1	-1	1	-1

The x-axis is the axis of symmetry.

[7] Similar handling of a degenerate E wave function gives a product that consists of the characters of a *reducible* representation. These characters are the sum of the characters of the component irreducible representations; if one of the latter is A or A_g, the transition is allowed.

Ans. $\Psi_{2,3}$, B_u; $\Psi_{2,4}$, A_g; $\Psi_{1,3}$, A_g; $\Psi_{1,4}$, B_u. $\Psi_0 \rightarrow \Psi_{2,3}$ and $\Psi_0 \rightarrow \Psi_{1,4}$ are allowed with y,z-polarization; both are $A_g \rightarrow B_u$. The other two transitions are forbidden.

Exercise. Show that all four transitions are allowed in *cis*-butadiene.

Forbidden transitions are sometimes found experimentally because the molecular symmetry may be altered in higher vibrational levels. Even in this event, the band generally has low intensity. $N \rightarrow Q$ transitions are generally of relatively low intensity for these reasons. Many such transitions involve transition of an electron from an unshared pair having only part s-character to an antibonding π-orbital. Only the s-part of the hybrid forms an allowed transition but of relatively high energy. Transition from a p-orbital is more favorable energetically but is normally forbidden by symmetry; hence transition occurs only when the local symmetry is distorted by appropriate vibrations.[8]

8.2 $\pi \rightarrow \pi^*$ Transitions in Polyene Hydrocarbons

Saturated hydrocarbons absorb in the far UV region <1700 A. The transitions involve firmly held σ-electrons and require high energy. The transitions form a Rydberg series and may be described as $N \rightarrow R$. Unconjugated olefins have overlapping bands which extend towards longer wavelengths and which are due to $\pi \rightarrow \pi^*$ transitions. Such transitions are usually highly allowed and are rather intense. If the π-system is extended, say by adding another double bond, the original bonding and antibonding levels split into two levels each, with a consequent reduction in the energy required for transition (Fig. 8.1). Associating the first intense absorption ($^1A \rightarrow {}^1B^\circ$ in Platt s terminology) with the transition of an electron from the highest occupied MO, ψ_m, to the lowest unoccupied MO, ψ_{m+1}, we look for a correlation between the observed frequency and the corresponding HMO energy difference. Since the energy levels of linear polyenes are given in analytic form (2.35), the corresponding energy difference is given as

$$\Delta E = 4\beta \sin \frac{\pi}{2(n + 1)} \tag{3}$$

in which n is the number of π-centers. The simple polyenes are known through octatetraene. Table 8.2 lists the frequency of the $^1A \rightarrow {}^1B^\circ$ absorption band and the HMO energy difference for each of these polyenes.

---

[8] For a review of $N \rightarrow Q$ transitions, see J. W. Sidman, *Chem. Rev.*, **58**, 689 (1958). For some recent simple MO discussions of such transitions see also B. Pullman and S. Diner, *J. chim. phys.*, **55**, 212 (1958), and S. F. Mason, *J. Chem. Soc.*, **1959**, 1240, 1247, and 1263.

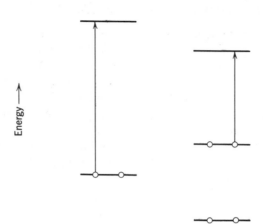

Fig. 8.1   Illustrating the decrease in energy for the first $\pi \to \pi^*$ transition from ethylene to butadiene.

TABLE 8.2

ELECTRONIC SPECTRA OF LINEAR POLYENES

	λ_{max} mμ	ν cm.$^{-1}$	HMO Δm (3)
Ethylene	162.5*	61500	2.000
Butadiene	217.0†	46080	1.236
Hexatriene	251‡	39750	0.890
Octatetraene	304§	32900	0.695

* J. R. Platt, H. B. Klevens, and W. C. Price, *J. Chem. Phys.*, **17**, 466 (1949). For a discussion on the assignment of this band see F. A. Matsen's paper in ref. 2, p. 647.

† W. C. Price and A. D. Walsh, *Proc. Roy. Soc.*, **A174**, 220 (1940).

‡ *Ibid.*, **A185**, 182 (1946).

§ G. F. Woods and L. H. Schwartzman, *J. Am. Chem. Soc.*, **71**, 1396 (1949).

The plot of these data in Fig. 8.2 shows a good linear relation; the slope corresponds to $\beta_0 = -2.62$ ev, or -60.5 kcal./mole.

This correlation is limited to only four compounds because of a paucity of data. If we include carotenoids, data for substituted olefins are available for systems with up to fifteen conjugated double bonds. Such compounds also give a good linear relation in a plot such as that in Fig. 8.2.[9] The polyene aldehydes serve as an additional example. We treat the aldehyde group as a double bond; to a first approximation the perturbation caused by the oxygen will be similar on both MO's and will tend to cancel in

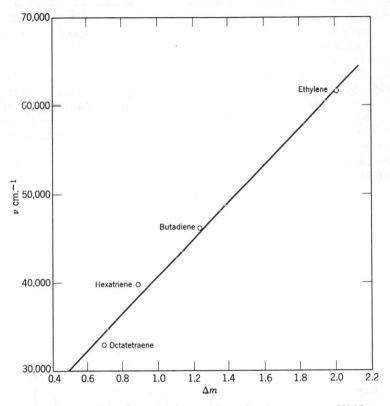

Fig. 8.2 Frequency of the first $\pi \rightarrow \pi^*$ transition of polyenes versus HMO energy difference.

taking the energy difference. The data in Table 8.3 are plotted in Fig. 8.3. An excellent linear correlation is again found whose slope corresponds to $\beta_0 = -3.08$ ev $= -71.1$ kcal./mole.

These amazingly good correlations become truly astonishing when we take a closer look at what we have done. The geometry of an excited state often differs from that of the ground state. Spectral bands are often characterized by the energy difference from the first vibrational level of the ground state to the first vibrational level of the excited state—the

[9] D. P. Stevenson, private communication.

TABLE 8.3

SPECTRA OF POLYENE ALDEHYDES

CH$_3$(CH=CH)$_{n-1}$CHO	n	λ_{max}* mμ	ν cm.$^{-1}$	HMO Δm
Crotonaldehyde	2	220	45500	1.236
2,4-Hexadienal	3	270	37000	0.890
2,4,6-Octatrienal	4	312	32100	0.695
2,4,6,8-Decatetraenal	5	343	29200	0.569
2,4,6,8,10-Dodecapentaenal	6	370	27000	0.482
2,4,6,8,10,12-Tetradecahexaenal	7	393	25400	0.418
2,4,6,8,10,12,14-Hexadecaheptaenal	8	415	24100	0.391

* In dioxane, except for crotonaldehyde, which is for an alcohol solution.
Data from K. W. Hausser, R. Kuhn, A. Smakula, and M. Hoffer, *Z. physik.
Chem.*, **B29,** 371 (1935) and E. R. Blout and M. Fields, *J. Am. Chem. Soc.*, **70,**
189 (1948).

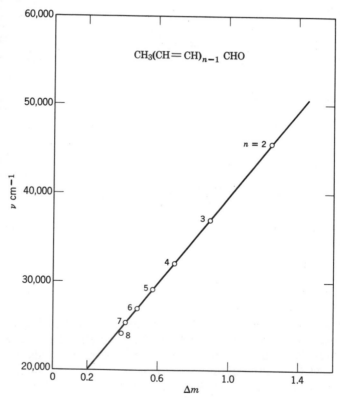

Fig. 8.3   Absorption spectra of polyene aldehydes and HMO energy differences.

so-called 0—0 transition. Since we use the same β-values for both states, we assume an identical geometry in our calculations; that is, we calculate a Franck-Condon transition which may correspond to a transition to a higher vibrational level in the excited state. The difference is illustrated in Fig. 8.4. This situation is related to that discussed with ionization

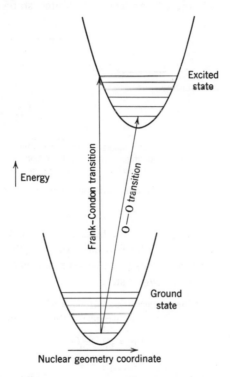

Fig. 8.4   Illustrating the difference between a Franck-Condon transition and a 0—0 transition.

potentials (Sec. 7.3) and is usually ignored.[10] Furthermore, we have neglected the alternation of single and double bonds; however, a more serious omission is our now familiar neglect of electronic interactions. We have noted earlier (Sec. 4.1) that one limitation of the HMO method is that it uses a single configuration wave function. The interaction of other configurations is a way of allowing for electron correlation. This neglect is already serious for the ground state in which other configurations correspond to excited states of much higher energy; however, the problem

[10] For a discussion of one example of the magnitude of this effect see D. S. McClure, *J. Chem. Phys.*, **22**, 1668 (1954).

is even more serious in describing the excited states themselves, since energy differences between excited states are often less than with the ground state. The smaller the difference in energy between two states, the more strongly they will interact, provided they have appropriate symmetry. In interacting, the energy levels "repel" each other as shown in Fig. 8.5. On an absolute scale the energy levels of excited states may differ widely from HMO models.

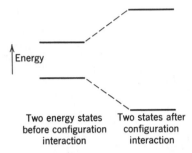

Fig. 8.5   Illustrating the effect of configuration interaction on single configuration energies.

Consider further the spin possibilities of an excited state in which two MO's are singly occupied. The two electrons involved may have either the same spins or opposed spins. The former case is that of a *triplet* state; the latter case is that of a *singlet* state.

There are actually four possibilities. If ψ_1 and ψ_2 are the singly occupied MO's and α and β are spin symbols, we derive the combinations

$$\Psi_1 = \psi_1(\alpha)\psi_2(\alpha)$$
$$\Psi_2 = \psi_1(\beta)\psi_2(\beta)$$

Because of electron exchange resulting from our inability to label electrons, we must take linear combinations of the two further distributions, $\psi_1(\alpha)\psi_2(\beta)$ and $\psi_1(\beta)\psi_2(\alpha)$.

$$\Psi_3 = \psi_1(\alpha)\psi_2(\beta) + \psi_1(\beta)\psi_2(\alpha)$$
$$\Psi_4 = \psi_1(\alpha)\psi_2(\beta) - \psi_1(\beta)\psi_2(\alpha)$$

Ψ_1, Ψ_2, and Ψ_3 are degenerate and constitute the triplet state. Ψ_4 is the singlet state.

In a triplet state the parallel spins of the electrons prevent them from approaching closely. With opposed spins, close approach is denied only by Coulombic forces which result in an increase in energy. Hence a triplet state corresponds to lower energy than a singlet state formed from the same orbitals; the energy difference is called singlet-triplet splitting. No distinction between singlet and triplet states is made in the HMO method.

We can associate the calculated HMO energy with the "center of gravity" of the singlet and triplet levels. Transition from a normal ground state to a triplet level involves a change in spin multiplicity and is forbidden. Hence observed spectra of normal intensities derive from singlet-singlet transitions. The observed correlations must mean that there is a high degree of proportionality within a family of systems between the effects of configuration interaction and singlet-triplet splitting and the over-all transition energy. The failure of the correlation lines to encounter the origin is a probable consequence of these effects. In applications of MO theory to spectral predictions it has usually been assumed in the literature that the origin must be a point on the line. We find that this assumption is not only not necessary but completely unwarranted.[11]

The systematic change produced in spectra by homologous conjugated systems has long been known and has served as the basis of numerous empirical and semiempirical correlations.[12] The α,ω-diphenylpolyenes constitute a further example. The data in Table 8.4 are plotted in Fig. 8.6.

TABLE 8.4

SPECTRA OF α,ω-DIPHENYLPOLYENES

$C_6H_5(CH{=}CH)_nC_6H_5$	n	λ_{max}* mμ in Benzene	ν cm.$^{-1}$	HMO Δm†
	0	251.5	39760	1.408
	1	319	31350	1.011
	2	352	28410	0.772
	3	377	26530	0.620
	4	404	24750	0.520
	5	424	23580	0.444
	6	445	22470	0.392

* Data of K. W. Hausser, R. Kuhn, and A. Smakula, *J. physik. Chem.*, **B29**, 384 (1935).

† Most of these values were obtained from G. J. Hoijtink and J. van Schooten, *Rec. trav. chim.*, **72**, 691 (1953).

The slope of the least squares correlation line corresponds to $\beta_0 = -2.02$ ev $= -46.5$ kcal./mole. This value differs from the polyene value probably because differences in β for single and double bonds have not been taken into account.

[11] This point was first made to the author by Dr. D. P. Stevenson.

[12] For further examples and a discussion in more classical terms see G. E. K. Branch and M. Calvin, *The Theory of Organic Chemistry*, Prentice-Hall, Englewood Cliffs, N.J. (1941); pp. 155–182.

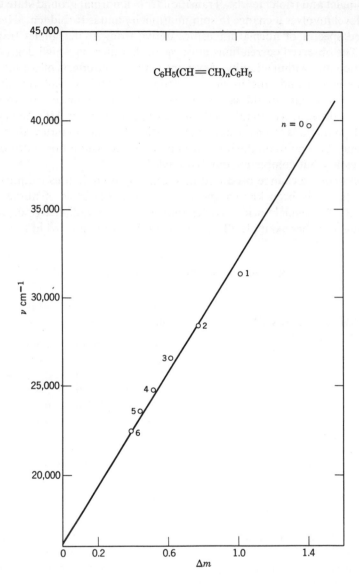

Fig. 8.6   Spectra of α,ω-diphenylpolyenes and HMO transition energies.

8.3 $\pi \rightarrow \pi^*$ Transitions in Aromatic Hydrocarbons

On an empirical basis Clar[13] has established a useful classification of absorption bands which form a consistent and recurrent pattern with polycyclic benzenoid hydrocarbons. The α-bands of log ϵ 2.5–3 are often the longest wavelength bands in these spectra. The *para*- or *p*-bands have log ϵ of about 4, and the shorter wavelength β-bands are generally very intense with log $\epsilon \geqslant 5$. The *p*-bands are generally more sensitive to variations in structure than the α-bands. For example, in naphthalene $\lambda_\alpha > \lambda_p$. As we go along the series, naphthalene, anthracene, tetracene, and pentacene, the α-bands are overtaken by the *p*-bands. In pentacene both bands are clearly distinguishable with $\lambda_\alpha < \lambda_p$. Clar has noted that the empirical relationship, $\lambda_\alpha / \lambda_\beta \cong 1.35$, holds for many hydrocarbons. The reason for this relationship is discussed below. Examples of the patterns of the bands are shown in Fig. 8.7.

The quantitative interpretation of these spectra has long been an important goal of molecular quantum mechanics. Rapid progress has been made in recent years with advanced MO methods, culminating in the theory of Pariser and Parr which has been applied successfully to a number of benzenoid hydrocarbons.[14] In this chapter we are concerned chiefly with the simple MO theory. The HMO theory itself with its normal neglect of overlap actually fell into disrepute rather early, although reasonably successful results have been obtained when the overlap integral is included for bonded atoms. This situation apparently arose because of the general demand that the origin be a necessary point in a correlation line. When this demand is removed, the simple HMO theory accounts rather well for the major features of the spectra of benzenoid hydrocarbons.

Theoretical excited states are symbolized by group representation notation, although the widely used symbolism developed by Platt[4] is based on a free electron model of the perimeters of hydrocarbons. This model has been further developed by Moffitt[15] and gives essential agreement with the MO results outlined as follows.

The *p*-band is associated with a transition from the highest occupied to the lowest unoccupied MO; that is, $\Psi_0 \rightarrow \Psi_{m,m+1}$. In the linear acenes (naphthalene, anthracene, and tetracene), this transition is ^1L_a in Platt's nomenclature[4] and $^1B_{2u}$ in group notation and is calculated to have y- (short

[13] E. Clar, *Aromatische Kohlenwasserstoffe*, second edition, Springer-Verlag, Berlin (1952).

[14] For a very readable account of these developments, see H. C. Longuet-Higgins, "Recent Developments in Molecular Orbital Theory" in I. Prigogine's *Advances in Chemical Physics*, Vol. I, Interscience Publishers, New York (1958).

[15] W. Moffitt, *J. Chem. Phys.*, **22**, 320 (1954).

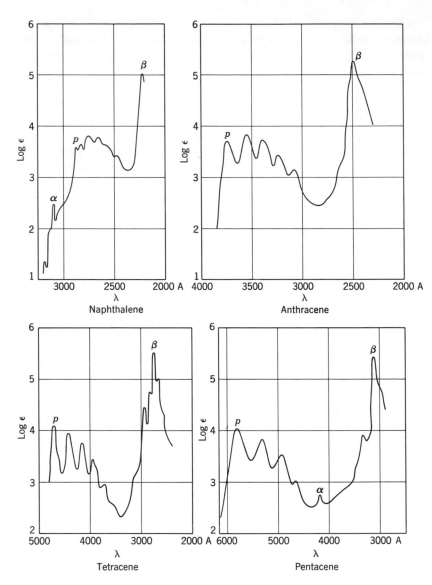

Fig. 8.7   Ultraviolet spectra of the linear acenes (reproduced with permission from ref. 13).

axis or transverse) polarization,[4,16,17] the direction found experimentally.[17,18] This agreement represents a substantial achievement for MO theory.

Table 8.5 lists the α-, p- and β-bands for a number of aromatic hydrocarbons.[19] A plot of the frequency of the p-bands against the HMO energy difference between highest occupied and lowest unoccupied MO's ($m_{m+1} - m_m$) in Fig. 8.8 shows an excellent linear correlation,

$$\nu\,(\text{cm.}^{-1}) = (19020 \pm 330)\,\Delta m + (10520 \pm 340) \qquad (4)$$

in which the slope corresponds to $\beta_0 = -54.4$ kcal./mole. It is interesting that the fluoranthene, II, derivatives which are non-AH also fit the correlation.

II

For AH's the transition energy is twice the HMO ionization potential or electron affinity. Watson and Matsen[20] pointed out the excellent linear correlation between the frequency of the p-band and the polarographic half-wave reduction potentials given by AH's. This relationship was developed empirically by Bergman.[21]

We next consider the transition of an electron from the highest occupied to the second lowest unoccupied MO ($\Psi'_0 \rightarrow \Psi'_{m,m+2}$) and from the penultimate occupied to the lowest unoccupied MO ($\Psi'_0 \rightarrow \Psi'_{m-1,m+1}$). For AH's these transitions have identical energy in the HMO approximation. The resulting excited states are not only degenerate but also belong to the same group representation (see the example for anthracene in Exercise 2, p. 206. The interaction between degenerate states has been called *first-order configuration interaction* by Moffitt[15] to emphasize its importance compared to the *second*-order configuration interaction between nondegenerate states.

[16] C. A. Coulson, *Proc. Phys. Soc.*, **60**, 257 (1948).

[17] D. S. McClure, *J. Chem. Phys.*, **22**, 1256, 1668 (1954).

[18] J. W. Sidman, *J. Chem. Phys.*, **25**, 115, 122 (1956).

[19] A list of transition energies with the corresponding Platt symbol has recently been compiled for a number of aromatic hydrocarbons by E. M. Layton, Jr., *J. Mol. Spect.*, **5**, 181 (1960).

[20] A. T. Watson and F. A. Matsen, *J. Chem. Phys.*, **18**, 1305 (1950).

[21] I. Bergman, *Trans. Faraday Soc.*, **50**, 829 (1954).

TABLE 8.5

ELECTRONIC SPECTRA OF BENZENOID HYDROCARBONS

Hydrocarbon	α-Band*		p-Band*		β-Band*		HMO ΔE†	
	λ_{max} mμ	$10^{-2}\nu$ cm.$^{-1}$	λ_{max} mμ	$10^{-2}\nu$ cm.$^{-1}$	λ_{max} mμ	$10^{-2}\nu$ cm.$^{-1}$	$m_{m+1} - m_m$	$m_{m+1} - m_{m-1}$
Benzene	264	379	206.8	484	179	559	2.000‡	
Naphthalene	314.5	318	288.5	347	221	452	1.236	1.618
Phenanthrene	344.5	290	294.5	340	254.7	392	1.210	1.374
Triphenylene	342.5	292	287	349	259.5	385	1.368‡	
Chrysene	360	278	319	313	267	375	1.040	1.312
3,4-Benzphenanthrene	372	269	315	317	281	356	1.135	1.230
3,4,5,6-Dibenzphenanthrene	395	253	329	304	310	323	1.071	1.192
Picene	376	266	328.5	305	286.5	349	1.004	1.182
Anthracene			378.5	264	251.5	397	0.828	1.414
Tetraphene	385	260	359	278	290	345	0.905	1.167
1,2,3,4-Dibenzanthracene	375	267	349	286	290	345	0.998	1.213
1,2,5,6-Dibenzanthracene	395	253	351	285	300	333	0.946	1.158
1,2,7,8-Dibenzanthracene	395	253	351	285	304	329	0.983	1.110
Pentaphene	423	237	359	279	317	316	0.874	0.958
2′,1′,2,1-Anthraanthracene			420	238	307	326	0.696	0.918
Tetracene			471	212	274	365	0.590	1.072
Pentacene			575.5	174	310	323	0.439	0.838
Perylene	428	234	439.5	228	251	398	0.695	1.347§
1,12-Benzperylene	337.5	296	387.5	258	303	330	0.878	1.124
Coronene	406.5	246	341.5	293	305	328	1.078‡	
Pyrene	428	234	337.2	297	274.1	365	0.890	1.324
4,5-Benzpyrene	371.2	269	331.5	302	289	346	0.994	1.212
1,2-Benzpyrene	388	258	384.5	260	296.5	337	0.742	1.173
1,2,4,5-Dibenzpyrene	403	248	402	248	332	301	0.797	1.062
1,2,7,8-Dibenzpyrene	454	221	397	252	332	302	0.684	1.024
Peropyrene	433	231	436.5	229	323.5	309	0.570	

Compound								
Anthanthrene	363		433	231	310	323	0.582	1.041
Fluoranthene			358.5	279	287	349	0.989	1.115 (1.523)‖
1,2-Benzfluoranthene		276	428	234	308	325	0.703	0.970 (1.296)‖
2,3-Benzfluoranthene	369	271	350	286	301	332	0.979	1.064 (1.340)‖
7,8-Benzfluoranthene			383	261	318	315	0.837	0.985 (1.294)‖
8,9-Benzfluoranthene			400	250	308	325	0.860	1.184 (1.038)‖
Periphenylenefluoranthene	317	315	410	244	292.5	342	0.766	0.804 (1.435)‖
1,2-Benzchrysene	371	270	334	299	286	350	1.063	1.243
5,6-Benzchrysene	386	259	320.5	312	292.5	342	1.100	1.152
1,2-Benztetraphene	401	249	372	369	308	325	0.837	1.079
1,2,3,4,5,6-Ditribenzanthracene	338	258	345	290	304	329	1.045	1.160
3,4-Benztetraphene	393	255	367	273	307	326	0.810	1.109
1,2-Benztetracene			452.5	221	319	314	0.654	1.015
1,2,7,8-Dibenztetracene			437.5	229	325.5	307	0.717	1.038
2,3,8,9-Dibenzperylene			433.5	231	303.5	330	0.711	1.052
Pyranthrene			462	217	354	283	0.516	0.934
3,4,8,9-Dibenzpyrene			451	222	313.5	319	0.605	1.096

* Taken from Clar.[13] Solvent is usually an alcohol or benzene. When available, the spectrum at −170° in alcohol was used.

† Most of the calculated energies were taken from C. A. Coulson and R. Daudel, *Dictionary of Values of Molecular Constants.* Some were obtained from O. Chalvet and J. Peltier, *J. chim. phys.,* **53,** 402 (1956) and from G. G. Hall, *Trans. Faraday Soc.,* **53,** 573 (1957). The remainder are calculations made in the author's research group and are available in A. Streitwieser, Jr., and J. I. Brauman, *Tables of Molecular Orbital Calculations,* Pergamon Press, New York (in press).

‡ Both levels are degenerate. See text.

§ m_{m-1} is degenerate.

‖ Values in parentheses are $m_{m+2} - m_m$. Since these compounds are nor-AH, these values differ from $m_{m+1} - m_{m-1}$.

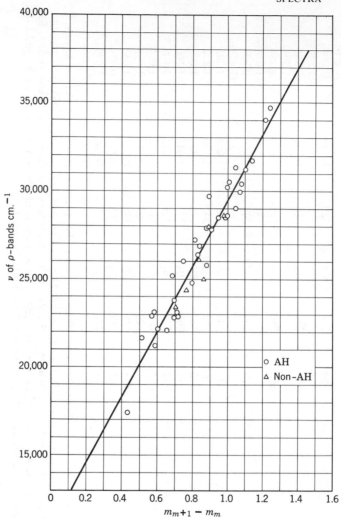

Fig. 8.8 p-Bands of benzenoid hydrocarbons and HMO transition energies for $\Psi_0' \rightarrow \Psi_{m,m+1}'$ transitions.

Linear combination gives two states, which Pariser[22] calls the *plus* and *minus* states and which we symbolize as Ψ'^+ and Ψ'^-:

$$\Psi'^+ = \Psi'_{m,m+2} + \Psi'_{m-1,m+1} \tag{5}$$
$$\Psi'^- = \Psi'_{m,m+2} - \Psi'_{m-1,m+1}$$

Transition to the plus state is highly allowed and gives rise to Clar's β-bands. Transition to the minus state is predicted to have much lower

[22] R. Pariser, *J. Chem. Phys.*, **24**, 250 (1956).

intensity and corresponds to Clar's α-bands which are comparatively weak.[23] In the linear acenes these transitions are $A_{1g}\rightarrow B_{3u}^+$ and $A_{1g}\rightarrow B_{3u}^-$ respectively, or in Platt's nomenclature,[4] $^1A\rightarrow^1B_b$ and $^1A\rightarrow^1L_b$, respectively. Both bands are predicted to have z- (long axis) polarization. This direction is found experimentally for the α-band of naphthalene.[17]

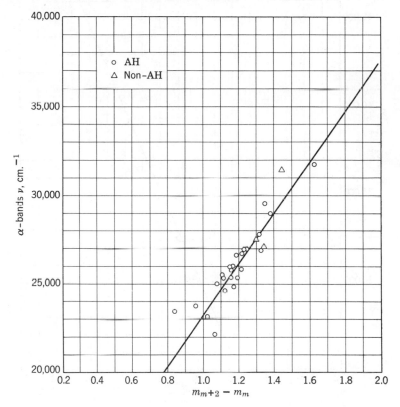

Fig. 8.9 α-Bands and $\Psi_0\rightarrow\Psi_{m,m+2}'$ transitions.

If the splitting due to configuration interaction is proportional to the over-all transition energy, we might expect correlations between $m_{m+2} - m_m$ and the positions of the α- and β-bands. The data in Table 8.5 are plotted in Figs. 8.9 and 8.10. Fair correlations are obtained. As expected, the slopes and intercepts of these correlations differ. Note that although configuration interaction is considered in our analysis it is actually not included in the correlation which involves only the usual HMO energy levels.

[23] The preceding analysis is essentially that of M. J. S. Dewar and H. C. Longuet-Higgins, *Proc. Phys. Soc.*, **A67**, 795 (1954).

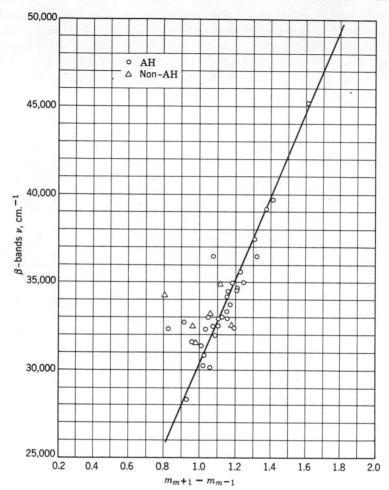

Fig. 8.10 β-Bands and $\Psi_1 \rightarrow \Psi_{m-1,m+1}$ transitions.

Benzene, the parent member of the series, is actually a rather complex case. Both the highest occupied and lowest unoccupied levels are doubly degenerate and lead to four possible transitions, all of the same energy. Electronic interaction splits these transition energies into three, one of which is doubly degenerate. It is perhaps too much to expect that the effects of such electronic interaction in benzene will exactly parallel the behavior of the less degenerate transitions in other aromatic hydrocarbons. The spectrum of benzene has been used to obtain values of β and γ, which were then used to predict spectra of other compounds.[24] We see that

[24] For one example see M. J. S. Dewar, *J. Chem. Soc.*, **1952**, 3532.

benzene is an a priori poor choice for this purpose. The same type of degeneracy results from all compounds with D_{3h} or D_{6h} symmetry, for example, triphenylene and coronene. Such compounds have been omitted from the correlations in Figs. 8.8 to 8.10. In practice, the deviations from the correlations are not serious. This agreement, which is shown in Table 8.6, must result from a similarity of electronic interaction effects that could not have been predicted.

TABLE 8.6

CALCULATED AND OBSERVED $\lambda_{\max}(m\mu)$ OF D_{3h} AND D_{6h} HYDROCARBONS

Hydrocarbon	p-bands		α-bands		β-bands	
	Calculated	Observed	Calculated	Observed	Calculated	Observed
Benzene	206	207	263	264	185	179
Triphenylene	274	287	349	343	255	260
Coronene	322	342	409	428	311	305

With non-AH's, $\Psi'_{m,m+2}$ and $\Psi'_{m-1,m+1}$ are no longer degenerate. Configuration interaction effects might well be different so that such compounds may no longer fit the α- and β-band correlations. Actually, fluoranthene derivatives, which may be regarded as perturbed benzenoid hydrocarbons, fit these correlations satisfactorily if the α-band is identified with the $\Psi'_0 \rightarrow \Psi'_{m,m+2}$ transition and the β-band is identified with the $\Psi'_0 \rightarrow \Psi'_{m-1,m+1}$ transition. With nonbenzenoid non-AH's the correlations do not hold. Such is the case for azulene and the benzazulenes; for example, the calculated and experimental $\lambda_{\max}$ for azulene are α: 381 mμ, 690 mμ; β: 270 mμ, 274 mμ. The α-band, which is responsible for the purple color of azulene, is far from the calculated value. When electronic interaction is explicitly considered, as in the Pariser-Parr advanced MO method, a good account of these bands is obtained.[25] Even for the azulenes, however, our simple MO correlation is in satisfactory agreement with the p-bands; for example, calculated and observed $\lambda_{\max}$ are azulene, 368 mμ; 339 mμ; 1,2-benzazulene, 424 mμ, 405 mμ; 4, 5-benzazulene, 401 mμ, 410 mμ; and 5, 6-benzazulene, 407 mμ, 390 mμ.[26,27]

[25] R. Pariser, J. Chem. Phys., 25, 1112 (1956).

[26] Observed absorption spectra were taken from E. Kloster-Jensen, E. Kovats, A. Eschenmoser, and E. Heilbronner, Helv. Chim. Acta, 39, 1051 (1956).

[27] MO data were obtained from G. Berthier, B. Pullman, and J. Baudet, J. chim. phys., 50, 209 (1953). Our assignments of transitions are consistent with those of Pariser[25] and agree with the experimental polarizations.[28] These assignments differ in part from those of the recent analysis of Heilbronner.[29]

[28] J. W. Sidman and D. S. McClure, J. Chem. Phys., 24, 757 (1956).

[29] E. Heilbronner, Chap. 5 of D. Ginsburg's Non-benzenoid Aromatic Compounds, Interscience Publishers, New York (1959), p. 171.

When overlap integrals are not neglected, the degeneracy between $\Psi'_{m,m+2}$ and $\Psi'_{m-1,m+1}$ for AH's disappears; in particular, the antibonding levels are spread. In the usual approximation the overlap integral for bonded atoms is taken as 0.25. The effect of this change on the MO levels of naphthalene is shown as an example in Fig. 8.11. In most quantitative

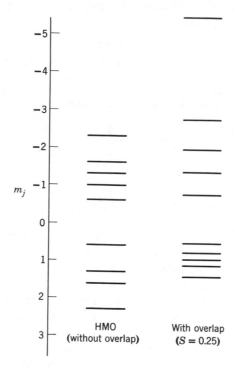

Fig. 8.11   MO energy levels of naphthalene in terms of β or γ.

applications of simple MO theory to spectra of aromatic systems overlap is included and γ is given a value of about 23,000 cm^{-1}.[30] A direct proportionality between frequency and ΔE in γ is normally assumed; that is, the origin is assumed to be a point on the line. This method has been applied with fair success to benzenoid systems,[16,30,31] cyclobutadiene

[30] H. B. Klevens and J. R. Platt, *J. Chem. Phys.*, **17**, 470 (1949).

[31] O. Chalvet and J. Peltier, *J. chim. phys.*, **53**, 402 (1956); R. Pauncz and F. Berencz, *Acta Chim. Acad. Sci. Hung.*, **3**, 361 (1953); *ibid.*, **4**, 333 (1954); R. Pauncz and I. Wilheim, *Acta Chim. Acad. Sci. Hung.*, **11**, 63 (1956); F. Berencz, W. Kolos and R. Pauncz, *Acta Univ. Szegediensis, Acta Phys. et Chem.* [N.S.] **2**, 3 (1956); J. I. F. Alonso, J. Mira, and M. Luisa, *Anales real soc. españ. fis. y quim. (Madrid)*, **53B**, 101 (1957); A. Pullman, B. Pullman, E. D. Bergmann, G. Berthier, E. Fischer, D. Ginsburg, and Y. Hirshberg, *Bull. soc. chim. France*, **1951**, 707.

derivatives,[32] azulene compounds,[26,33] and other non-AH's.[34] Dewar[35] has described a simple procedure based on perturbation theory that is applicable to AH's and that yields rather good agreement with observed p-bands. This method has the important advantage that the calculations necessary are trivial—no solution of secular equations is necessary. In this procedure the π-system of an AH is considered to be synthesized by bonding atoms r, s, $\cdots$, of one odd-AH π-system to atoms t, u, $\cdots$, of another odd-AH π-system of approximately equal extent. The formation of these bonds is a perturbation that splits the NBMO's of the two odd-AH radicals into the highest occupied and lowest vacant MO's of the complete AH. The magnitude of the splitting corresponds to the energy of the p-band and is given approximately by $2\beta\Sigma a_r b_t$; a_r and b_t are the coefficients of the newly bound atoms r and t in the NBMO's of the divided radicals (p. 54). Dewar finds an excellent proportionality between the summation term and the observed energies of the p-bands:

$$\lambda_p = \frac{208}{\Sigma a_r b_t} \tag{6}$$

The proportionality constant is obtained from benzene.

Some examples may illustrate the convenience and success of the method. The π-system of anthracene is made by combination of the π-systems of two benzyl radicals:

$\lambda = 208/(1/\sqrt{7} \cdot 2/\sqrt{7} + 2/\sqrt{7} \cdot 1/\sqrt{7}) = 364$ mμ. The observed value (Table 8.5) is 378.5 mμ.

[32] V. A. Crawford, *Can. J. Chem.*, **30**, 47 (1952); M. A. Silva and B. Pullman, *Compt. rend.*, **242**, 1888 (1956).

[33] D. E. Mann, J. R. Platt, and H. B. Klevens, *J. Chem. Phys.*, **17**, 481 (1949); G. Berthier, B. Pullman, and J. Baudet, *J. chim. phys.*, **50**, 209 (1953).

[34] C. Sandorfy, *Compt. rend.*, **229**, 653 (1949); B. Pullman and G. Berthier, *Compt. rend.*, **229**, 717 (1949); A. Pullman, B. Pullman, E. D. Bergmann, G. Berthier, and J. Pontis, *Bull. soc. chim. France*, **1951**, 681; E. D. Bergmann, D. Ginsburg, Y. Hirshberg, M. Payot, A. Pullman, and B. Pullman, *Bull. soc. chim. France*, **1951**, 697; B. Pullman, A. Pullman, E. D. Bergmann, H. Berthod, E. Fischer, Y. Hirshberg, D. Lavie, and M. Mayot, *Bull. soc. chim. France*, **1952**, 73; G. Berthier, B. Pullman, and J. Baudet,

Phenanthrene is generated by a different combination of benzyl radicals: $\lambda = 208/(1/\sqrt{7} \cdot 1/\sqrt{7} + 2/\sqrt{7} \cdot 2/\sqrt{7}) = 291$ mμ. The experimental value (Table 8.5) is 294.5 mμ.

8.4 $\pi\rightarrow\pi^*$ Transitions in Carbonium Ions, Carbanions, and Radicals

In cations of odd-AH the lowest transition is associated with passage of an electron from the highest filled MO to the vacant NBMO [ψ_m; $m = (n + 1)/2$]. In the corresponding anions the NBMO is doubly occupied

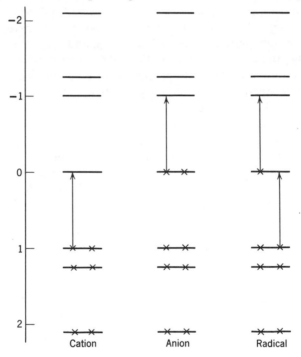

Fig. 8.12   Electronic transitions in benzyl cation, anion, and radical.

and transition is from this orbital to the lowest antibonding orbital. Both transitions are of identical energy; hence anions and cations of odd-AH should, in the HMO approximation, have their longest wavelength bands in the same position. These relationships are illustrated in Fig. 8.12. It would not be surprising if configuration interaction effects and singlet-triplet splitting differed in these cases from the $\Psi_0\rightarrow\Psi_{m,m+1}$ transition of

---

J. chim. phys., **49**, 641 (1952); B. Pullman, A. Pullman, G. Berthier, and J. Pontis, *J. chim. phys.*, **49**, 20 (1952); R. D. Brown, *J. Chem. Soc.*, **1951**, 2391.

[35] M. J. S. Dewar, *J. Chem. Soc.*, **1952**, 3532.

even-AH's such that the same correlation curve would not apply. Actually triphenylmethyl cation and anion, for example, absorb in the same region in the visible, 420–500 mμ, but at longer wavelengths than calculated by using Fig. 8.8. The highest occupied orbital of triphenylmethyl cation and the lowest unoccupied MO of the anion are multiply degenerate and are expected to give rise to the splitting of first-order configuration interaction. This effect may be the reason for the small difference in λ_{max} between triphenylmethyl cation, 430 mμ[36,37,38,39,40] and the anion, 495 mμ.[36,38] Because of symmetry, triphenylmethyl, benzhydryl, and benzyl have the same energies for ψ_{m+1}.[41] Hence, except for differences in configuration interaction, these ions should possess a common long wavelength band. Benzhydryl cation and anion (III) both have λ_{max} 441 mμ[38,40] in good agreement with the triphenylmethyl ions. Although the spectra of the benzyl ions are not known, advanced MO calculations give λ_{max} 410–460 mμ.[42] For ions of non-AH's, ψ_{m+1} and ψ_{m-1} differ in energy and ψ_m is no longer necessarily a NBMO. The spectra of the corresponding anions and cations are now expected to differ. The transition energy of fluorenyl anion, IV, 0.993β, is close to the 1.000β of benzhydryl anion,

III   IV

and its λ_{max}, 480 mμ,[38] also agrees. The transition energy of fluorenyl cation, however, is only 0.524β and its λ_{max}, 655 mμ,[38] is correspondingly at longer wavelengths. Meuche, Strauss, and Heilbronner[43] have shown that the HMO energies for $\Psi_0 \rightarrow \Psi_{m, m-1}$ transitions give an excellent correlation with the experimental long wavelength spectra of several non-AH cations. The $\Psi_0 \rightarrow \Psi_{m+1, m}$ transitions of anions also fit the same

[36] L. C. Anderson, *J. Am. Chem. Soc.*, **57**, 1673 (1935).

[37] T. L. Chu and S. I. Weissman, *J. Chem. Phys.*, **22**, 21 (1954).

[38] I. V. Astaf'ev and A. I. Shatenshtein, *Optics and Spectroscopy* (English translation) **6**, 410 (1959).

[39] N. C. Deno, P. T. Groves, and G. Saines, *J. Am. Chem. Soc.*, **81**, 5790 (1959).

[40] S. F. Mason and G. Grinter in G. W. Gray's *Steric Effects in Conjugated Systems*, Academic Press, New York (1958), p. 52.

[41] This effect is thoroughly discussed by Mason and Grinter[40] and is also treated by Deno, Groves, and Saines.[39]

[42] W. Bingel, *Z. Naturforsch.*, **10A**, 462 (1955); H. C. Longuet-Higgins and J. A. Pople, *Proc. Phys. Soc.*, **68A**, 591 (1955); see also F. A. Matsen in W. West's *Chemical Applications of Spectroscopy*, Interscience Publishers, New York (1956), p. 691.

[43] D. Meuche, H. Strauss, and E. Heilbronner, *Helv. Chim. Acta*, **41**, 57 (1958).

TABLE 8.7
Long Wavelength Spectra of Anions and Cations

System	Cations λ_{max} mμ	$m_m - m_{m-1}$	Anions λ_{max} mμ	$m_{m+1} - m_m$
Triphenylmethyl	431*	1.000	495*	1.000
	430†			
Tri-p-biphenylylmethyl	510†	0.705		
Diphenyl-p-biphenylylmethyl	474†	0.751		
Phenyl-di-p-biphenylylmethyl	500†	0.705		
Tri-m-biphenylylmethyl	500†	0.705		
Benzhydryl	441†	1.000	440†	1.000
	440*‡			
Di-p-biphenylylmethyl	555†	0.705		
Fluorenyl	655*‡	0.524	480*	0.993
Cinnamyl			420*	1.000
α-Naphthylmethyl			560*	0.799
Tropylium	275§	1.692‖		
Benztropylium	426‖¶	1.028‖		
1,2,4,5-Dibenzotropylium	540**	0.824‖		
Naphtho(1,2)tropylium	450‖	0.980‖		
Tribenzotropylium	560††	0.805‡‡		

    * Ref. 38.

    † Ref. 40.

    ‡ N. C. Deno, J. J. Jaruzelski, and A. Schriesheim, *J. Org. Chem.*, **19**, 155 (1954).

    § W. E. Doering and L. H. Knox, *J. Am. Chem. Soc.*, **76**, 3203 (1954).

    ‖ Ref. 43.

    ¶ H. H. Rennhard, G. DiModica, W. Simon, E. Heilbronner, and A. Eschenmoser, *Helv. Chim. Acta*, **40**, 957 (1957).

    ** G. Berti, *J. Org. Chem.*, **22**, 230 (1957).

    †† M. Stiles and A. J. Libbey, *J. Org. Chem.*, **22**, 1243 (1957).

    ‡‡ D. Meuche, H. Strauss, and E. Heilbronner, *Helv. Chim. Acta*, **41**, 414 (1958).

correlation. The AH anions and cations also fit fairly well. The data in Table 8.7 are plotted in Fig. 8.13. The slope of the line gives $\beta_0 = -2.06$ ev $= 47.7$ kcal./mole. This type of correlation should be rather useful, inasmuch as the average deviation of experimental λ_{max} from that calculated from Fig. 8.13 is ± 26 mμ, even though some of these ions are not coplanar.

In odd-AH radicals the NBMO is singly occupied. Longuet-Higgins

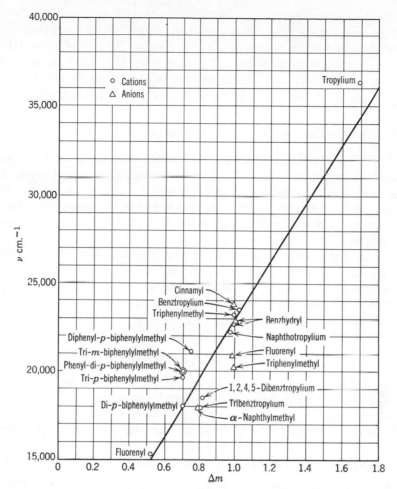

Fig. 8.13   Long wavelength absorption of anions and cations and HMO transition energies.

and Pople[44] have pointed out that the two singly excited states, Ψ'_{m+1} and Ψ'_{m-1}, are degenerate (see Fig. 8.12 for an example) and lead to plus and minus states much as in the related situation for the degenerate excited states of even-AH. Transition to the plus state occurs at lower wavelengths and is of low expected intensity. Transition to the minus state is highly allowed. Triarylmethyl radicals show a doublet with a relatively weak long wavelength band[36,37] in excellent agreement with the predictions of Longuet-Higgins and Pople.[44]

[44] H. C. Longuet-Higgins and J. A. Pople, *Proc. Phys. Soc.*, **68A,** 591 (1955).

The electronic spectra of a number of aromatic hydrocarbon radical cations and anions have recently been measured.[45,46] For AH the transition energies for the cations are the same as for the anions in the HMO approximation with its neglect of overlap integrals. The spectra of the cations are actually very similar to those of the anions.[46] This result of the simple HMO theory holds even in more rigorous calculations.[47]

8.5   Effect of Heteroatoms on $\pi \to \pi^*$ Transitions

In principle, the UV spectra of heteroatom π-systems should be calculable if appropriate h_X and k_{CX} parameters (Sec. 5.1) are known. On the other hand, the observed spectra could be used to determine these parameters. Although some limited work has been done along these lines,[48] no extensive correlations of heteroatom systems and UV spectra have been attempted.

Studies have been made of the qualitative and semiquantitative effect on spectra of inductive and resonance effects. A simple model of a substituent that exerts an inductive effect only is that in which the attached carbon of the parent π-system is assigned an appropriate altered α-value. If a change, h_r, is made on atom r, to a first approximation, the change in transition energy, ΔE, is given by [derived from (4.29)][49,50]

$$\Delta E \cong (c_{m+1, r}^2 - c_{m,r}^2)h_r \tag{7}$$

In an AH system $c_{m+1,r}^2$ and $c_{m,r}^2$ are identical; hence, to the first order in such a system, an inductive substituent should cause no change in the spectrum. This conclusion is affirmed by a more detailed study of Murrell and Longuet-Higgins[51] who find that only the α-band should be appreciably affected by an inductive substituent. The inclusion of overlap or taking the series expansion to a higher order, however does cause some changes.[49,50]

[45] D. E. Paul, D. Lipkin, and S. I. Weissman, *J. Am. Chem. Soc.*, **78**, 119 (1956); P. Balk, G. J. Hoijtink, and J. W. H. Schreurs, *Rec. trav. chim.*, **76**, 813 (1957); N. S. Hush and J. R. Rowlands, *J. Chem. Phys.*, **25**, 1076 (1956); A. I. Aalbersberg, G. J. Hoijtink, E. L. Mackor, and W. P. Weijland, *J. Chem. Soc.*, **1959**, 3055; G. J. Hoijtink and P. H. van der Meij, *Z. Physik. Chem.*, NF., **20**, 1 (1959).

[46] G. J. Hoijtink and W. P. Weijland, *Rec. trav. chim.*, **76**, 836 (1957); W. I. Aalbersberg, G. J. Hoijtink, E. L. Mackor, and W. P. Weijland, *J. Chem. Soc.*, **1959**, 3049; G. J. Hoijtink, *Mol. Phys.*, **2**, 85 (1959).

[47] See, for example, A. D. McLachlan, *Mol. Phys.*, **2**, 271 (1959).

[48] For a few examples, see F. A. Matsen, *J. Am. Chem. Soc.*, **72**, 5243 (1950); W. W. Robertson and F. A. Matsen, *ibid.*, **72**, 5248, 5250, 5252 (1950); L. Goodman and H. Shull, *J. Chem. Phys.*, **27**, 1388 (1957).

[49] C. A. Coulson, *Proc. Phys. Soc.*, **A65**, 933 (1952).

[50] H. C. Longuet-Higgins and R. G. Sowden, *J. Chem. Soc.*, **1952**, 1404.

[51] J. N. Murrell and H. C. Longuet-Higgins, *Proc. Phys. Soc.*, **A68**, 328 (1955).

It has long been known that the salts of aromatic amines possess UV spectra that are remarkably similar to the spectra of the parent hydrocarbons.[52] The $-NH_3^+$ group may be considered as purely inductive;[53] the results provide a significant confirmation of the simple theory. The addition of a vinyl group, on the other hand, contributes a conjugative effect that always lowers the energy of the excited state more than that of the ground state; hence such a substituent is associated with a bathochromic shift (to longer wavelengths).[49,54]

The combination of inductive and conjugative influence on spectra has been given a thorough MO treatment for methyl substituents, stimulated by Plattner's observations of hypsochromic shifts by methyl groups at the 2, 4- and 6-positions in azulene.[55] MO calculations of all of the monomethylazulenes have given complete qualitative agreement with the direction and magnitude of the spectral shifts caused by methyl substituents.[56] As expected from our analysis of inductive and conjugative influences, the hypsochromic shifts of methyl groups are caused almost entirely by an overriding inductive effect[49,50] and can occur only in non-AH. The normal bathochromic shifts of methyl substituents in AH's are in good semiquantitative accord with various MO calculations.[50,57] These effects in neutral hydrocarbons are usually small and rarely exceed 10 mμ. In methyl or methylene substituted carbonium ions the shifts are frequently much larger. Strong hypsochromic and bathochromic shifts are known which follow nicely the principles previously stated. If we treat the methyl group as exerting an inductive effect only, we note that c_{m+1}^2 ($m + 1$ is now the NBMO) vanishes at inactive positions (Sec. 2.8). Since h_r is negative for methyl groups, (7) says that methyl groups at inactive positions should be bathochromic. The pentamethylbenzenium ion, V, absorbs at 23 mμ longer wavelength than does mesitylenium ion, VI.[58] If the methylene group is neglected, both ions are methyl substituted pentadienyl cations in which V has methyl groups at inactive positions.

[52] Several examples have been summarized by D. Peters, *J. Chem. Soc.*, **1957**, 4182.

[53] Because of the formal positive charge, the bonding electrons are expected to be held too tightly for hyperconjugation to be significant.

[54] This statement is true only if carbon α-values remain constant. Application of the ω-technique to non-AH's shows that phenyl substituents can in some cases produce a hypsochromic (to shorter wavelengths) shift; cf. B. Pullman, G. Berthier, and J. Baudet, *J. Chem. Phys.*, **21**, 188 (1953), *J. chim. phys.*, **50**, 69 (1953).

[55] P. A. Plattner, *Helv. Chim. Acta*, **24**, 283E (1941); P. A. Plattner and E. Heilbronner, *Helv. Chim. Acta*, **31**, 804 (1948). The experimental and theoretical results have recently been reviewed thoroughly by E. Heilbronner.[29]

[56] B. Pullman, M. Mayot, and G. Berthier, *J. Chem. Phys.*, **18**, 257 (1950).

[57] B. Pullman and G. Berthier, *J. chim. phys.*, **52**, 114 (1955); G. M. Badger and I. S. Walker, *J. Chem. Soc.*, **1954**, 3238; D. Peters, *J. Chem. Soc.*, **1957**, 646, 4182.

[58] G. Dallinga, E. L. Mackor, and A. A. V. Stuart, *Mol. Phys.*, **1**, 123 (1958).

For the active positions of pentadienyl cation $c_{m+1}^2 > c_m^2$; hence methyl groups at these positions should be hypsochromic. The longest wavelength absorption of benzenium, VII, toluenium, VIII, and mesitylenium, VI, cations, are, respectively, 417 mμ, 400 mμ, and 390 mμ,[59] in complete agreement with the simple theory.[60]

The π-system of pyridine is usually treated in the HMO method by altering α of one carbon atom so that it corresponds to a nitrogen (Sec. 5.2). This change is the same perturbation as that discussed for the inductive effect of a substituent on a benzene ring. Following the same argument, we expect relatively little change in spectra for heterocyclic derivatives of AH's. The spectra of nitrogen heterocycles actually do closely resemble their benzenoid hydrocarbon counterparts. Relatively little has been done with the HMO theory and the UV spectra of such heterocycles; more work has been done with advanced MO techniques often founded in HMO origins.[61] The shifts caused by methyl substituents on pyridine have been treated by simple MO theory and show good qualitative agreement with experiment.[62]

Heteroatom substituents attached to an aryl ring as in aniline, chlorobenzene, or phenol, for example, cause important changes in the UV spectra. Such spectra have been discussed in simple MO terms by Matsen and Robertson[63] and more recently by Peters[64] and Mason.[65] These

[59] C. Reid, *J. Am. Chem. Soc.*, **76**, 3264 (1954).

[60] For more extensive calculations of ions of this type cf. N. Muller, L. W. Pickett, and R. S. Mulliken, *J. Am. Chem. Soc.*, **76**, 4770 (1954); A. A. V. Stuart and E. L. Mackor, *J. Chem. Phys.*, **27**, 826 (1957); T. Morita, *J. Chem. Phys.*, **25**, 1290 (1956), *Bull. Chem. Soc. Japan*, **31**, 322 (1958); and ref. 58.

[61] For examples, see D. W. Davies, *Trans. Faraday Soc.*, **51**, 449 (1955); R. McWeeny and T. E. Peacock, *Proc. Phys. Soc.*, **70A**, 41 (1957); R. McWeeny, *Proc. Phys. Soc.*, **70A**, 593 (1957); J. N. Murrell, *Mol. Phys.*, **1**, 384 (1958); T. Anno, *J. Chem. Phys.*, **29**, 1161 (1958); N. Mataga, *Bull. Chem. Soc. Japan*, **31**, 453, 463 (1958); T. E. Peacock, *J. Chem. Soc.*, **1959**, 2308.

[62] A. K. Chandra and S. Basu, *J. Chem. Soc.*, **1959**, 1623.

[63] F. A. Matsen, *J. Am. Chem. Soc.*, **72**, 5243 (1950); W. W. Robertson and F. A. Matsen, *J. Am. Chem. Soc.*, **72**, 5248, 5250, 5252 (1950).

[64] D. Peters, *J. Chem. Soc.*, **1957**, 1933.

[65] S. F. Mason, *J. Chem. Soc.*, **1959**, 1253; **1960**, 219.

authors found that satisfactory results could be obtained even with the use of approximation methods. One problem that arises with these types of compounds is the identification of bands; the α-bands are frequently as intense as the p-bands, and both bands run together to give diffuse spectra.

Dyes are another important class of heterosubstituted compounds. Rather few HMO applications have been attempted so far, but these few have been successful for triarylmethane and cyanine dyes both in correlating experimental spectra and in accounting for changes in absorption with structure.[66] The simple theory has also been applied to the spectra of azo dyes,[67] azines,[68] porphyrines,[69] phthalocyanines,[70] and quinones.[71]

8.6   Infrared Spectra

Absorption in the infrared region is due usually to transitions between vibrational levels. The complex vibrations of a polyatomic molecule can be dissected into sets of vibrations of the individual atoms called *normal modes of vibration*. In a normal mode each atom reaches its maximum amplitude of vibration simultaneously. Although the vibrations of many atoms in the molecule contribute to every normal mode, to a close approximation many transitions behave as though they were properties of individual groups of atoms. This effect leads to the characteristic infrared absorption bands within relatively narrow regions for such functions as the methyl group, vinyl hydrogen, nitrile group, and different types of carbonyl groups.[72]

For a π-system we can consider that the energy for vibration has σ- and π-bond components. A change in bond distance will change the β-value for the bond; because of the effect of a change in β on the π-energy, we

[66] C. Sandorfy, *Compt. rend.*, **232**, 633 (1951); M. J. S. Dewar, *J. Chem. Soc.*, **1950**, 2329; D. A. Brown and M. J. S. Dewar, *J. Chem. Soc.*, **1954**, 2134; M. J. S. Dewar in G. W. Gray's *Steric Effects in Conjugated Systems*, Academic Press, New York, (1958), p. 46.

[67] O. Chalvet, *Compt. rend.*, **239**, 1135 (1954); F. Gerson and E. Heilbronner, *Helv. Chim. Acta*, **41**, 1444, 2332 (1958), **42**, 1877 (1959).

[68] O. Polansky, *Monatsh. Chem.*, **88**, 670 (1957).

[69] H. C. Longuet-Higgins, C. W. Rector, and J. R. Platt, *J. Chem. Phys.*, **18**, 1174 (1950); G. R. Seely, *J. Chem. Phys.*, **27**, 125 (1957).

[70] S. Basu, *Indian J. Phys.*, **28**, 511 (1954).

[71] B. Pullman, A. Pullman, G. Berthier, and J. Pontis, *Bull. soc. chim. France*, **18**, 271 (1952); L. Paoloni, *J. chim. phys.*, **51**, 385 (1954); B. Pullman and S. Diner, *J. chim. phys.*, **55**, 212 (1958).

[72] For more extensive treatments see A. B. F. Duncan (Chap. 3) and R. N. Jones and C. Sandorfy (Chap. 4) in W. West's *Chemical Applications of Spectroscopy*, Interscience Publishers, New York (1956), and L. J. Bellamy, *Infrared Spectra of Complex Molecules*, J. Wiley and Sons, New York, second edition (1958).

might expect the π-component of the vibration energy to be a function of the bond order, p (cf. Sec. 4.3). A detailed analysis by Coulson and Longuet-Higgins[73] indicates that the force constant for a bond is a function of both the bond order and the bond polarizability. Nevertheless, for empirical correlations of a family of systems the use of p alone apparently suffices for satisfactory results. Berthier, Pullman, and Pontis[74] have found that the characteristic frequencies of open-chain or six-membered ring aldehydes and ketones give an excellent correlation with the bond order

TABLE 8.8

INFRARED FREQUENCIES AND BOND ORDERS OF CARBONYL
COMPOUNDS*

Compound	Bond Order $p_{C=O}$	Carbonyl Frequency cm.$^{-1}$
Formaldehyde	0.958	1744
Glyoxal	0.937	1730
Benzaldehyde	0.905	1708
Acrolein	0.895	1700
Benzophenone	0.857	1664
Naphthophenone	0.820	1644
p-Benzoquinone	0.856	1667
9,10-Anthraquinone	0.860	1679
1,2,5,6-Dibenz-9,10-anthraquinone	0.858	1663
Diphenoquinone	0.819	1626
1,8-Pyrenequinone	0.833	1645
o-Benzoquinone	0.879	1669
9,10-Phenanthraquinone	0.885	1683

* Ref. 74.

of the carbonyl group. They used the values $k_{CO} = 1.2$, $h_O = 2$ for the carbonyl parameters, but the nature of the correlation should not be sensitive to a change in these values. Their bond orders and the experimental frequencies are listed in Table 8.8; the corresponding linear correlation is shown in Fig. 8.14. It should be mentioned that this correlation does not apply to rings of other sizes—five- and seven-membered ring ketones tend to have carbonyl frequencies higher than predicted by the correlation. A change in the σ-bond contribution to the force constants is probably responsible—cyclopentanone has a carbonyl frequency 30 cm.$^{-1}$

[73] C. A. Coulson and H. C. Longuet-Higgins, *Proc. Roy. Soc.*, **A193**, 456 (1948); see also H. C. Longuet-Higgins and F. H. Burkitt, *Trans. Faraday Soc.*, **48**, 1077 (1952).
[74] G. Berthier, B. Pullman, and J. Pontis, *J. chim. phys.*, **49**, 367 (1952).

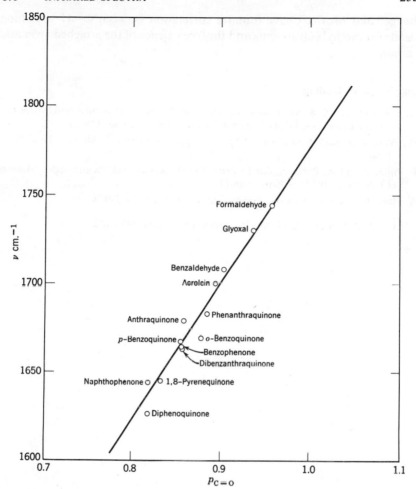

Fig. 8.14   Infrared frequencies and bond orders for carbonyl groups.

higher than that of cyclohexanone. Similar correlations have been extended
to additional carbonyl compounds.[75]

Other types of correlations with IR spectra have been reported. Mason[76]
noted a correlation between the force constant for N—H stretching vibrations in primary aromatic amines and the electron density on nitrogen.

[75] I. Estellés and J. I. F. Alonso, *Anales Real Soc. Españ. de fis. y quim.* (Madrid),
B50, 151 (1954); M. L. Josien and J. Deschamps, *J. chim. phys.,* **52,** 213 (1955); J.
Baudet, G. Berthier and B. Pullman, *J. chim. phys.,* **54,** 282 (1957); cf. also E. Scrocco
and P. Chiorboli, *Atti accad. nazl. Lincei, Rend. classe sci. fis., mat. e. nat.,* **8,** 248 (1950);
C.A., **44,** 7147 (1950).

[76] S. F. Mason, *J. Chem. Soc.,* **1958,** 3619.

Badger and Moritz[77] have found a correlation between C—H stretching bands for methyl substituents and the free valence of the attached aromatic carbon.

Supplemental Reading

A. E. Gillam and E. S. Stern, *An Introduction to Electronic Absorption Spectroscopy in Organic Chemistry*, Edward Arnold, London, second edition (1957).

W. West, *Chemical Applications of Spectroscopy*, Interscience Publishers, New York (1956).

B. Pullman and A. Pullman, *Les Théories electroniques de la chimie organique*, Masson et Cie, Paris (1952); Chaps. 6 and 8.

A. Maccoll, "Colour and Constitution," *Quart. Revs.*, **1**, 16 (1947).

[77] G. M. Badger and A. G. Moritz, *Spectrochim. Acta*, **1959**, 672.

9 Resonance energy

9.1 Empirical Resonance Energies

In VB theory the energy of a structure, say A, is found in principle by solving the wave equation for the corresponding approximate VB wave function. Addition of further structures as linear combinations to the wave function generally results in a lowering of the calculated energy (Secs. 1.4, 1.6). If an infinite number of appropriate structures is used, the true energy of the molecule could be determined. If A is that single structure of lowest energy, the difference between this energy and the true energy of the molecule is known as the *resonance energy*, or, more exactly, the vertical resonance energy (Sec. 9.3).

Since calculation of these energies from first principles is impractical at present, empirical energy values for individual structures are estimated by analogy with other compounds; comparison with the experimental energies of actual molecules then gives *empirical resonance energies*. Procedures for evaluating the empirical energies have been developed with various levels of sophistication. A simple method is that of the average bond energy. The energy required to convert methane into its five constituent isolated atoms (*atomization energy*) is 398.0 kcal./mole; hence we may say that each C—H bond has an average energy of 99.5 kcal. If we assume that the same values apply to the C—H bonds in ethane, the atomization energy of 676.2 kcal./mole yields the value 79.2 kcal. for the energy of the C—C bond. Similarly, the atomization energy of ethylene, 539.3 kcal./mole, yields 141.3 kcal. as the energy of a C=C bond. We next assume that these same values may be applied to one Kekulé structure of benzene, I, and calculate that this structure should have an atomization energy of 1258.5 kcal./mole. The experimental value for benzene is 1322.9 kcal./mole; therefore, benzene is more stable than a single Kekulé structure by 64.4 kcal./mole according to this method. The value, 64.4 kcal., may be identified as an *empirical resonance energy*.

The foregoing procedure is rather crude; in actual practice, average

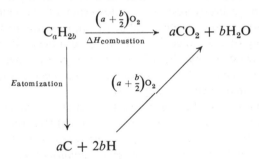

I

bond energies are derived to give best results for a large number of compounds and differ somewhat from the values used above.[1] Because of the implicit and incorrect premise that average bond energies are independent of environment, such energies can be used to give only a rough measure of total atomization energy.

For most purposes it is also more convenient to deal directly with heats of combustion that may be simply related to atomization energies:

$$C_aH_{2b} \xrightarrow[\Delta H_{combustion}]{\left(a + \frac{b}{2}\right)O_2} aCO_2 + bH_2O$$

$E_{atomization}$ $\left(a + \frac{b}{2}\right)O_2$

$$aC + 2bH$$

In the same way, we may speak of average contributions of bonds to heats of combustion. The extensive list of such contributions developed by Klages[2] includes corrections for different environments of bonds.[3] Part of this list is given in Table 9.1, which calculates the heat of combustion for benzene as 825.1 kcal./mole; the experimental value is 789.1 kcal./mole.[4]

[1] For a convenient table of a number of average bond energies, see J. Hine, *Physical Organic Chemistry*, McGraw-Hill Book Co., New York (1956), p. 34.

[2] F. Klages, *Ber.*, **82**, 358 (1949).

[3] An improved list of bond contributions and an extensive discussion may be found in G. W. Wheland, *Resonance in Organic Chemistry*, J. Wiley and Sons, New York (1955), p. 86.

[4] F. D. Rossini, K. S. Pitzer, W. J. Taylor, J. P. Ebert, J. E. Kilpatrick, C. W. Beckett, M. G. Williams, and H. G. Werner, *Selected Values of Properties of Hydrocarbons*, U.S. Government Printing Office, Washington, D.C. (1947); F. D. Rossini, K. S. Pitzer, R. L. Arnett, R. M. Braun, and G. C. Pimentel, *Selected Values of Physical and Thermodynamic Properties of Hydrocarbons and Related Compounds*, Carnegie Press, Pittsburgh (1953).

TABLE 9.1

SOME BOND CONTRIBUTIONS TO HEATS OF COMBUSTION
ACCORDING TO KLAGES*†

Bond	Contribution, kcal.
C—H	54.0
C—C	49.3
C=C	121.2‡
	117.4§
	112.0¶

Correction for six-membered ring: +1.0

* Ref. 2.
† Taken from ref. 3.
‡ For ethylene.
§ For *cis*-1,2-disubstituted ethylenes in six-membered rings.
¶ For tetrasubstituted ethylenes.

The empirical resonance energy is 36.0 kcal. Such resonance energies have been determined for a number of conjugated and aromatic hydrocarbons (Table 9.2).[3,5]

In a related method of Franklin[6] consideration is given to groups of atoms that are treated as units from one compound to another. Resonance energies derived by this method are generally comparable to those of Klages.[3]

An alternative method for arriving at resonance energies is based on heats of hydrogenation. The hydrogenation of cyclohexene to cyclohexane liberates 28.6 kcal./mole. A Kekulé structure of benzene might be expected to liberate three times as much heat, or 85.8 kcal./mole. The observed heat of hydrogenation, 49.8 kcal./mole, indicates that benzene is more stable by 36 kcal./mole than expected for structure I; this value, which may be identified as a resonance energy, is identical with that derived from heats of combustion.[7] Unfortunately, heats of hydrogenation are available for relatively few compounds; resonance energies for polycyclic aromatic hydrocarbons are presently available only from heats of combustion.

9.2 Delocalization Energies

In Sec. 2.4 the delocalization energy, DE, was defined as the calculated additional bonding energy which results from delocalization of electrons originally constrained to isolated double bonds. In practice, the reference

[5] A. Magnus, H. Hartmann, and F. Becker, *Z. physik. chem.*, **197**, 75 (1951).
[6] J. L. Franklin, *Ind. Eng. Chem.*, **41**, 1070 (1949); see discussion in ref. 3.
[7] G. W. Wheland, *Resonance in Organic Chemistry*, John Wiley and Sons, New York (1955), p. 80.

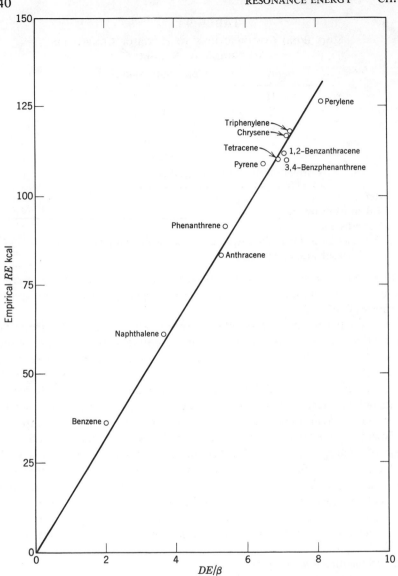

Fig. 9.1 Empirical resonance energies and HMO DE of aromatic hydrocarbons.

compound is treated as a set of double bonds, each of which contributes a π-energy of 2β to the total energy. Cyclohexatriene, I, for example, as a set of three isolated double bonds, has a π-energy of 6β. Benzene has a π-energy of 8β; hence DE for benzene is 2β. In this manner DE-values have been obtained for a large number of compounds.

The set of isolated double bonds corresponds clearly to a Kekulé structure; the delocalized structure used as a model for the HMO calculations corresponds to the resonance hybrid. Thus the DE corresponds to the resonance energy. For a number of aromatic hydrocarbons a good correlation has been demonstrated between HMO DE and VB resonance energies (Table 9.2).[8] An excellent correlation also exists between the

TABLE 9.2

RESONANCE AND DELOCALIZATION ENERGIES OF AROMATIC
HYDROCARBONS

Hydrocarbon	VB RE in α*	HMO DE in β*	α/β	Empirical RE kcal.†
Benzene	1.106	2.000	1.81	36.0
Naphthalene	2.040	3.683	1.80	61.0
Anthracene	2.951	5.314	1.80	83.5
Phenanthrene	3.019	5.448	1.80	91.3
Tetracene		6.932		110.0
1,2-Benzanthracene		7.101		111.6
Chrysene		7.190		116.5
Triphenylene		7.275		117.7
3,4-Benzphenanthrene		7.187		109.6
Pyrene		6.506		108.9‡
Perylene		8.245		126.3‡

* From B. Pullman and A. Pullman, *Les Théories electroniques de la chimie organique*, Masson and Cie, Paris (1952), p. 226.

† G. W. Wheland, *Resonance in Organic Chemistry*, John Wiley and Sons, New York (1955), p. 98.

‡ Ref. 2.

theoretical delocalization or resonance energies and the observed empirical resonance energies RE. Since the latter may vary, depending on the method used to estimate the energies of the reference structures, values derived by Wheland[9] using the method of Klages are listed in Table 9.2 and plotted in Fig. 9.1.

The correlation line is assumed to pass through the origin; its slope corresponds to $\beta = -16$ kcal. (0.69 ev)—a value far smaller than values derived by other correlations. This difference is discussed below. The value of β derived from empirical resonance energies depends in part on the way the resonance energies have been calculated. An older so-derived

[8] G. W. Wheland, *J. Chem. Phys.*, **2**, 474 (1934).
[9] Ref. 7, p. 98.

β-value, -20 kcal.,[8,10,11,12] has been used extensively in subsequent applications of HMO theory. As we have emphasized repeatedly, such a procedure is completely invalid. Unless otherwise independently demonstrated, this value of β should be used for only one purpose: the calculation of empirical resonance energies—and then only for benzenoid aromatic hydrocarbons (*vide infra*).

Several methods have been reported for approximating HMO *DE* values. Some are completely empirical and depend on such things as the number of doubly, triply, or quadruply substituted double bonds in a Kekulé structure,[13] although others are based on perturbation methods for building up aromatic hydrocarbons from simpler systems.[14]

The hydrocarbons used for the correlation in Fig. 9.1 are all planar polycyclic benzenoid systems. Empirical resonance energies are available for other types of compounds that may be divided into three classes: polyenes, noncoplanar benzenoid hydrocarbons, and nonbenzenoid aromatic systems.

Empirical resonance energies for substituted olefins and polyenes are compared with HMO *DE*-values in Table 9.3. The average value of β derived from these compounds, -6 kcal., is considerably smaller than the value derived from the aromatic hydrocarbons. In part, the difference is due to the alternation of bond distances that occurs in polyenes, but for the most part the discrepancy is undoubtedly due to the effects of electron correlation; the conjugation π-energy between adjacent double bonds in a polyene is less than that given by HMO theory (*vide infra*).

In hydrocarbons such as biphenyl conjugation energy between the rings is less than that calculated by HMO theory, not only because of electron correlation effects but also because the rings are twisted from coplanarity. The results in Table 9.4 show that there is only negligible conjugation energy between rings in compounds of this type, probably for both the foregoing reasons.

Theoretical delocalization energies have been used frequently for

[10] E. Hückel, *Z. Elektrochem.*, **43**, 752 (1937).

[11] J. Syrkin and M. E. Dyatkina, *Acta Physicochim. U.R.S.S.*, **21**, 641 (1946), derive $\beta = -17.2$ kcal. from empirical resonance energies.

[12] Magnus, Hartmann, and Becker[5] give $\beta = -18.5$ kcal. from empirical resonance energies. The empirical resonance energies derived by H. J. Dauben, Jr. (private communication), differ slightly from those in Table 9.2 and lead to $\beta = -16.5$ kcal./mole.

[13] V. M. Tatevskiĭ, *Zhur. Fiz. Khim.*, **25**, 241 (1951); *C.A.*, **45**, 5988 (1951); C. Vroelant, *Compt. rend.*, **235**, 958 (1952); A. L. Green, *J. Chem. Soc.*, **1956**, 1886.

[14] M. J. S. Dewar in J. W. Cook's, *Progress in Organic Chemistry*, Vol. 2, Academic Press, New York (1953), p. 1; M. J. S. Dewar and R. Pettit, *J. Chem. Soc.*, **1954**, 1617; D. Peters, *J. Chem. Soc.*, **1958**, 1023.

TABLE 9.3

RESONANCE AND DELOCALIZATION ENERGIES FOR OLEFINS

Hydrocarbon	RE kcal.	DE/β	RE/DE kcal./β
1,3-Butadiene	3.5,* 3†	0.472	7.4, 6.4
1,3,5-Cycloheptatriene	6.7*	0.988†	6.8
Styrene	0.9,* 2.1†‡	0.424§	2.1, 5.0
trans-Stilbene	7.0,* 4.9†‡	0.878§	8.0, 5.6
1,4-Diphenyl-1,3-butadiene	10.7*‡	1.401§	7.6

* From heats of hydrogenation; G. W. Wheland, *Resonance in Organic Chemistry*, John Wiley and Sons, New York (1955), p. 80.
† From heats of combustion, *ibid.*, p. 98.
‡ 36.0 kcal. has been subtracted for each benzene ring.
§ 2.00β has been subtracted for each benzene ring.

TABLE 9.4

EXTRA RESONANCE ENERGIES OF BIARYLS

Compound	$\Delta'RE$* kcal.	$\Delta DE/\beta$†	ΔDE kcal.¶
Biphenyl	−1.0	0.383	6.1
1,3,5-Triphenylbenzene	4.9	1.15‡	18.4
9,10-Diphenylanthracene	−3.9	0.86§	13.8
9,9′-Bianthryl	−0.4	0.58§	9.3

* Resonance energies of component rings have been subtracted; from ref. †, Table 9.2.
† DE for aromatic rings have been subtracted; calculated for coplanar systems. For rings twisted by 90°, $\Delta DE = 0$.
‡ Ref. 8.
§ Ref. 5.
¶ $\beta = -16.0$ kcal.

estimating the stabilities of unusual nonclassical or nonbenzenoid aromatic hydrocarbons. Examples are cyclobutadiene and azulene. Such compounds are treated in greater detail in Chap. 10. For the present, however, a number of empirical resonance energy values are available for comparison with DE-values (Table 9.5). For the most part, the experimental values are considerably lower than the theoretical estimates. A substantial part of the discrepancy is apparently due to ring strain effects that should be considered in estimating the energy of the reference structure. Dauben[15]

[15] H. J. Dauben, Jr., private communication.

TABLE 9.5

RESONANCE ENERGIES OF NONCLASSICAL "AROMATIC"
HYDROCARBONS*

Compound	RE kcal.	DE/β	DE kcal.†	Strain Energy kcal.§	Net RE Calculated kcal.
Fulvene II	11	1.466	23	10	13
Heptafulvene III	14	1.994	32	7	25
Fulvalene IV	20	2.779	44	19	25
Heptafulvalene V	29	4.004	64	14	50
Cycloöctatetraene	4–5	1.657‡	27	23	4
Azulene	30–32	3.364	54	16	38
Biphenylene VI	10	4.506	72	63	9
Acepleiadylene VII	77	5.997	96	33	63

* Adapted from a compilation of H. J. Dauben, Jr. (ref. 15).

† Using $\beta = -16$ kcal.; Dauben used $\beta = -16.5$ kcal.

‡ Calculated for a coplanar ring. For double bonds completely orthogonal $DE = 0$.

§ Evaluated by the procedure of H. J. Dauben, Jr. (ref. 15).

has developed procedures for estimating these strain energies. When corrected for strain, the calculated net resonance energies agree in several cases rather well with the experimental quantities. In others, particularly for the compounds best considered as cyclic polyenes, the calculated values are still too high. At least part of this remaining difference may be attributed to the use of the same β for the alternating single and double bonds in the HMO calculations.

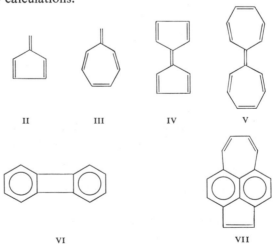

II   III   IV   V

VI   VII

9.3   Vertical Resonance Energies

The energy of a reference Kekulé structure of benzene is estimated from empirical bond contributions derived from alkanes and alkenes. This energy corresponds, therefore, to a cyclohexatriene, VIII, with alternating bond lengths corresponding to single and double bonds. A valence-bond

VIII   IX

resonance structure of benzene must possess the same geometry as the total system; hence the energy estimated for cyclohexatriene, VIII, does not correspond to a Kekulé structure, IX, for benzene. To convert VIII into IX we must stretch the double bonds and compress the single bonds until all are 1.397 A in length; IX must have higher energy than VIII.

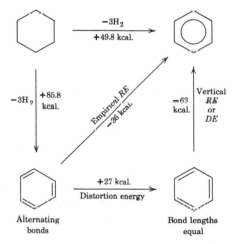

Fig. 9.2   Energy cycle showing the various resonance energies of benzene.

The resonance energy referred to IX is the *vertical resonance energy* and corresponds more exactly to the delocalization energy; this energy must be greater than the resonance energy referred to VIII; that is, IX is less stable than VIII. The distortion energy for VIII → IX has been estimated as 27 kcal./mole by Coulson and Altmann.[16] Since the vertical resonance energy must overcome this strain and still give an observed empirical resonance energy of 36 kcal., the vertical *RE* is 63 kcal. (Fig. 9.2).

[16] C. A. Coulson and S. L. Altmann, *Trans. Faraday Soc.*, **48**, 293 (1952).

Mulliken and Parr[17] have derived a vertical RE of 73 kcal. for benzene by using an independent estimate of 37 kcal./mole for the distortion energy. This value of the distortion energy was also used by Glockler[18] who obtains 111.5 kcal. for the vertical RE of benzene. Each of these derived values has its own merits, and it is not our point here to choose among them. Values for distortion energies are only rough estimates, but this quantity is clearly of important magnitude compared to the empirical resonance energies. The various values for the vertical RE of benzene correspond to $\beta = -31.5$ to -55.8 kcal. (-1.4 to -2.4 ev); these estimates are closer to those derived from other correlations. The observed correlation of HMO DE's with empirical RE's for benzenoid aromatics must require a remarkable proportionality between distortion energies and empirical RE-values.

Finally, for a discussion of the resonance energies of some radicals, see Chap. 13.

9.4 Further Comments

A critical assumption which is implied in all of the empirical resonance energies previously derived is that a single bond between sp^3-carbons has the same bond energy as that between two sp^2-carbons. As an example of the dependence of empirical RE-values on this assumption, let us consider that C_{sp^2}—C_{sp^2} bonds are 10 kcal. more stable than C_{sp^3}—C_{sp^3} bonds; that is, the contribution of the former bonds to heats of combustion is is 39.3 kcal. Use of this value gives an empirical RE for benzene of only 6 kcal! By using bond energies for different types of bonds based on not unreasonable assumptions, Dewar and Schmeising[19] have derived a vertical RE of benzene of only 13 kcal.!

A similar situation arises with heats of hydrogenation. In the hydrogenation of ethylene a double bond is converted to a single bond and two C—H bonds, but, in addition, four C_{sp^2}—H bonds are converted to C_{sp^3}—H bonds. Any energy difference associated with this change has been generally ignored in deriving empirical RE's from heats of hydrogenation.[20] In essence, the problem is simply that energies of known systems are dissected into components in order to estimate the energies of unknown systems. The dissection can be carried out in any number of

[17] R. G. Parr, *J. Chem. Phys.*, **19**, 799 (1951); R. S. Mulliken and R. G. Parr, *J. Chem. Phys.*, **19**, 1271 (1951).

[18] G. Glockler, *J. Chem. Phys.*, **21**, 1249 (1953).

[19] M. J. S. Dewar and H. N. Schmeising, *Tetrahedron*, **5**, 166 (1959), **11**, 96 (1960).

[20] See the discussions by R. S. Mulliken, *Tetrahedron*, **6**, 68 (1959) and W. F. Yates, *J. Phys. Chem.*, **65**, 185 (1961).

different ways to yield different values for the unknown energies. Furthermore, the energy of a structural feature is required for which suitable analogies are rare. This feature is a single bond between two double bonds with no π-delocalization. Most of the differences between various estimates of Kekulé energies hinge on the respective treatments of this structural unit.

Perhaps the best source of the π-conjugation energy across such a single bond is the barrier height for rotation about the 2-3 bond in 1,3-butadiene. This energy is 5 kcal./mole[21] and after consideration of several factors has been identified by Mulliken[20] as being close to the desired π-conjugation energy.[22] We may conclude that π-delocalization stabilization exists in conjugated systems[22] but that its magnitude is somewhat greater in benzenoid aromatic hydrocarbons (cf. Sec. 9.6). With estimates varying from 13 to 112 kcal. for the vertical resonance energy of benzene, the folly of the general use of $\beta = -20$ kcal. could not be clearer.

Thus the *empirical resonance energies*, as normally computed from experimental thermodynamic quantities, may contain σ-bond energy terms. To emphasize this point, the derived energies are sometimes called *stabilization energies*.

9.5   Hyperconjugation

The heats of hydrogenation of alkyl-substituted ethylenes decrease with the degree of substitution (Table 9.6). The observed heats can be derived from the value for ethylene by adding -2.7 kcal. for each alkyl group and applying corrections of $+1.5$ kcal. for each *cis*-dialkyl group and $+0.7$ kcal. for each 1,1-dialkyl group. These corrections undoubtedly have a steric origin; the primary effect of alkyl substituents has traditionally been regarded as a hyperconjugation energy resulting from delocalization of σ-electrons into the double bond π-system. However, in the same manner as that alluded to in Sec. 9.4, this energy change may be a σ-bond stabilization; that is, a C_{sp^2}—C_{sp^3} bond may be sufficiently more stable than a C_{sp^3}—C_{sp^3} bond to account for this difference.[19,20,23] At least part of the alkyl stabilization is probably due to σ-bond energy differences; the hyperconjugation energy from π-delocalization is probably rather small. Various models of a methyl substituent in

[21] J. G. Aston, G. Szasz, H. W. Woolley, and F. C. Brickwedde, *J. Chem. Phys.*, **14**, 67 (1946); cf. also C. M. Richards and J. R. Nielsen, *J. Opt. Soc. Amer.*, **40**, 438 (1950).

[22] Even this conclusion may be wrong. Internal dispersion or Van der Waals forces between the double bonds of butadiene may contribute substantially to this barrier. See W. T. Simpson, *J. Am. Chem. Soc.*, **73**, 5363 (1951).

[23] R. B. Turner, *Tetrahedron*, **5**, 127 (1959).

TABLE 9.6

HEATS OF HYDROGENATION OF SIMPLE OLEFINS

Olefin	Heat of Hydrogenation kcal./mole*
$CH_2{=}CH_2$	31.0
$RCH{=}CH_2$	28.0–28.9
$RCH{=}CHR$ *cis*	26.7–27.3
$RCH{=}CHR$ *trans*	25.4–26.0
$R_2C{=}CH_2$	25.8–26.4
$R_2C{=}CHR$	25.0–25.1
$R_2C{=}CR_2$	24.9

* At absolute zero; ref. 4.

the HMO method give π-hyperconjugation stabilizations that are undoubtedly far too high (Sec. 5.7); advanced MO techniques generally predict smaller magnitudes of such hyperconjugation energy for neutral systems, and the effect of electron correlation would be to make these energies smaller still. Hence here also we must be careful to distinguish between a theoretical hyperconjugation energy and an experimental *stabilization energy*.

9.6 Delocalization Energies and Equilibria

Despite the limited theoretical foundation for empirical resonance energy quantities, they and their theoretical HMO DE counterparts are nevertheless useful in a number of applications involving relative equilibrium constants and reaction rates. Many of these applications are discussed in succeeding chapters; a few are mentioned at this time.

When 6,13-dihydropentacene, X, is heated to 250°, it is converted to about 5% of 5,14-dihydropentacene, XI;[24] σ-bond energies strain energies, etc., may be expected to be so much the same in both compounds that we may take the principal energy difference to be that of delocalization. If

we neglect the π-hyperconjugation of the methylene groups, the delocalization energy of X is that of two naphthalenes, 7.366β, whereas the DE of

[24] E. Clar, *Aromatische Kohlenwasserstoffe*, Springer-Verlag, Berlin (1952), p. 58.

XI is that of anthracene plus benzene, 7.314β. X is more stable by 0.052β, in qualitative agreement with Clar's results.

Similarly, 5,16-dihydrohexacene, XII, is substantially converted to the 6,15-dihydro-isomer, XIII, by merely refluxing in xylene. At 300°, XII is converted 98% into XIII.[24] The lesser amount of XII at equilibrium is consistent with the greater difference in DE, 0.065β, compared to the previous case.

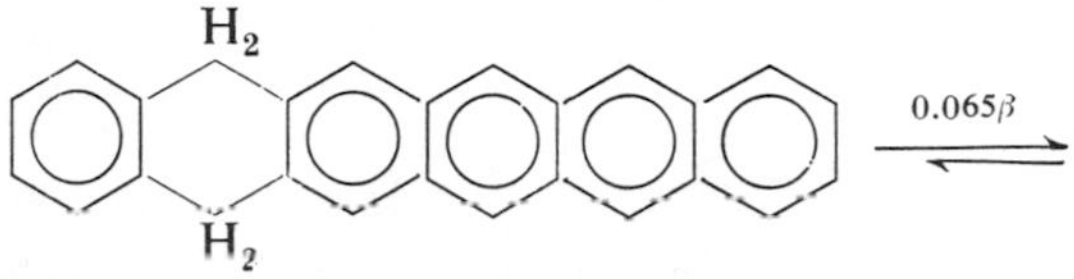

XII

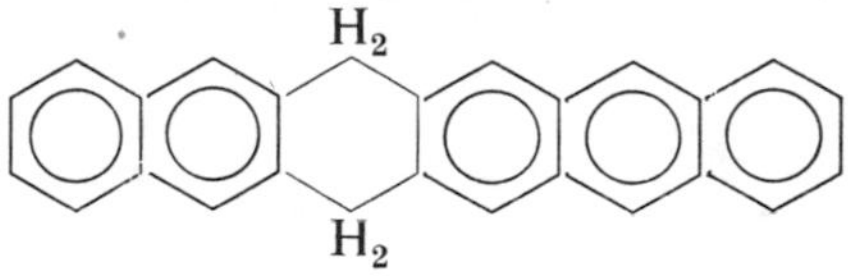

XIII

5,18-Dihydrohydroheptacene, XIV, is completely converted to about a 50 : 50 mixture of its isomers, XV and XVI, by refluxing in nitrobenzene.[25] The difference in DE between XIV and XV is 0.071β; between XV and XVI, 0.013β, in qualitative agreement with these observations.

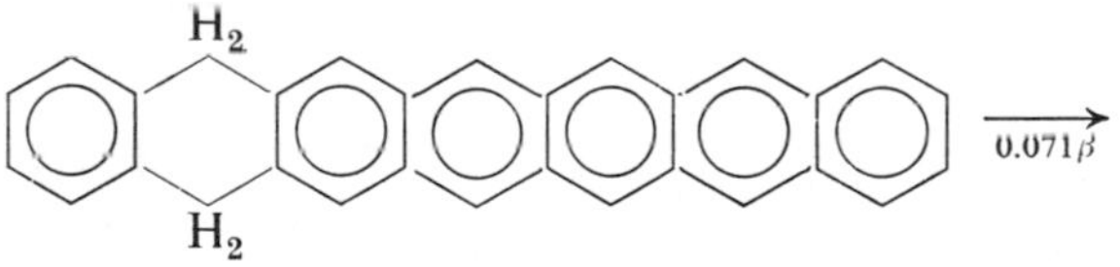

XIV

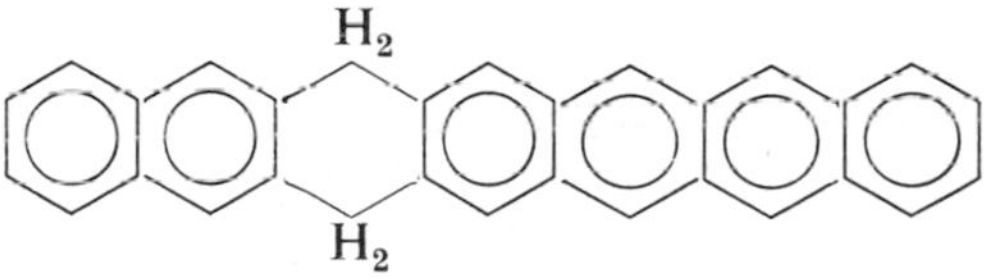

XV

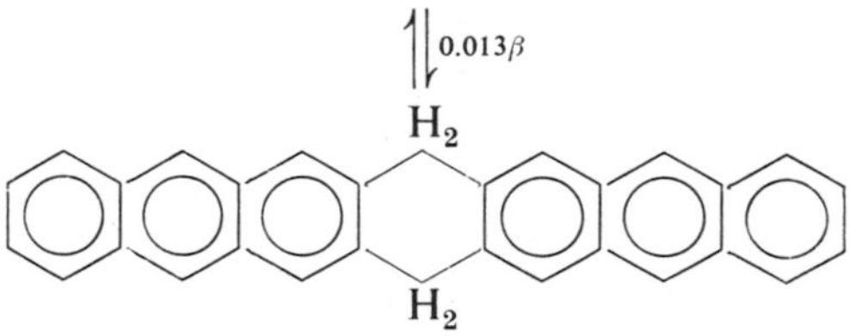

XVI

[25] C. Marschalk, *Bull. soc. chim. France*, **5**, 306 (1938); **8**, 354 (1941).

A related example is found in the methylacene-methylene-dihydroacene equilibria. The energy difference between the isomers may be dissected into a σ-bond energy difference and a π-delocalization energy difference. The first quantity may be expected to be relatively constant for a series of compounds. For the conversion of toluene to methylenedihydrobenzene

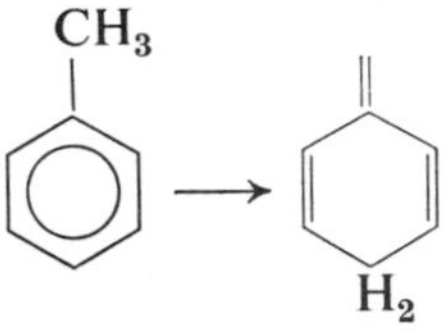

we neglect π-hyperconjugation and calculate the change in energy from a benzene π-system to a methylenepentadiene π-system:

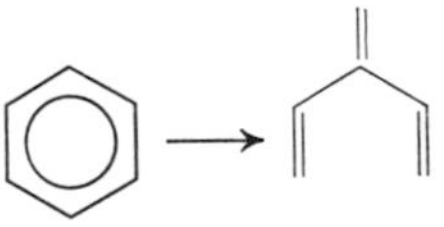

$$\Delta DE = -1.10\beta$$

XVII   XVIII
2.00β 0.90β

In this conversion we lose 1.10β of DE. In the similar reaction of 6-methylpentacene, XIX, the loss in DE is only 0.38β.[26] In this case the equilibrium lies on the side of the methylene-dihydroisomer, XX.[27]

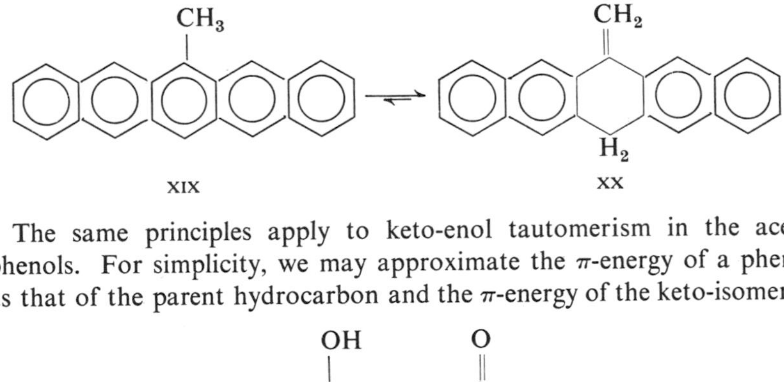

XIX   XX

The same principles apply to keto-enol tautomerism in the acene phenols. For simplicity, we may approximate the π-energy of a phenol as that of the parent hydrocarbon and the π-energy of the keto-isomer as

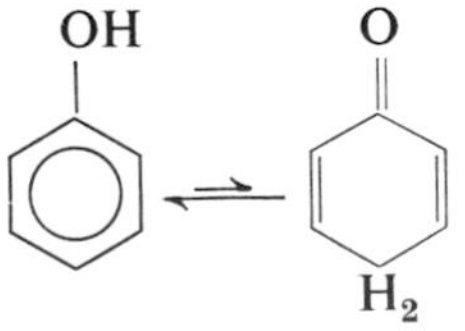

XXI   XXII

[26] B. Pullman and A. Pullman, *Les Théories electroniques de la chimie organique*, Masson and Cie, Paris (1952), p. 250.
[27] Ref. 24, p. 62.

that of the corresponding methylene derivative; for example, the π-energy change for the equilibrium between phenol, XXI, and 2,5-cyclohexadienone, XXII, is calculated as that for XVII $\rightleftharpoons$ XVIII, -1.10β. In the pentacene series 6-hydroxypentacene, XXIII, is unknown, the stable form being 6-keto-6,13-dihydropentacene, XXIV.[27]

| XXIII | XXIV |

Even the 9-anthrol-anthrone equilibrium, with a DE difference of -0.50β in this approximation, lies in the anthrone direction.

The application of resonance energies to quinone-hydroquinone redox potentials has long been known. Two decades ago Branch and Calvin[28] demonstrated an amazing correlation between these potentials and the relative numbers of Kekulé resonance structures in the quinone and hydroquinone. The MO equivalent has been treated by Dyatkina and Syrkin[29] in terms of two simplified MO models of a quinone. As in the foregoing cases, the σ-bond change for the equilibrium is considered substantially the same for a series of quinones and hydroquinones. The observed variations of potential are considered to be due entirely to DE differences. In both of Dyatkina and Syrkin's approximations the π-energy of the hydroquinone is taken as that of the parent acene; that is, the conjugation of a phenolic hydroxyl with an aromatic ring is ignored. In the first approximation the carbonyl groups of the quinone are considered to be completely localized; that is, conjugation of the carbonyl groups with the rest of the π-system is ignored. In this approximation the difference in DE for 1,4-naphthalenediol and 1,4-naphthoquinone, for example, is obtained as

$$\Delta DE = 1.68\beta$$

This approximation was also suggested by Evans.[30] A summary of results with a number of quinone-hydroquinone pairs is listed in Table 9.7;

[28] G. E. K. Branch and M. Calvin, *The Theory of Organic Chemistry*, Prentice-Hall, Englewood Cliffs, N.J. (1941), p. 305.

[29] M. E. Dyatkina and J. Syrkin, *Acta Physicochim. U.R.S.S.*, **21**, 921 (1946).

[30] M. G. Evans, *Trans. Faraday Soc.*, **42**, 113 (1946).

TABLE 9.7

CORRELATIONS OF REDOX POTENTIALS OF QUINONES-HYDROQUINONES*

No. in Figs. 9.3 and 9.4	Quinone	Reduction Potential Volts	DE of Parent Acene in β	First Approximation		Second Approximation	
				DE of Quinone in β	$\Delta DE/\beta$	DE of Quinone in β	$\Delta DE/\beta$
1	o-Benzoquinone	0.794	2.00	0.47	1.53	1.96	0.04
2	1,2-Naphthoquinone	0.576	3.68	2.42	1.26	3.79	−0.11
3	1,2-Anthraquinone	0.490	5.31	4.10	1.21		
4	9,10-Phenanthrenequinone	0.460	5.45	4.38	1.07	5.72	−0.27
5	1,2-Phenanthrenequinone	0.660	5.45	4.12	1.32	5.52	−0.07
6	3,4-Phenanthrenequinone	0.621	5.45	4.10	1.35		
7	1,2-Benz-3,4-anthraquinone	0.424	7.10	6.08	1.02		
8	4,5-Pyrenequinone	0.468†	6.51	5.45	1.06		
9	1,2-Benz-4,5-pyrenequinone	0.435†	8.21	7.19	1.02		
10	5,6-Chrysenequinone	0.47	7.19	6.08	1.11		
11	p-Benzoquinone	0.715	2.00	0	2.00	1.92	0.08
12	1,4-Naphthoquinone	0.484	3.68	2.00	1.68	3.80	−0.12
13	9,10-Anthraquinone	0.154	5.31	4.00	1.31	5.68	−0.37
14	1,4-Anthraquinone	0.401	5.31	3.68	1.63	5.48	−0.17
15	1,4-Phenanthrenequinone	0.530	5.45	3.68	1.77		
16	1,2-Benz-9,10-anthraquinone	0.230	7.10	5.68	1.42		
17	1,2,5,6-Dibenz-9,10-anthraquinone	0.268	8.88	7.36	1.52		
18	p-Diphenoquinone,	0.954	4.38	1.92	2.46	4.12	0.26
19		0.854	4.88	2.48	2.40	4.65	0.23

* G. M. Badger and H. A. McKenzie, Nature, 172, 458 (1953).

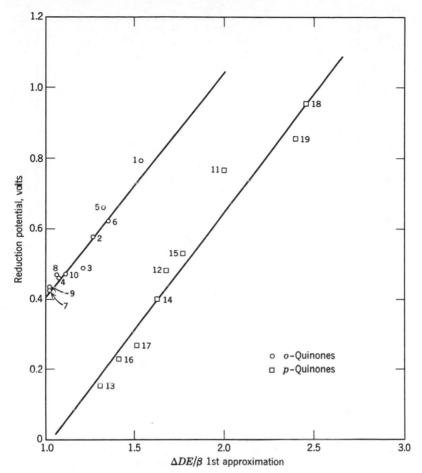

Fig. 9.3   Reduction potentials of quinones and Dyatkina and Syrkin's first approximation. Numbers refer to compounds in Table 9.7.

the correlation with redox potentials found for the first approximation is plotted in Fig. 9.3. Considering the simplicity of the approach, the observed correlations are amazingly good. It should not be unexpected that *ortho*- and *para*-quinones give different correlations for the changes in σ-bond energies are different. In *para*-quinones four bonds between carbonyl carbons and sp^2-carbons are converted into aromatic bonds, whereas only two such bonds as well as a bond between carbonyl groups are so involved in *ortho*-quinones. The slopes of the two correlations are similar.

In Dyatkina and Syrkin's second approximation the quinone carbonyl

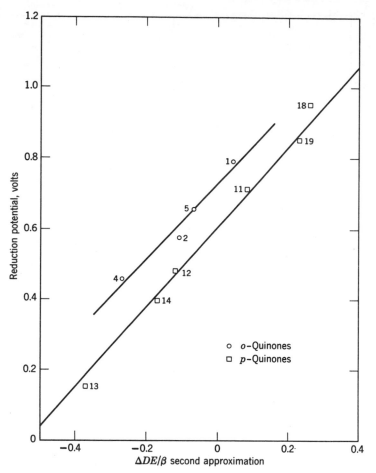

Fig. 9.4 Reduction potentials of quinones and Dyatkina and Syrkin's second approximation. Numbers refer to compounds in Table 9.7.

group is treated as a double bond. The reduction potential of 1,4-naphthoquinone, for example, is related to the delocalization energy change:

$$\Delta DE = -0.12\beta$$

In effect, the quinone is calculated with $k_{C=O} = 1$, $h_O = 0$; the hydroquinone is calculated with $k_{C-O} = 0$. The results are summarized in

Table 9.7 and are plotted in Fig. 9.4. Good correlations are obtained again.

Of course, a closer approximation to the π-energy change could be obtained by including suitable values for the oxygen parameters. Various calculations of this type have been reported.[31,32] As might be expected from the success of the foregoing simple approaches, the goodness of fit of the correlations is not sensitive to the particular parameter values used. With certain parameter values, the difference between the correlation lines for *ortho*- and *para*-quinones vanishes; however, because of the possible difference in σ-bond energy changes between these systems, this merging is not necessarily significant.

[31] M. G. Evans, J. Gergely, and J. De Heer, *Trans. Faraday Soc.*, **45**, 312 (1949); M. G. Evans and J. De Heer, *Trans. Faraday Soc.*, **47**, 801 (1951); *Quart. Revs.*, **4**, 94 (1950).

[32] V. Gold, *Trans. Faraday. Soc.*, **46**, 109 (1950).

10 Aromaticity and the $4n + 2$ rule

10.1 Hückel's $4n + 2$ Rule

The aromatic sextet has long been a part of organic chemistry; most aromatic and heterocyclic compounds can be divided into rings which share six electrons per ring.[1] A question of long-standing puzzlement was "why *six* electrons?" Part of the answer lies in the six-membered ring. The internal angle of a plane hexagon, 120°, is exactly the bond angle of a trigonal carbon. Strainless six-membered ring systems can be built from such carbons. Angle strain is not the whole answer, however. The cyclopentadienyl anion, I, with six electrons is relatively stable, whereas the corresponding cation, II, with four electrons, is not. Cycloheptatrienyl

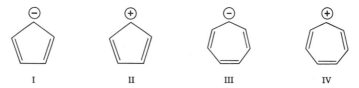

| I | II | III | IV |

anion, III, with eight electrons is relatively unstable, whereas the cation, IV, with six electrons is comparatively stable (*vide infra*). These examples, and others, were recognized decades ago by Hückel, who found a satisfying explanation in simple MO theory.[2] This theory leads to a rule, usually referred to as the $4n + 2$ rule, which may be stated as follows: *those*

---

[1] For historical discussion see G. W. Wheland, *Advanced Organic Chemistry*, John Wiley and Sons, New York, second edition (1949), p. 102; C. K. Ingold, *Structure and Mechanism in Organic Chemistry*, Cornell University Press, Ithaca (1953), p. 156; W. Hückel, *Theoretical Principles of Organic Chemistry*, Vol. 1, Elsevier Publishing Co., Amsterdam (1955), Chapter IX.

[2] E. Hückel, *Z. Physik.*, **70**, 204 (1931); **76**, 628 (1932); International Conference on Physics, London, 1934, Vol. II, The Physical Society, London (1935), p. 9; *Z. Elektrochem.*, **43**, 752 (1937).

monocyclic coplanar systems of trigonally hybridized atoms which contain $4n + 2$ *π-electrons will possess relative electronic stability.* This rule is inherent in the form of the MO's which can be written in analytic form for such systems (pp. 47 and 50). The results of these equations may be expressed graphically in the following manner:[3] to obtain the MO energy levels for a cyclic system of k-atoms, inscribe a k-fold regular polygon within a circle of radius 2β such that one apex is at the lowest point. The distance of each apex from a horizontal mid-line then represents an energy level in units of β. Examples for $k = 5$, 6, and 7 are given in Fig. 10.1.

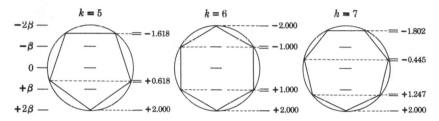

Fig. 10.1   Graphic depiction of HMO energies of some monocyclic systems.

Each set of energy levels may be thought of as a *shell*. The lowest-lying MO is always single and can harbor two electrons. The remaining low-lying orbitals occur in pairs, each MO of which can accommodate two electrons; hence each low-lying shell above the first requires four electrons to become *filled*; that is, $4n + 2$ electrons yield filled shell configurations. This feature also comes out of free electron MO theory[4] which ascribes a particular significance to the doubly degenerate levels. In this theory each successive level corresponds to an increase in orbital angular momentum. The lowest level has zero angular momentum. For succeeding levels a degeneracy arises, for the angular momentum can be ascribed a clockwise or counterclockwise direction around the ring. Hence these levels occur as degenerate pairs.[4a] In HMO theory the degeneracy arises from the equivalence of orthogonal descriptions of the nodal planes.

The state of occupancy of the shells seems to have more fundamental significance than delocalization energy. Cycloöctatetraene, for example, has 1.657β *DE*, a substantial quantity. The instability of this compound may be associated with the half-filled shell of two NBMO's (*vide infra*).

Guided by the $4n + 2$ rule, we can evaluate much of the chemistry of monocyclic ring systems and their derivatives in a qualitative fashion.

[3] A. A. Frost and B. Musulin, *J. Chem. Phys.*, **21**, 572 (1953); cf. also H. C. Longuet-Higgins, *Proc. Chem. Soc.*, **1957**, 157.

[4] (a) J. R. Platt, *J. Chem. Phys.*, **17**, 484 (1949); (b) H. H. Jaffé, *J. Chem. Phys.*, **20**, 1646 (1952).

In succeeding sections various rings are discussed and compared with experimental observations (Secs. 10.2 to 10.8). The extension of the concepts to polycyclic systems (Secs. 10.9 and 10.10) is then considered. HMO *DE* values by themselves do not suffice to estimate aromatic character in these systems, but with ancillary criteria reliable predictions can apparently be made. These predictions were developed by theoretical chemists; it has remained for the synthetic organic chemist to test the theories. In recent years a number of remarkable compounds have been synthesized in a brilliant and fascinating chapter of organic chemistry.

10.2   Cyclopropenyl

The cyclopropenyl system has the HMO energies: $\epsilon_1 = \alpha + 2\beta$, $\epsilon_2 = \epsilon_3 = \alpha - \beta$ (Fig. 10.2). The cyclopropenyl cation, V, with two

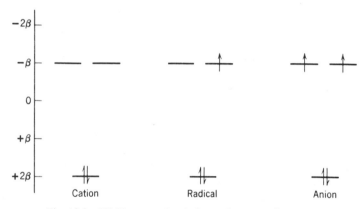

Fig. 10.2   HMO energy levels for cyclopropenyl systems.

electrons, obeys the $4n + 2$ rule ($n = 0$) and should be stable. The cyclopropenyl radical and anion both have electrons in antibonding MO's; they also have unfilled shells and should be relatively unstable in theory.

V

The parent cation has not been isolated, although cyclopropene is reported to undergo facile hydride exchange with triphenylmethyl cation.[5] The triphenyl derivative, VI, on the other hand, has been synthesized and is found to be relatively stable.[6] Both the fluoborate and bromide of VI

[5] K. B. Wiberg, Abstracts of 131st Meeting of American Chemical Society, Miami, April 7–12, 1957, p. 39–O.
[6] R. Breslow and C. Yuan, *J. Am. Chem. Soc.*, **80**, 5991 (1958).

are saltlike compounds that are insoluble in most organic solvents. The salts of VI dissolve in alcohol with the formation of the covalent ether, VII, but the cation is regenerated by treatment with acid. This cation is

VII   VI

clearly more stable than triphenylmethyl cation. HMO calculations of VI have been reported.[7,8] The cation has the same filled-shell characteristic of the parent cyclopropenyl cation and a total DE of 9.19β, which exceeds the sum of the DE's of three benzene rings by 3.19β (the DE of triphenyl-methyl cation, for comparison, exceeds that of three benzene rings by only 1.80β). Relative DE's are not always a reliable measure of stability; nevertheless, the stability of VI is in qualitative accord with HMO theory (Sec. 12.2) and represents a remarkable achievement for the theory.

   Diphenylcyclopropenone, VIII, has also been prepared.[9,10] This remarkably stable ketone requires heating to 130° to lose carbon monoxide.[9] In addition, VIII has a dipole moment of $5.08 D$,[10] a high value compared to other ketones (benzophenone, $2.97 D$; acetone, $2.8 D$) that undoubtedly reflects the substantial contribution of structure IX to the resonance hybrid.

VIII   IX

This conclusion is reinforced by the basicity of VIII which forms a stable crystalline hydrobromide salt, X.[10]

   [7] D. A. Bochvar, I. V. Stankevich, and A. L. Chistyavkov, *Bull. Acad. Sci. U.S.S.R.* (Engl. Trans.), **1958**, 775; *J. Russ. Phys. Chem.* **33**, 593 (1959).
   [8] A. Streitwieser, Jr., *J. Am. Chem. Soc.*, **82**, 4123 (1960).
   [9] R. Breslow, R. Haynie, and J. Mirra, *J. Am. Chem. Soc.*, **81**, 247 (1959).
   [10] M. E. Vol'pin, Y. D. Koreshkov and D. N. Kursanov, *Izvest. Akad. Nauk. S.S.S.R. Otdel. Khim. Nauk.*, **1959**, 560.

The relative stability of these cations contrasts with the instability of the corresponding radicals. The odd electron in triphenylcyclopropenyl radical is in an antibonding orbital which, furthermore, is doubly degenerate. The Jahn-Teller theorem[11] states that if an electronic state is degenerate forces will persist in the symmetrical configuration, and

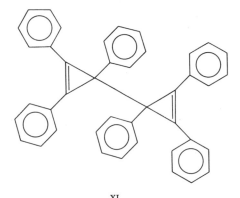

X

equilibrium in this state is impossible. A triphenylcyclopropenyl radical would distort to destroy the symmetry that leads to the degeneracy. Such distortion would represent a further destabilizing influence. The dimer, bis-triphenylcyclopropenyl, XI, has also been prepared and shows no tendency to dissociate even under vigorous conditions.[12]

XI

These qualitative observations have received firm quantitative support by application of a polarography technique that gives -1.13 v for the reversible one-electron reduction potential of VI in acetonitrile compared to only -0.09 v for triphenylmethyl cation.[12a]

The Jahn-Teller theorem does not apply to a cyclopropenyl anion that has both degenerate antibonding orbitals singly occupied (Fig. 10.2).

[11] H. A. Jahn and E. Teller, *Proc. Roy. Soc.*, **A161**, 220 (1937); W. L. Clinton and B. Rice, *J. Chem. Phys.*, **30**, 542 (1959).

[12] R. Breslow and P. Gal, *J. Am. Chem. Soc.*, **81**, 4747 (1959).

[12a] R. Breslow, W. Bahary, and W. Reinmuth, *J. Am. Chem. Soc.*, **83**, 1763 (1961).

This diradical has an unfilled shell and should be comparatively unstable. Little is known about the acidity of cyclopropene;[13] however, the ester, XII, has been prepared and has been found to exchange the α-hydrogen for deuterium with potassium *t*-butoxide in *t*-butanol. This exchange reaction, however, is much slower than the comparable reaction of a cyclopropane ester and suggests a destabilizing influence of the double bond.[15]

COOBu-*t*

XII

10.3   Cyclobutadiene[16]

The cyclobutadiene ring system has fundamental significance as the first cyclic polyene for which Kekulé structures, XIII, can be written. The simple MO predictions for this molecule are straightforward. Square

XIII   XIV   XV

cyclobutadiene, XIV, has four electrons and does not fit the $4n + 2$ rule. A pair of degenerate NBMO's is each singly occupied and the molecule should be a diradical having a DE of 0 (Fig. 10.3). The degeneracy is destroyed for a rectangular cyclobutadiene, XV, but the compound still has zero delocalization energy (Fig. 10.3). More detailed considerations suggest that a square cyclobutadiene should distort to the more stable rectangular configuration.[17,18] Both models have no resonance stabilization, and this strained ring system should theoretically be an exceedingly unstable substance.

[13] In a preliminary report this hydrocarbon was reported not to form a Grignard derivative under conditions that convert cyclopentadiene to the Grignard reagent.[14]

[14] J. D. Roberts, A. Streitwieser, Jr., and C. M. Regan, *J. Am. Chem. Soc.*, **74**, 4579 (1952).

[15] R. Breslow and M. Battiste, *Chem. and Ind.*, **1958**, 1143.

[16] For a recent review see W. Baker and J. F. W. McOmie in D. Ginsburg's *Non-Benzenoid Aromatic Compounds*, Interscience Publishers, New York (1959), p. 43.

[17] J. E. Lennard-Jones and J. Turkevich, *Proc. Roy. Soc.*, **A158**, 297 (1937).

[18] A. D. Liehr, *Z. physik. Chem.*, *NF* **9**, 338 (1956).

In complete contrast, simple VB theory predicts that square cyclo-butadiene is a singlet with substantial resonance energy.[19] Advanced MO theory with configuration interaction suggests that the molecule is not a diradical and has but little resonance energy.[20,21] A more complete VB

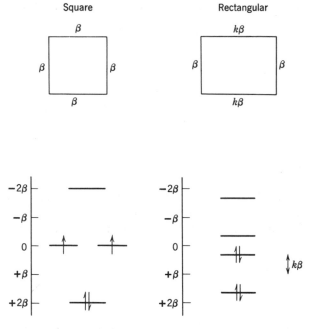

Fig. 10.3   Comparison of MO energy levels for square and rectangular cyclobutadiene.

calculation gives the same result.[22] These calculations do not rule out the possibility that the molecule is a diradical but their prediction of a non-totally symmetric ground state[20,22] would speak for a low order of stability for cyclobutadiene.   Further discussion is presented in Sec. 10.8.

The cyclobutadiene ring structure has eluded a number of ingenious attempts at synthesis; these clever experiments provide definite evidence of but little resonance energy in this system.   The diene, XVI, has been pre-pared and shows no observable tendency to rearrange to methylcyclo-butadiene, XVII.[23]   The ketones, XVIII and XIX, cannot be converted

[19] G. W. Wheland, *J. Chem. Phys.*, **2**, 474 (1934); *Proc. Roy. Soc.*, **A164**, 397 (1938).

[20] D. P. Craig, *Proc. Roy. Soc.*, **A202**, 498 (1950); *J. Chem. Soc.*, **1951**, 3175.

[21] Cf. also W. E. Moffitt and J. Scanlan, *Proc. Roy. Soc.*, **A220**, 530 (1953).

[22] R. McWeeny, *Proc. Roy. Soc.*, **A227**, 288 (1955).

[23] D. R. Howton and E. R. Buchman, *J. Am. Chem. Soc.*, **78**, 4011 (1956);   D. E. Applequist and J. D. Roberts, *J. Am. Chem. Soc.*, **78**, 4012 (1956).

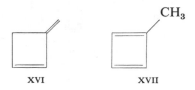

XVI   XVII

into the corresponding enols which are cyclobutadiene derivatives.[24] The
parent ketone, XX, has also been prepared recently and shows no enol

XVIII   XIX   XX

content in the infrared.[25] Phenylcyclobutadienoquinone, XXI, apparently cannot be converted to the hydroquinone, a cyclobutadiene
derivative.[26]

XXI

Blomquist and Meinwald[27] conceived an elegant approach to cyclo-
butadienes by Diels-Alder reactions of diphenyldimethylenecyclobutene,
XXII; however, although 1,2-dimethylenecyclobutane, XXIII, reacts
readily with maleic anhydride to form a cyclobutene derivative, XXIV,[28]
the triene, XXII, does not form an adduct. A product was obtained with
the reactive dienophile, tetracyanoethylene, but this was shown to be the
1,2-adduct, XXV, and not a cyclobutadiene derivative.

[24] R. B. Woodward and G. Small, *J. Am. Chem. Soc.*, **72**, 1297 (1950); J. D. Roberts,
G. B. Kline, and H. E. Simmons, *J. Am. Chem. Soc.*, **75**, 4765 (1953).
[25] E. Vogel, private communication.
[26] E. J. Smutny and J. D. Roberts, *J. Am. Chem. Soc.*, **77**, 3420 (1955); E. J. Smutny,
M. C. Caserio, and J. D. Roberts, *J. Am. Chem. Soc.*, **82**, 1793 (1960).
[27] A. T. Blomquist and Y. C. Meinwald, *J. Am. Chem. Soc.*, **81**, 667 (1959).
[28] A. T. Blomquist and J. A. Verdol, *J. Am. Chem. Soc.*, **77**, 1806 (1955).

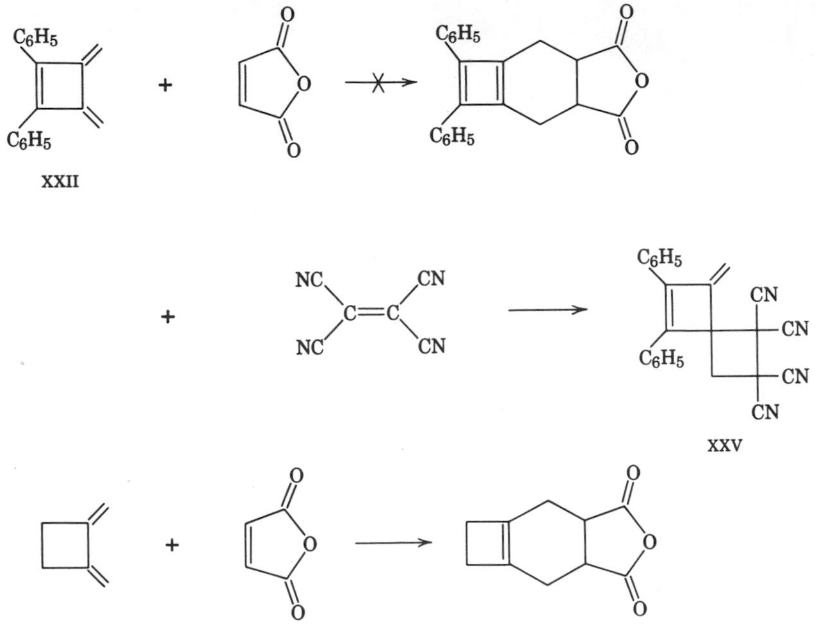

XXII

XXV

XIV

Cram and Allinger[29] have prepared the cyclic diacetylene, XXVI, and have found no evidence for a valence tautomerism to the cyclobutadiene, XXVII.

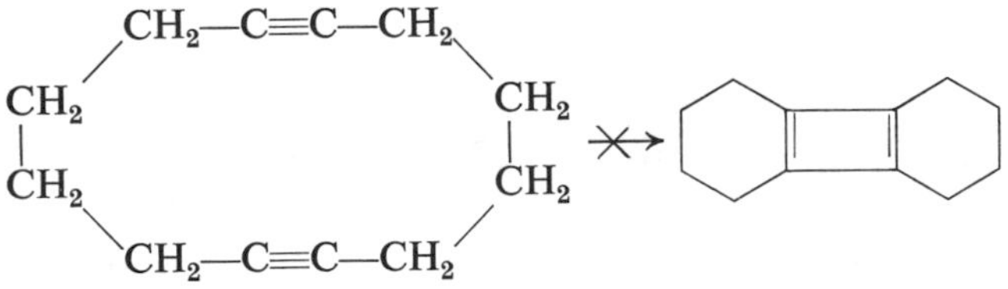

XXVI   XXVII

Other more or less classical approaches to cyclobutadiene by dehydrohalogenation[30] or by Hofmann exhaustive methylation[31] have also failed.

Nevertheless, other recent work teaches that cyclobutadiene is capable of existence at least transiently. The production of 20% butadiene by Hofmann elimination of the bis-quaternary hydroxide, XXVIII, was

[29] D. J. Cram and N. L. Allinger, *J. Am. Chem. Soc.*, **78**, 2518 (1956).
[30] R. Willstätter and W. von Schmaedel, *Ber.*, **38**, 1992 (1905).
[31] E. R. Buchman, M. Schlatter, and A. D. Reims, *J. Am. Chem. Soc.*, **64**, 2701 (1942).

interpreted by Nenitzescu et al.[32] as arising from the reduction of intermediate cyclobutadiene:

N(CH₃)₃⁺OH⁻

XXVIII

Similarly, the production of small amounts of butadiene from the thermal decomposition of the Diels-Alder adduct, XXIX, of cyclooctatetraene and dimethyl acetylenedicarboxylate was interpreted in terms of the intermediate formation of cyclobutadiene:[32]

XXIX

However, the first really concrete evidence for a cyclobutadiene intermediate comes from the observation by Criegee and Louis[33] that 3,4-dichloro-1,2,3,4-tetramethylcyclobutene, XXX, with lithium amalgam yields the tricyclic hydrocarbon, XXXI. Longuet-Higgins and Orgel[34] have suggested that Reppe's synthesis of cyclooctatetraene from acetylene and nickelous cyanide[35] involves an intermediate nickel complex of cyclobutadiene, $(C_4H_4)Ni(CN)_2$. Criegee and Schröder[36] have found that the reaction of the dichloride, XXX, with nickel carbonyl yields a stable red-violet compound, $C_8H_{12}NiCl_2$, which seems to be a complex of the type, XXXIII.[37]

---

[32] M. Avram, C. D. Nenitzescu, and E. Marica, *Ber.*, **90**, 1857 (1957).

[33] R. Criegee and G. Louis, *Ber.*, **90**, 417 (1957).

[34] H. C. Longuet-Higgins and L. E. Orgel, *J. Chem. Soc.*, **1956**, 1969.

[35] Cf. M. H. Bigelow and J. W. Copenhaver, *Acetylene and Carbon Monoxide Chemistry*, Reinhold Publishing Corp., New York (1949).

[36] R. Criegee and G. Schröder, *Ann.* **623**, 1 (1959).

[37] X-Ray analysis shows that the compound has a sandwich structure in which the nickel is bound to a central position of the ring. Private communication from R. Criegee reported by M. Avram, D. Dinu, G. Matecscu, and C. D. Nenitzescu, *Ber.*, **93**, 1789 (1960).

The parent molecule cyclobutadiene has been found to behave in a similar fashion in Nenitzescu's laboratory.[38] 1,2,3,4-Tetrabromobutane with lithium amalgam gives a mercury derivative that on treatment with

silver nitrate yields a crystalline complex, $C_4H_4AgNO_3$, whose infrared spectrum indicates a polymeric 1:1 complex between cyclobutadiene moieties and silver ions.[39] On treatment with steam, the complex yields a cyclobutadiene dimer apparently related to XXXI in structure; the dimer reverts to the original complex on treatment with silver nitrate.

XXXIII

Simple molecular orbital calculations on benzocyclobutadiene, XXXIV, indicate that this compound is not a diradical; it has substantial de-localization energy (2.38β; 0.38β more than benzene) but should be very reactive.[14] This compound has not been isolated, but it seems to be an intermediate in several reactions. Cava and Napier[40] found the dehalogenation of dibromo- or diiodobenzocyclobutene, XXXV, with zinc to give a good yield of the hydrocarbon, XXXVI, which can be formulated as a dimer product of an intermediate benzocyclobutadiene (see p. 267).

In the presence of cyclopentadiene, the adduct XXXVII has been iso-lated.[41,42] The reaction of the dibromide, XXXV (X = Br), with potas-sium hydroxide yields 3-bromo-1,2-benzobiphenylene, XXXVIII, which

[38] M. Avram, E. Marica, and C. D. Nenitzescu, *Ber.*, **92**, 1088 (1959); M. Avram, G. Matecscu, I. G. Dinulescu, E. Marica, and C. D. Nenitzescu, *Tetrahedron Letters*, No. 1, 21 (1961).

[39] H. P. Fritz, J. F. W. McOmie, and N. Sheppard, *Tetrahedron Letters*, No. 26, 35 (1960).

[40] M. P. Cava and D. R. Napier, *J. Am. Chem. Soc.*, **79**, 1701 (1957).

[41] C. D. Nenitzescu, M. Avram, and D. Dinu, *Ber.*, **90**, 2541 (1957).

[42] M. P. Cava and M. J. Mitchell, *J. Am. Chem. Soc.*, **81**, 5409 (1959).

XXXV   XXXIV

XXXVII

XXXVI

can also be formulated as a dimerization product of an intermediate benzocyclobutadiene.[43]

XXXVIII

Similarly, the reaction of the tetrabromide, XXXIX, with anhydrous base apparently yields a dibromobenzocyclobutadiene, XL, which immediately dimerizes.[44]

XXXIX   XL

Dibenzocyclobutadiene or biphenylene is relatively normal according to HMO studies.[14,45] It has a relatively high DE of 4.506β, 0.506β more than

[43] M. P. Cava and J. F. Stucker, *J. Am. Chem. Soc.*, **79**, 1706 (1957).

[44] F. R. Jensen and W. E. Coleman, *Tetrahedron Letters* No. 20, 7 (1959).

[45] R. D. Brown, *Trans. Faraday Soc.*, **46**, 146 (1950); J. I. F. Alonso and F. Perade-jordi, *Anales real soc. españ. fis. y quim. (Madrid)*, **50B**, 253 (1954); J. I. F. Alonso and R. Domingo, *Anales real soc. españ. fis. y quim. (Madrid)*, **51B**, 447 (1955).

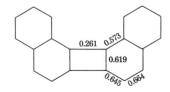

Biphenylene

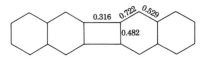

1,2,7,8-Dibenzobiphenylene

2,3,6,7-Dibenzobiphenylene

Fig. 10.4   Bond orders for some biphenylenes.

two benzene rings. The bond orders as given in Fig. 10.4 correspond best to the Kekulé structure XLI in which a cyclobutadiene structure is absent.

XLI

This hydrocarbon was first prepared by Lothrop[46] by the distillation of 2,2′-diiodobiphenyl with cuprous oxide. This synthesis has been improved[47] and several other syntheses are now available.[48] According to X-ray[49] and electron diffraction[50] studies, the six-membered rings are normal benzene rings and the remaining four-membered ring bonds are sp^2—sp^2 single bonds of 1.46 A length.[50a] These bond distances are in

[46] W. C. Lothrop, *J. Am. Chem. Soc.*, **63**, 1187 (1941).

[47] W. Baker, M. P. V. Boarland, and J. F. W. McOmie, *J. Chem. Soc.*, **1954**, 1476.

[48] J. Collette, D. McGreer, R. Crawford, F. Chubb, and R. B. Sandin, *J. Am. Chem. Soc.*, **78**, 3819 (1956); W. S. Rapson and R. G. Shuttleworth, *J. Chem. Soc.*, **1941**, 487; W. S. Rapson, R. G. Shuttleworth, and J. N. van Niekerk, *J. Chem. Soc.*, **1943**, 326; G. Wittig and W. Herwig, *Ber.*, **87**, 1511 (1954); G. Wittig and L. Pohmer, *Ber.*, **89**, 1334 (1956); H. Heaney, F. G. Mann, and I. T. Millar, *J. Chem. Soc.*, **1957**, 3930.

[49] J. Waser and C.-S. Lu, *J. Am. Chem. Soc.*, **66**, 2035 (1944).

[50] J. Waser and V. Schomaker, *J. Am. Chem. Soc.*, **65**, 1451 (1943).

[50a] A more recent X-ray study gives 1.52 A for this bond distance: T. C. W. Mark and J. Trotter, *Proc. Chem. Soc.*, **1961**, 163.

complete accord with the HMO bond orders. The heat of combustion of biphenylene gives an empirical *RE* of 20 kcal.[51] The rather low value is undoubtedly caused by ring strain of the small ring. Aromatic substitution reactions with this compound are of especial interest and are discussed in Chap. 11.

A number of benzo- and dibenzobiphenylenes have been prepared. These include 1,2-benzobiphenylene, XLII,[52] 2,3-benzobiphenylene, XLIII,[44] 1,2,7,8-dibenzobiphenylene, XLIV,[52,53] and 2,3,6,7-dibenzobiphenylene, XLV.[54] HMO calculations of some of these compounds

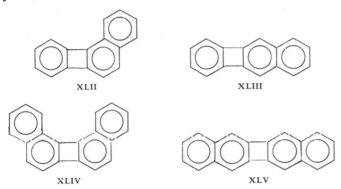

have been published.[55,56] Interestingly, XLIV is far less stable than XLV in agreement with the difference in *DE* values, 7.82β and 7.98β, respectively.[56a] Furthermore, the bond orders[56a] (Fig. 10.4) show that XLIV has more cyclobutadiene character than XLV in agreement with the suggestion of Cava and Stucker[53] for explaining the difference in the chemistry of the two isomers. Silva and Pullman[56a] also point out that the energy difference between the highest occupied and the lowest vacant MO's is much smaller for XLIV than for XLV in approximate agreement with the color difference: XLIV is deep red, XLV is pale yellow.

10.4   Cyclopentadienyl

The cyclopentadienyl cation has four electrons and an unfilled bonding shell (Fig. 10.1). We expect the cation to be electronically unstable or,

[51] R. C. Cass, H. D. Springall, and P. G. Quincey, *J. Chem. Soc.*, **1955**, 1188.

[52] M. P. Cava and J. F. Stucker, *J. Am. Chem. Soc.*, **77**, 6022 (1955).

[53] M. P. Cava and J. F. Stucker, *Chem. and Ind.*, **1955**, 446.

[54] R. F. Curtis and G. Viswanath, *Chem. and Ind.*, **1954**, 1174; *J. Chem. Soc.*, **1959**, 1670; E. R. Ward and B. D. Pearson, *J. Chem. Soc.*, **1959**, 1676.

[55] V. A. Crawford, *Can. J. Chem.*, **30**, 47 (1952).

[56] (*a*) M. A. Silva and B. Pullman, *Compt. rend.*, **242**, 1888 (1956). (*b*) M. A. Ali and C. A. Coulson, *Tetrahedron*, **10**, 41 (1960).

more precisely, we anticipate that the cation will be more reactive than otherwise expected for such a structure. The parent cation has never been observed in solution, but there are several indirect pieces of evidence of its relative instability. A reluctance of the cyclopentadienyl system to give up electrons carries through to highly substituted derivatives. The dissociation constant for the ionization of 9-chloro-9-phenylfluorene, XLVI, in liquid sulfur dioxide is less than 0.0001 of that of the analogous triphenylmethyl chloride, XLVII; that is,

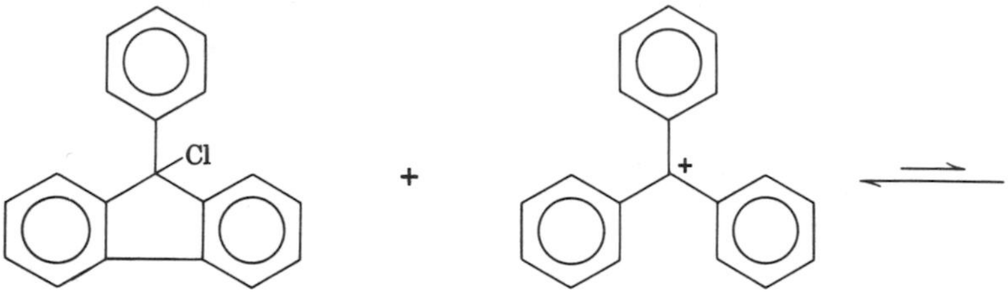

XLVI   +

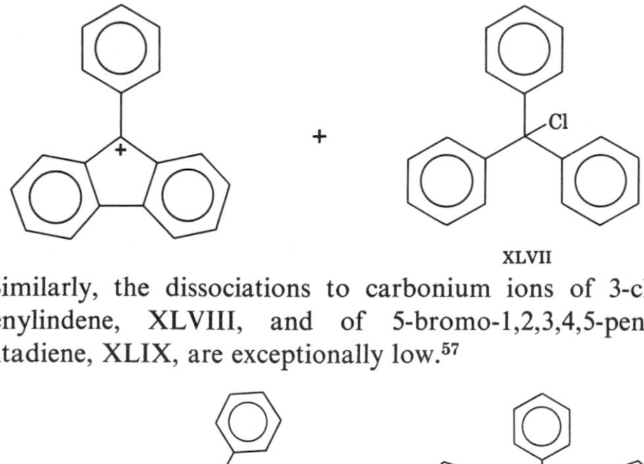

XLVII

Similarly, the dissociations to carbonium ions of 3-chloro-1,2,3-triphenylindene, XLVIII, and of 5-bromo-1,2,3,4,5-pentaphenylcyclopentadiene, XLIX, are exceptionally low.[57]

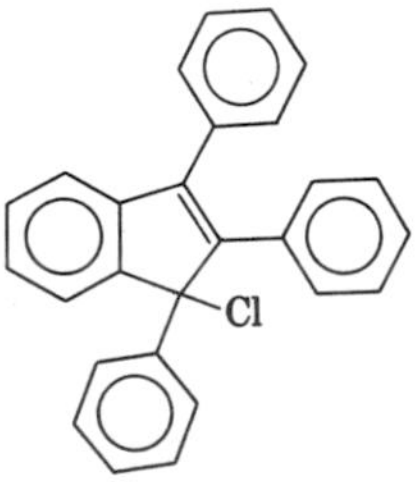

XLVIII 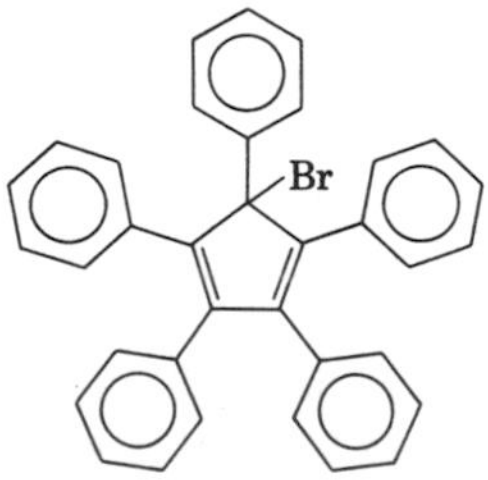   XLIX

The pseudo-basic character of carbinols follows a similar vein:

$$ROH + H^+ \rightleftharpoons R^+ + H_2O$$

[57] K. Ziegler and H. Wollschitt, *Ann.*, **479**, 104 (1930).

9-Fluorenol, L, is substantially less basic than benzhydrol, LI;[58] on the basis of simple resonance theory, we would have expected the reverse to hold, especially since the fluorenyl cation is planar, whereas benzhydryl cation probably is not (for a quantitative discussion, cf. Sec. 12.2).

Because the carbon in a carbonyl group has carbonium ion character,

$$\diagdown C{=}O \leftrightarrow \diagdown C^{+}{-}O^{-},$$ the properties of cyclopentadienone derivatives are of

particular significance in the present connection. The most striking characteristic of these compounds is their pronounced tendency toward dimerization. It has long been known that the condensation of cyclopentadiene with ethyl nitrite yields not cyclopentadienone oxime but its dimer, LII.[59]

Recent preparations of the parent ketone, LIII, also yield the dimer, LIV.[60,61] HMO calculations of cyclopentadienone theoretically confirm this easy dimerization.[62]

Tetraphenylcyclopentadienone (tetracyclone), LV, is readily available from the condensation of dibenzyl ketone with benzil[63] and exists as an

[58] N. C. Deno, J. J. Jaruzelski, and A. Schriesheim, *J. Am. Chem. Soc.*, **77**, 3044 (1955).
[59] J. Thiele, Ber., **33**, 669 (1900).
[60] K. Alder and F. H. Flock, *Ber.*, **87**, 1916 (1954).
[61] C. H. DePuy and C. E. Lyons, *J. Am. Chem. Soc.*, **82**, 631 (1960).
[62] R. D. Brown, *J. Chem. Soc.*, **1951**, 2670.
[63] J. R. Johnson and O. Grummit, *Organic Synthesis*, Coll. Vol. III, John Wiley and Sons, New York (1955), p. 806.

intensely colored monomer. At least three substituents are required for cyclopentadienones to exist as monomers.[64]

LV

Indenone, LVI, a compound that has recently been prepared,[65] polymerizes with extreme ease. A combination of MO and VB techniques is convenient for interpreting this chemistry. The instability of these ketones

LVI

may be associated with the relatively high energy of structures of the type of LVII; hence such structures cannot contribute so effectively to the overall resonance hybrid, and the ketones are less stabilized than their analogues without five-membered rings. In complete contrast are the cycloheptatrienones, which are more unsaturated and far more stable (*vide infra*).

LVII

The same argument has been cleverly applied by DePuy[66] to the relatively high reactivity of maleic anhydride, LVIII, in Diels-Alder reactions. Structures such as LVIIIa that would normally be expected to stabilize an anhydride function cannot in the present instance participate so effectively because such structures are isoelectronic with cyclopentadienones!

[64] C. F. H. Allen and J. A. Van Allen, *J. Am. Chem. Soc.*, **72**, 5165 (1950).

[65] C. S. Marvel and C. W. Hinman, *J. Am. Chem. Soc.*, **76**, 5435 (1954).

[66] C. H. DePuy, private communication.

Consequently, maleic anhydride is less stable or more reactive than it would be otherwise.

LVIII   LVIIIa

These examples refer to an instability of cyclopentadienone rings toward reaction. Suitable data on thermodynamic stability would be valuable, but they are rare. One available example pertains to fluorenone in which the cyclopentadienone character is "diluted" by the two benzo rings; this ketone has a lower reduction potential (less readily reduced) than benzophenone,[67] a direction opposite that expected from the foregoing considerations alone.

The fifth electron in cyclopentadienyl radical can be placed in one of two degenerate and partially occupied bonding orbitals. Further considerations indicate that the regular pentagon will distort to produce a more stable radical with lower symmetry.[18] The circumstance that the odd electron is in a bonding MO is reflected in the ionization potential of cyclopentadienyl radical, 8.69 ev.[68] This value is higher than that of any other hydrocarbon radical measured with the single exception of methyl radical.[68]

Cyclopentadienyl radical has high DE (1.85β) and may be comparatively stable compared with alkyl radicals. Further stabilization of the radical is achieved by phenyl substituents. Thus pentaphenylcyclopentadienyl, LIX, exists only as a purple monomer and shows no tendency to dimerize.[69]

LIX

[67] The equilibrium, fluorenone + diphenylmethanol = fluorenol + benzophenone, lies towards the left; with equimolar starting materials, the fluorenone/benzophenone ratio at equilibrium is 60/40. R. H. Baker and H. Adkins, *J. Am. Chem. Soc.*, **62**, 3305 (1940).

[68] A. G. Harrison, L. R. Honnen, H. J. Dauben, Jr., and F. P. Lossing, *J. Am. Chem. Soc.*, **82**, 5593 (1960).

[69] K. Ziegler and B. Schnell, *Ann.*, **445**, 266 (1925).

Cyclopentadienyl anion has a full complement of six π-electrons; the bonding shells are filled and the $4n + 2$ rule is satisfied. The relatively high stability of this system is reflected in the great acidity of cyclopentadiene. This hydrocarbon is predicted by HMO considerations[70] (Sec 14.1) to have a pK_a of 17, a value consistent with qualitative observations; for example, cyclopentadiene reacts with potassium in the cold to form the potassium salt of the anion.[71] This property also carries through to substituted derivatives. Fluorene is at least ten powers of ten more acidic than diphenylmethane.[72]

The ability of the fully unsaturated five-membered ring to attract electrons is also manifest in the dipole moment of fulvenes. The direction of the moment has been shown to be toward the five-membered ring (Sec. 6.1), which proves conclusively that structure LXa contributes far more to the resonance hybrid of fulvene than does the cyclopentadienyl cation counterpart, LXb.

Heterocyclic derivatives, such as pyrrole, LXI, and furan, LXII, all have "aromatic character" by virtue of the "aromatic sextet" of electrons; these compounds are isoelectronic with cyclopentadienyl anion.

10.5 Benzene

In benzene the two bonding shells that comprise three MO's are completely filled by six electrons. The aromatic character of benzene may be associated with π-electronic stability in a strainless six-membered ring. Polycyclic benzenoid hydrocarbons may be built from such units in a familiar way.

Changing the coulomb and bond integrals by the introduction of heteroatoms will distort the energy levels, but the basic stabilization of the six

[70] A. Streitwieser, Jr., *Tetrahedron Letters*, No. 6, 23 (1960); cf. also, G. W. Wheland, *J. Chem. Phys.*, **2**, 474 (1934).

[71] J. Thiele, *Ber.*, **34**, 68 (1901); R. E. Dessy (personal communication) finds an experimental pK of 16 for cyclopentadiene.

[72] W. K. McEwen, *J. Am. Chem. Soc.*, **58**, 1124 (1936).

H
|
B⁻
H   N⁺   N⁺   H
    ⁻B   B⁻
H   N   H
    |
    H⁺

LXIII

π-electrons largely holds force still. The borazole structure, LXIII, for example, is stable.[73] Similarly, stable derivatives of 9-aza-10-boraphenanthrenes, LXIV, have been prepared and characterized.[74]

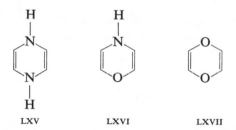

LXIV

Any additional electrons added to benzene must be stored in highly antibonding MO's. Benzene has a rather low electron affinity (Sec. 7.1). Replacement of one or more carbons by more electronegative elements ameliorates this situation somewhat but still leaves substantial electronic instability. The eight-electron ring compounds, LXV, LXVI and LXVII, are examples whose chemistry is completely consistent with HMO theory.

LXV   LXVI   LXVII

The MO energy levels of 1,4-dihydropyrazine, LXV, may be sufficiently well approximated by using $k_{C-N} = 1$, $h_{\ddot{N}} = 1$ (Sec. 5.2). We find that

[73] For a recent review of borazole chemistry see J. C. Sheldon and B. C. Smith, *Quart. Revs.*, **14**, 200 (1960).

[74] M. J. S. Dewar, V. P. Kubba, and R. Pettit, *J. Chem. Soc.*, **1958**, 3073.

the last two electrons must still be placed in an antibonding orbital (Fig. 10.5).   Alkyl and phenyl substituted 1,4-dihydropyrazines are known. These compounds hydrolyze easily as expected for enamines.   They are

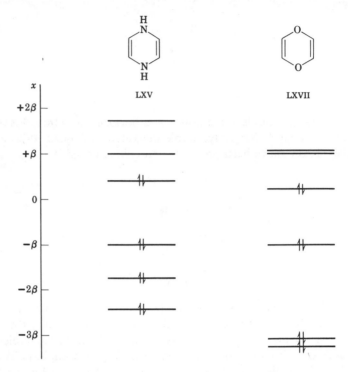

Fig. 10.5   MO energy levels for models of 1,4-dihydropyrazine, LXV, and dioxadiene, LXVII.

oxidized readily to the corresponding pyrazines.[75] 5,10-Dihydrophenazines, LXVIII, turn green in air and are readily oxidized.[76]   Neither the simple

LXVIII

[75] Y. T. Pratt and R. C. Elderfield, *Heterocyclic Compounds*, John Wiley and Sons, New York (1957), Vol. 6, Chap. 9, p. 414.
[76] H. McIlwain, *J. Chem. Soc.*, **1937**, 1704.

1,4-oxazines, LXVI,[77] nor the monobenzo derivatives[78] seem to be known. However, the dibenzo derivative, phenoxazine, LXIX, is well known. The two benzene rings provide substantial stabilization, yet this compound gives a radical readily by loss of an electron.[79]

LXIX

Oxygen is more electronegative than nitrogen; yet our simplified model for dioxadiene, LXVII ($h_O = 2$, $k_{C-O} - 1$) still finds two electrons in an antibonding orbital. This compound is known and shows the properties of a typical unsaturated aliphatic ether.[80]

These compounds clearly have no aromatic character in good qualitative agreement with our expectations from HMO theory of eight-electron monocycles.

10.6   Cycloheptatrienyl[81]

The cycloheptatrienyl or tropylium π-system has three bonding MO's in two shells (Fig. 10.1). These shells are completely filled by six electrons; hence tropylium cation should be relatively stable. Tropylium bromide is saltlike in character and seems to have the ionic structure $C_7H_7^+ Br^-$. The stability of tropylium cation is manifested by its various preparations. That of Doering and Knox[82,83] involves the thermal dehydrobromination of the addition product, LXX, of cycloheptatriene and bromine. Tropylium isocyanate is formed by heating norcaradiene carboxazide, LXXI.[84]

[77] N. H. Cromwell, *Heterocyclic Compounds*, John Wiley and Sons, New York, (1957), Vol. 6, Chap. 11, p. 499.

[78] R. C. Elderfield, W. H. Todd, and S. Gerber, *Heterocyclic Compounds*, John Wiley and Sons, New York (1957), Vol. 6, Chap. 12, p. 588.

[79] D. E. Pearson, *Heterocyclic Compounds*, John Wiley and Sons, New York (1957), Vol. 6, Chap. 14.

[80] C. B. Kremer and L. K. Rocher, *Heterocyclic Compounds*, John Wiley and Sons, New York (1957), Vol. 6, p. 33.

[81] For a recent complete review, including tropones and tropolones, see T. Nozoe in D. Ginsburg's *Non-Benzenoid Aromatic Compounds*, Interscience Publishers, New York, (1959), Chap. 7.

[82] W. E. Doering and L. H. Knox, *J. Am. Chem. Soc.*, **76**, 3203 (1954).

[83] This preparation was apparently accomplished much earlier by G. Merling, *Ber.*, **24**, 3108 (1891), who did not recognize the significance of his discovery.

[84] M. J. S. Dewar and R. Pettit, *Chem. and Ind.*, **1955**, 199; *J. Chem. Soc.*, **1956**, 2021, 2026.

Especially revealing is the hydride exchange method of Dauben et al.[85] The reaction of cycloheptatriene with triphenylmethyl cation in acetonitrile

or liquid sulfur dioxide gives tropylium cation and triphenylmethane. Tropylium cation has also been suggested as a product of mass spectral cracking of toluene, cycloheptatriene, and other compounds.[86]

Tropylium cation in water is in equilibrium with the corresponding carbinol:

$$C_7H_7^+ + H_2O \rightleftharpoons C_7H_7OH + H^+; \qquad K = \frac{[C_7H_7OH][H^+]}{[C_7H_7^+]} = 1.8 \times 10^{-5}$$

From the equilibrium constant it is apparent that tropylium cation liberates about as much hydrogen ion as does acetic acid.[82]

A number of benzo derivatives of tropylium cation have been prepared; for example, LXXII[87] and LXXIII.[88] As expected, these cations hydrolyze

more readily than the parent tropylium cation. The dibenzo-[89] and tribenzo-[90,91] derivatives, LXXIV and LXXV, are especially unstable,

[85] H. J. Dauben, Jr., F. A. Gadecki, K. M. Harmon, and D. L. Pearson, *J. Am. Chem. Soc.*, **79**, 4557 (1957).

[86] P. N. Rylander, S. Meyerson, and H. M. Grubb, *J. Am. Chem. Soc.*, **79**, 842 (1957); S. Meyerson and P. N. Rylander, *J. Chem. Phys.*, **27**, 901 (1957); V. Hanus, *Nature*, **184**, 1796 (1959).

[87] H. H. Rennhard, E. Heilbronner, and A. Eschenmoser, *Chem. and Ind.*, **1955**, 415.

[88] H. Fernholz, *Ann.*, **568**, 63 (1950).

[89] M. J. S. Dewar and C. R. Ganellin, *J. Chem. Soc.*, **1959**, 3139.

[90] M. Stiles and A. J. Libbey, *J. Org. Chem.*, **22**, 1243 (1957).

[91] D. Meuche, H. Strauss, and E. Heilbronner, *Helv. Chim. Acta*, **41**, 414 (1958).

apparently because of steric interference of neighboring benzene hydrogens in planar structures.

The absorption spectrum of tropylium cation has been compared with results of advanced MO calculations.[92] This spectrum and those of several benzo-derivatives have been shown by Meuche, Strauss, and Heilbronner[93] to give an excellent correlation with the HMO energy differences between highest occupied and lowest vacant orbitals (cf. Sec. 8.4). These authors also showed that the simple theory gives a good account of the pK-values of the corresponding carbinols (cf. Sec. 12.2).

The electronic stability of the tropylium cation is also reflected in the properties of cycloheptatrienone or tropone, LXXVI. The pioneering

LXXVI   LXXVIa

syntheses of tropone by Doering and Detert[94] and by Dauben and Ringold[95] were soon followed by several other general and specific methods.[96] The relative basicity and other properties of tropone indicate that the "cyclo-heptatrienylium oxide"[94] structure, LXXVIa, contributes importantly to the resonance hybrid. This conclusion also derives from the relatively high dipole moment of tropone, $4.17D$.[97] As in the tropylium cation itself, fusion of a benzene ring to the seven-membered ring reduces

LXXVII

the gain in delocalization energy in "going" to the dipolar structure. 2,3-Benzotropone, LXXVII, for example, behaves as a typical dienone.[98]

The odd electron in tropylium radical must be placed in an antibonding orbital; this species would be expected to oxidize easily. The ionization potential of this radical has recently been measured by the electron impact

[92] J. N. Murrell and H. C. Longuet-Higgins, *J. Chem. Phys.*, **23**, 2347 (1955).
[93] D. Meuche, H. Strauss, and E. Heilbronner, *Helv. Chim. Acta*, **41**, 57 (1958).
[94] W. E. Doering and F. L. Detert, *J. Am. Chem. Soc.*, **73**, 876 (1951).
[95] H. J. Dauben, Jr., and H. J. Ringold, *J. Am. Chem. Soc.*, **73**, 876 (1951).
[96] Summarized by T. Nozoe.[81]
[97] M. Kubo, T. Nozoe, and Y. Kurita, *Nature*, **167**, 688 (1951).
[98] G. L. Buchanan and D. R. Lockhart, *J. Chem. Soc.*, **1959**, 3586.

method,[68] 6.60 ev. This value is extremely low but agrees well with simple LCAO calculations by the ω-technique.[99]

The eighth electron of tropylium anion must also be placed in an anti-bonding MO and still leaves an incomplete shell. In consequence, cyclo-heptatriene is not markedly acidic. It does not undergo base catalyzed condensations with ketones or ethyl oxalate under conditions that give facile condensation of fluorene.[100] According to the HMO calculations, cycloheptatriene should have a pK_a of about 31[70] (Sec. 14.1).

Introduction of a heteroatom into the tropylium anion π-lattice alters the degeneracy of the first antibonding MO's and, for electronegative heteroatoms, lowers the energy. If the heteroatom is sufficiently electro-negative ($h_X > 1$ for $k_{CX} = 1$), the fourth MO is bonding; however, even in this case the highest occupied orbital has little bonding character and oxidation should be comparatively easy. We conclude that the π-system of a heterocyclic structure, LXXVIII, has less stability than would be expected from a simple consideration of resonance structures such as LXXVIIIa. Although some of these compounds have been assigned

LXXVIII   LXXVIIIa

trivial names (azepine, $X = N$; oxepine, $X = O$; thiapine, $X = S$), the parent molecules are still unknown. An aminoazepine, LXXIX, has been postulated as an intermediate in the decomposition of phenylazide in aniline, but, if formed, it tautomerizes to the amidine structure, LXXX, which must be more stable.[101]

LXXIX   LXXX

A series of benz derivatives, LXXXI, has been reported by Dimroth et al. by condensations of phthalaldehyde with the diesters, LXXXIII

[99] A. Streitwieser, Jr., *J. Am. Chem. Soc.*, **82**, 4123 (1960).

[100] J. Thiele, *Ann.*, **319**, 226 (1900); *Ber.*, **33**, 851 (1900).

[101] W. E. Doering and R. Odum, unpublished results; cf. R. Huisgen and M. Appl, *Ber.*, **91**, 12 (1958).

(X = NCH$_3$,[102] NC$_6$H$_5$,[103] O,[104] S[105]). Although some of the free acids are unstable and resinify readily, the diesters of LXXXI are reported to be stable. This remarkable difference in behavior is additionally striking

LXXXII   LXXXI

because of the observation that the esters are recoverable unchanged after prolonged standing in concentrated sulfuric acid solution.[102–105,105a]

The behavior of the benzodiazepines, LXXXIII, is more in accordance with theoretical expectations. These compounds rearrange readily to the diimino form, LXXXV, and, in general, show no aromatic character.[106]

LXXXIII   LXXXIV

The dibenzo compounds, LXXXV (X = O,[107] NR,[108] S[108]), have been reported but their chemistry has not been extensively studied.

In contrast to these eight-electron-ring compounds, a derivative, LXXXVI, of the six-electron-ring borepin system has been prepared and appears to be relatively stable.[109]

[102] K. Dimroth and H. Freyschlag, *Ber.*, **89**, 2602 (1956).

[103] K. Dimroth and H. Freyschlag, *ibid.*, **90**, 1628 (1957).

[104] K. Dimroth and H. Freyschlag, *ibid.*, **90**, 1623 (1957).

[105] K. Dimroth and G. Lenke, *Ber.*, **89**, 2608 (1956).

[105a] The NMR spectra of LXXXI (X = O) and several derivatives are consistent with the assigned structures (M. J. Jorgenson, personal communication).

[106] J. A. Barltrop, C. G. Richards, D. M. Russell, and G. Ryback, *J. Chem. Soc.*, **1959**, 1132.

[107] R. H. F. Manske and A. E. Ledingham, *J. Am. Chem. Soc.*, **72**, 4797 (1950); F. A. L. Anet and P. M. G. Bavin, *Can. J. Chem.*, **35**, 1084 (1957).

[108] R. Huisgen, E. Laschtuvka, and F. Bayerlin, *Ber.*, **93**, 392 (1960); E. D. Bergmann and M. Rabinowitz, *J. Org. Chem.*, **25**, 827, 828 (1960).

[109] E. E. van Tamelen, G. Brieger, and K. G. Untch, *Tetrahedron Letters*, No. 8, 14 (1960).

LXXXV   LXXXVI

The interesting compounds, LXXXVII (X = NCH$_3$, O, S), have been
reported recently and appear to be relatively stable.[110] The noteworthy
feature of these compounds is that the central ring has ten electrons and
fits the $4n + 2$ rule.

LXXXVII

10.7   Cyclooctatetraene and Larger Ring Systems

Cyclooctatetraene with eight π-electrons does not fit the $4n + 2$ rule;
the last two electrons occupy an unfilled shell. HMO theory predicts that
the planar molecule is unstable. The hydrocarbon synthesized by Willstätter[111] behaved as a highly unsaturated polyene and certainly had no
aromatic character. The synthesis was confirmed much later by Cope and
Overberger;[112] cyclooctatetraene is now available in quantity by Reppe's
synthesis from acetylene.[113]

The extensively developed chemistry of cyclooctatetraene is that of a
cyclic polyene;[114] the hydrocarbon has the nonplanar "tub" structure,
LXXXVIII, with alternating single and double bonds.[115] The unsaturated
character of the hydrocarbon is due to the negligible conjugation between
the almost orthogonal double bonds in this nonplanar structure; confirmation is found not only in MO calculations[116] but in the low empirical

[110] N. L. Allinger and G. A. Youngdale, *Tetrahedron Letters*, No. 9, 10 (1959).

[111] R. Willstätter and E. Waser, *Ber.*, **44**, 3423 (1911); R. Willstätter and M. Heidel-
berger, *Ber.*, **46**, 517 (1913).

[112] A. C. Cope and C. G. Overberger, *J. Am. Chem. Soc.*, **70**, 1433 (1948).

[113] W. Reppe, O. Schlichting, K. Klager, and T. Toepel, *Ann.*, **560**, 1 (1948).

[114] For reviews see L. E. Craig, *Chem. Rev.*, **49**, 103 (1951); W. Baker and J. F. W.
McOmie in J. W. Cook's *Progress in Organic Chemistry*, Vol. 3, Academic Press, New
York (1955), p. 44; R. A. Raphael in D. Ginsburg's *Non-Benzenoid Aromatic Compounds*,
Interscience Publishers, New York (1959), p. 465; B. C. L. Weedon in J. W. Cook's
Progress in Organic Chemistry, Vol. 1, Academic Press, New York (1952), p. 134.

[115] W. B. Person, G. C. Pimentel, and K. S. Pitzer, *J. Am. Chem. Soc.*, **74**, 3437 (1952).

[116] I. Tanaka, *Tokyo Kôgyô Daigaku Gakuhô, Ser. A.*, **1954**, No. 1, 1.

RE, 4.8 kcal./mole, from the heat of combustion.[117] The nonplanarity alone is not sufficient evidence that the eight π-electrons of the planar molecule represent an unstable configuration, for the steric energy requirements of a planar eight-membered ring are expected to be substantial. This

LXXXVIII

strain energy has recently been estimated to be 23.4 kcal./mole.[118] Consequently, we are left with the question, "Does planar cycloöctatetraene represent a π-system of electronic instability or are its properties the result of an overriding strain energy?" The answer has apparently emerged in recent studies of the electron affinity of cycloöctatetraene. The hydrocarbon is reduced at the dropping mercury electrode (Sec. 7.1) at a relatively low half-wave potential, -1.5 volts versus SCE.[119] Two electrons are involved in the reduction, and the half-wave potential is independent of *pH*, which means that no hydrogens take part in the reduction step. The reaction may be formulated as

$$C_8H_8 + 2e = C_8H_8^=$$

The earlier reports that cycloöctatetraene forms a di-salt with alkali metals[114] have been confirmed recently not only by isolation of such salts but also by ESR and NMR studies.[120] These studies show that alkali metals in ethers produce the cycloöctatetraene dianion in equilibrium with a small amount of radical anion. In contrast, the corresponding equilibria with aromatic hydrocarbons are in the radical-anion direction (Sec. 6.5). Both cycloöctatetraene radical-anion and di-anion appear to be planar; the observed hyperfine splitting constant (cf. Sec. 6.3) of the radical, 3.21 gauss, compares well with the HMO value calculated from Fig. 6.7, 3.6 gauss.

These results leave no doubt that cycloöctatetraene is nonplanar because of π-electronic instability and not because of steric strain. This phase of organic chemistry joins others in affirming the significance of many of the conclusions of simple LCAO theory even in its simplest and most naïve form.

[117] H. D. Springall, T. R. White, and R. C. Cass, *Trans. Faraday Soc.*, **50**, 815 (1954).
[118] H. J. Dauben, Jr., private communication.
[119] R. M. Elofson, *Anal. Chem.*, **21**, 917 (1949).
[120] T. J. Katz and H. L. Strauss, *J. Chem. Phys.*, **32**, 1873 (1960); T. J. Katz, *J. Am. Chem. Soc.*, **82**, 3784, 3785 (1960).

Various benzo-derivatives of cycloöctatetraene are known. The isolated double bonds of LXXXIX and XC are typically olefinic in character.[121] The tetrabenzo compound, tetraphenylene, XCI, is known to have the tub structure.[122] MO calculations of this hydrocarbon give only 0.01β *DE* over that of the four benzene rings.[123]

LXXXIX   XC   XCI

The problem of the strain of the planar version occurs also in larger cyclic polyenes. From scale models with normal van der Waals radii, Mislow[124] concluded that not until $C_{30}H_{30}$ could a planar cyclopolyene be obtained. In many such compounds the carbon skeleton can assume a relatively strain-free planar arrangement; the strain develops from crowding of interior hydrogens. An example is cyclodecapentaene, XCII,

XCII

which should have π-electronic stability in fitting the $4n + 2$ rule but has the destabilizing influence of steric strain between the two indicated hydrogens that could force a nonplanar configuration.

This hydrocarbon has never been isolated, but its transitory existence seems probable in some experiments of Nenitzescu et al.[32] In addition to the products mentioned earlier, pyrolysis of the Diels-Alder adduct of cycloöctatetraene and dimethyl acetylenedicarboxylate, XXIX, gave substantial quantities of dimethyl 2,6-naphthalenedicarboxylate, XCIII,

[121] W. S. Rapson, R. G. Shuttleworth, and J. N. van Niekerk, *J. Chem. Soc.*, **1943** 326; L. F. Fieser and M. M. Pechet, *J. Am. Chem. Soc.*, **68**, 2577 (1946); A. C. Cope and S. W. Fenton, *J. Am. Chem. Soc.*, **73**, 1668, 1673 (1951); G. Wittig, H. Tenhaeff, W. Schoch, and G. Koenig, *Ann.*, **572**, 1 (1951).

[122] I. L. Karle and L. O. Brockway, *J. Am. Chem. Soc.*, **66**, 1974 (1944).

[123] S. S. Pérez, *Anales real soc. españ. fis. y quim.* (*Madrid*), **51B**, 263 (1955).

[124] K. Mislow, *J. Chem. Phys.*, **20**, 1489 (1952).

and a dihydro derivative formulated as XCIV. The authors explain the formation of these products by the following reaction sequence which involves a derivative of XCII as an intermediate:*

XXIX

A similar path is used to explain the related formation of 1,4-dimethoxy-phenanthrene, XCV, by pyrolysis of XCVI.[32]

XCVI   XCV

Cyclododecahexaene with 12 π-electrons does not fit the $4n + 2$ rule; the only experimental clue to the properties of the parent hydrocarbon is the report by Wilke[125] of the preparation from the triene, XCVII, of an air sensitive orange compound, $C_{12}H_{12}$. The tetrabenzo, XCVIII, and

[125] G. Wilke, *Angew. Chem.*, **69**, 397 (1957).

* Added note: XXIX is incorrect; cf. M. Avram, G. Mateescu and C. D. Nenitzescu, *Ann.*, **636**, 174 (1960).

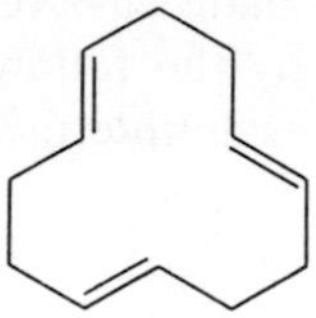

XCVII

hexabenzo, XCIX, derivatives are known and are not planar.[126] Sworski[127] has postulated that acetylenic hydrocarbons such as C might be stable

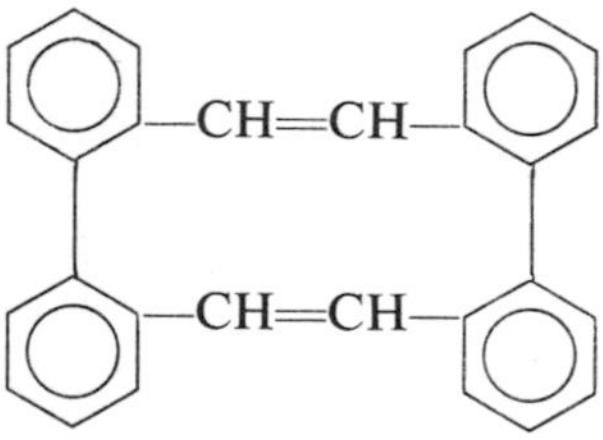

XCVIII

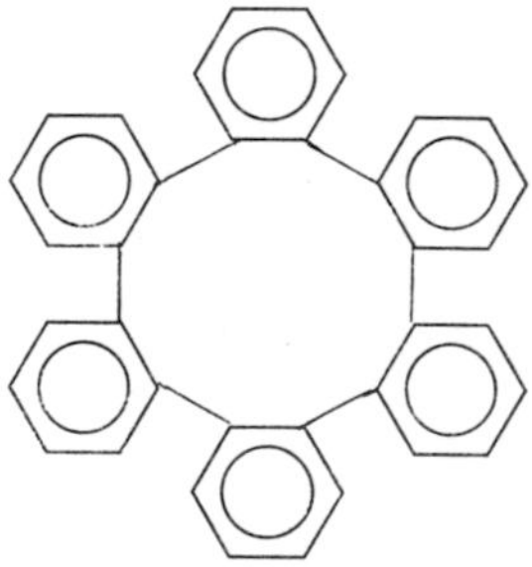

XCIX

because of the lack of interior hydrogens. The hydrocarbon, CI, has been prepared as an unstable yellow crystalline compound.[128]

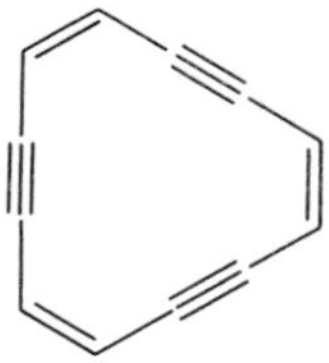

C

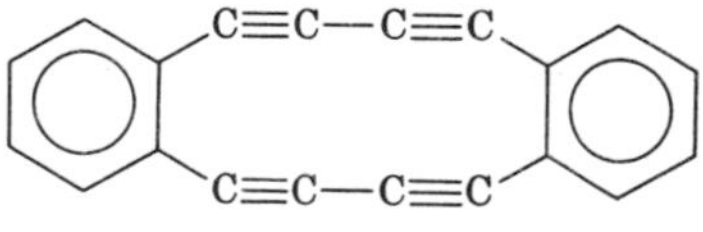

CI

An important contribution to this field has been made by Sondheimer and his co-workers who have found that the oxidation of $\alpha\omega$-diacetylenes with cupric acetate in pyridine gives cyclic dimers, trimers, etc.[129] Partial

---

[126] G. Wittig, G. Koenig, and K. Clausz, *Ann.*, **593**, 127 (1955); G. Wittig and G. Lehmann, *Ber.*, **90**, 875 (1957).

[127] T. J. Sworski, *J. Chem. Phys.*, **16**, 550 (1948).

[128] G. Eglinton and A. R. Galbraith, *Proc. Chem. Soc.*, **1957**, 350; W. K. Grant and J. C. Speakman, *Proc. Chem. Soc.*, **1959**, 231; O. M. Behr, G. Eglinton, and R. A. Raphael, *Chem. and Ind.*, **1959**, 699.

[129] F. Sondheimer, Y. Amiel, and R. Wolovsky, *J. Am. Chem. Soc.*, **78**, 4178 (1956); *Ibid.*, **79**, 4247, 6263 (1957).

hydrogenation of the eighteen-membered ring triyne, CII, gave cycloöcta-
decanonaene, CIII, as a relatively stable, brown-red crystalline compound.[130]
The fascinating question of whether this remarkable hydrocarbon has
alternating single and double bonds or whether the carbon-carbon bonds
have equal length will undoubtedly be answered in the near future. Already
the spectrum has been interpreted as indicative of alternating single and
double bonds.[131] An HMO treatment has been published.[132] The inference
from recent calculations of the strain energy is that the molecule cannot be
planar and that its resonance energy is relatively small.[133]

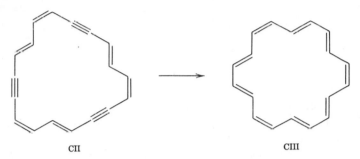

CII   CIII

It is difficult to believe that the alternation or nonalternation of bond
lengths in a large polyene should depend on whether the total system has
$4n$ or $4n + 2$ π-electrons, that is, that the local situation in a large molecule
should depend so much on the nature of the large system.[134] A more
detailed study which considers σ-bond compression indicates that in very
large cyclic polyenes alternate single and double bonds should obtain
regardless of the presence of $4n$ or $4n + 2$ electrons.[135]

Sondheimer's group has recently reported the synthesis of a dark blue
cyclotetracosadodecaene,[136] which, with 24 π-electrons, does not fit the
$4n + 2$ rule, and a dark red cyclotriacontapentadecaene,[137] which, with
30 π-electrons, does fit the rule; both are rather unstable substances.[138]

[130] F. Sondheimer and R. Wolovsky, *Tetrahedron Letters*, No. 3, 3 (1959).

[131] M. Gouterman and G. Wagniere, *Tetrahedron Letters*, No. 11, 22 (1960).

[132] D. W. Davies, *Tetrahedron Letters*, **8**, 4 (1959).

[133] C. A. Coulson and A. Golebiewski, *Tetrahedron*, **11**, 125 (1960).

[134] H. C. Longuet-Higgins in *Theoretical Organic Chemistry*, The Kekulé Symposium,
Butterworths Scientific Publications, London (1959), p. 9.

[135] Y. Ooshika, *J. Phys. Soc. Japan*, **12**, 1238, 1246 (1957); H. C. Longuet-Higgins and
L. Salem, *Proc. Roy. Soc.*, **A251**, 172 (1959).

[136] F. Sondheimer and R. Wolovsky, *J. Am. Chem. Soc.*, **81**, 4755 (1959).

[137] F. Sondheimer, R. Wolovsky, and Y. Gaoni, *J. Am. Chem. Soc.*, **82**, 755 (1960).

[138] For a further review of cyclic polyenes cf. W. Baker and J. F. W. McOmie in
D. Ginsburg's *Non-Benzenoid Aromatic Compounds*, Interscience Publishers, New York
(1959), Chap. 9.

10.8   Aromaticity and Pseudoaromaticity[139]

The success of the $4n + 2$ rule prompts its application to polycyclic systems, but even a casual review of polycyclic aromatic compounds spots a number of exceptions. Both pyrene, CIV, and fluoranthene, CV, for example, are stable aromatic hydrocarbons which possess sixteen π-electrons.

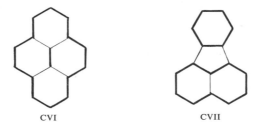

CIV   CV

A more successful approach is to examine the periphery of the molecule.[140] Many aromatic molecules may be formed by starting with a cyclic polyene and introducing cross-links. If the cross links are regarded as small perturbations, the derived aromatic compounds are directly related[141] to the original cyclic polyenes and the $4n + 2$ rule may be applied. In this way pyrene may be regarded as concentric 14 and 2 systems connected by cross-links, CVI, and fluoranthene may be looked on as a 6 and a 10 system connected together, CVII. This approach can be very helpful,

CVI   CVII

particularly in predicting possible new aromatic systems. Unfortunately, the cross links are not small perturbations and exceptions should be expected. One such example is 3,4-(*peri*-phenylene) fluoranthene, CVIII, which has a periphery of 20 atoms as shown (CVIII*a*). This compound is a normal aromatic hydrocarbon. However, we should note that this hydrocarbon may also be regarded as two 6's and a 10 united as in CVIII*b*.

[139] For a recent review see M. E. Vol'pin, *Russ. Chem. Rev.*, **29**, 129 (1960).

[140] J. R. Platt, *J. Chem. Phys.*, **22**, 1448 (1954).

[141] M. J. S. Dewar, *J. Am. Chem. Soc.*, **74**, 3345 (1952); M. J. S. Dewar and R. Pettit, *J. Chem. Soc.*, **1954**, 1617; D. Peters, *J. Chem. Soc.*, **1958**, 1023.

HMO calculations almost always predict substantial delocalization energies for polycyclic hydrocarbons undoubtedly because additional stabilizing β's are incorporated without any compensating destabilizing terms for electron repulsion effects. Symmetry factors rarely change the

CVIIIa   CVIIIb

result that the simple theory generally predicts thermodynamic stability for polycyclic systems. This limitation has led to proposals for additional criteria for stability.

The magnitude of the DE per π-electron has occasionally been used as such a criterion, the thought being that this quantity should be close to that in benzene or naphthalene ($\sim$0.3) for reasonable aromatic character. This criterion may be suggestive but it is not sufficient—a number of still unknown compounds fulfill this criterion.

Ring strain undoubtedly can be of major importance in some types of compounds, but detailed calculations show that such strain is only moderate in many hydrocarbons that contain only five-, six-, and seven-membered rings.[118]

The term stability as ordinarily used is ambiguous.[142] In one sense we use the term as a thermodynamic stability and imply a relative heat of formation from the elements or, equivalently, a heat of dissociation into separated atoms. In another widely used sense we mean stability in terms of relative reactivity; for example, a stable compound is one that does not polymerize, react with air, decompose, or rearrange. DE-values alone do not help in judging this kind of stability and other MO quantities have been suggested.

The free valence index, F (Sec. 2.9), has been used in this connection. A high F-value implies a greater reactivity and may provide a rough measure of the ease of reaction with radicals (Sec. 13.2) or in polymerization. For example, p-xylylene, CIX, has substantial DE, 1.924β or 0.25β per electron, but F at the methylene positions has the rather high value

[142] D. Peters, *J. Chem. Soc.*, **1960**, 1274.

of 0.97 (other values for reference are α-position of benzyl radical, 1.04; ethylene, 0.73; benzene, 0.23).[14] The hydrocarbon exists in the gas phase but polymerizes readily—it is *unstable*.[143] A high F-value at a position probably does indicate a tendency toward facile reaction at the position and should give warning in an attempted synthesis. On the other hand, a low F-value does not guarantee freedom from reaction. A conspicuous example is benzocyclobutadiene, XXXIV, which has $F = 0.62$ at a cyclobutadiene position, a value lower than that for ethylene; yet, this compound apparently dimerizes with extreme ease (Sec. 10.3).

CIX

Bush[144] has proposed that the energies of the lowest vacant and highest occupied MO's may provide a qualitative indication of stability, since the corresponding electron affinity and ionization potential, respectively, are related to oxidation-reduction potentials (Chap. 7). Pyracylene, CX, for example, has a vacant NBMO (Fig. 10.6) and may be expected to reduce with exceptional ease. A recent attempted preparation of this hydrocarbon failed.[145]

For a closely related criterion, we carry over from monocyclic systems the closed shell concept.[146] This concept suggests that neutral hydrocarbons will be thermodynamically stable only if they have no electrons in antibonding MO's and have all bonding MO's completely occupied. According to this criterion alone, pyracylene should be unstable. Similarly, the tricyclic hydrocarbons, acepentylene, CXI, with a vacant bonding MO, and pleiaheptalene, CXII, with an occupied antibonding MO, should be unstable (see Fig. 10.6).

A recent and important criterion for stability or "aromaticity" arises from the symmetry properties of wave functions. In HMO theory all compounds that have electrons paired in orbitals have totally symmetric

[143] The chemistry of CIX has been reviewed recently: L. A. Errede and M. Szwarc, *Quart. Revs.*, **12**, 301 (1958).

[144] J. B. Bush, private communication.

[145] A. G. Anderson and R. G. Anderson, *J. Org. Chem.*, **23**, 517 (1958).

[146] The significance of closed shells has recently been emphasized by M. E. Dyatkina and E. M. Shustorovich, *Proc. Acad. Sci. U.S.S.R.* (*English Translation*), **117**, 797 (1957); M. E. Dyatkina, S. N. Dobriakov, and E. M. Shustorovich, *Proc. Acad. Sci. U.S.S.R., Sec. Chem.* (*English Translation*), **123**, 783 (1958); D. A. Bochvar, N. P. Gambaryan, I. V. Stankevich, and A. L. Chistyakov, *Zhur. Fiz. Khim.*, **32**, 2797 (1958).

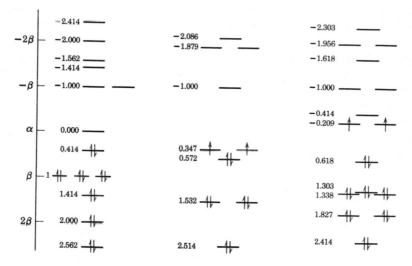

Fig. 10.6   HMO energy levels for some polycyclic hydrocarbons.

ground states (Sec. 8.1). The only exceptions arise when there are insufficient electrons to fill degenerate orbitals. In such cases the various possible wave functions generally belong to different group representations.

The wave functions used in these symmetry classifications are the complete antisymmetrized wave functions that include spin functions and permute over the available electrons (Sec. 16.1). For this purpose the correct character table for the molecule is used instead of the abbreviated tables that normally suffice for simplifying HMO calculations.

Although for most compounds the totally symmetric function normally represents the ground state, there is no reason in principle why this need necessarily be so. For cyclobutadiene advanced MO calculations with inclusion of configuration interaction indicate that the ground state is not totally symmetric; that is, the ground-state wave function changes sign on applying some symmetry operations. Because VB theory deals with structures and not with electron pairs in orbitals, totally symmetric wave functions are a less automatic outcome of this method.

Craig has proposed a set of rules, based on VB and other calculations of cyclobutadiene,[147] pentalene, CXIII, and, heptalene, CXIV,[148] which has become known as Craig's rules.[147,149] These rules predict whether the VB wave function is totally symmetric, and, although not part of a simple MO treatment, they are easy to apply to appropriate structures and serve as a useful adjunct to HMO calculations. The rules apply only to hydrocarbons in which at least two π-centers lie on a symmetry axis that converts one Kekulé structure to another. The π-centers are labeled with the spin functions, α and β, such that the ends of each double bond have different spins in both Kekulé structures and as few as possible of like spins adjoin. The sum is then taken of the number (f) of symmetrically related π-centers not on the symmetry axis and the number (g) of interconversions of α and β by rotation about the axis. If this sum, $f + g$, is even, the VB ground state is symmetric and the compound should be normally aromatic. If the sum is odd, the VB ground state is nontotally symmetric and the compound should not have aromatic stability; that is, the compound should be *pseudoaromatic*.[147] Examples of the application of these rules follow:

Pentalene.

CXIIIa   CXIIIb

Rotation about the vertical axis of symmetry (dotted; the horizontal axis could have been used just as well) converts structure CXIIIb into CXIIIa. The pairs of centers, 1-6, 2-5, 3-4, are related by the symmetry axis; $f = 3$. Upon rotation, no interconversions of α and β occur; $g = 0$. Hence, $f + g = 3$, an odd number, and pentalene should be pseudoaromatic (*vide infra*).

These rules were advanced as an empirical criterion for aromaticity.

---

[147] D. P. Craig, *J. Chem. Soc.*, **1951**, 3175.

[148] D. P. Craig and A. Maccoll, *J. Chem. Soc.*, **1949**, 964.

[149] For further discussion cf. D. P. Craig in *Theoretical Organic Chemistry*, The Kekulé Symposium, Butterworths Scientific Publications, London (1959), p. 20; and Chap. 1 in D. Ginsburg's *Non-Benzenoid Aromatic Compounds*, Interscience Publishers, New York (1959).

However, Longuet-Higgins[134] has shown that a correspondence between the symmetry characteristics of certain normal modes of vibration and

Acenaphthylene.

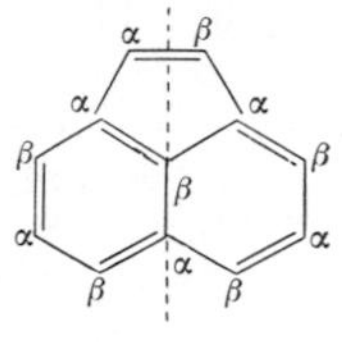

$f = 5$
$g = 1$
$f + g = 6$; aromatic

Heptalene.

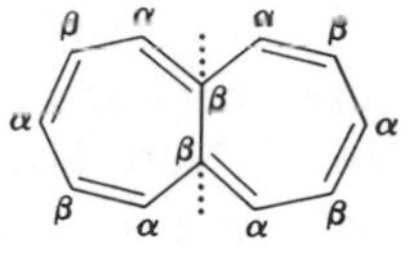

$f = 5$
$g = 0$
$f + g = 5$; pseudo-
              aromatic

CXIV

Pyrene.

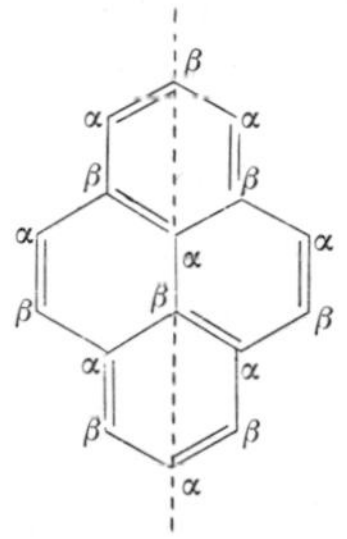

$f = 6$
$g = 0$
$f + g = 6$; aromatic

that of the ground state can lead to complete decomposition. The distortion in pentalene, for example, is predicted to lead in the limit to acetylene and diacetylene:

Until recently no aromatic compounds were known that were exceptions to Craig's rule, although numerous attempts have been recorded to synthesize such examples. In 1958, however, Hafner and Schneider[150] reported the preparation of the red aceheptylene derivative, CXV, a reasonably stable hydrocarbon that has aromatic properties. CXVIa has allowable labeling and the π-system is seen to be pseudoaromatic ($f + g = 7$). The structure CXVIb also has allowable labeling and corresponds to an aromatic system ($f + g = 8$). These structures, consequently, belong to different group representations and both cannot contribute to the same

[150] K. Hafner and J. Schneider, *Angew. Chem.*, **70**, 702 (1958); *Ann.*, **624**, 37 (1959).

linear combination VB resonance hybrid.[151] This type of ambiguity occasionally occurs in the application of Craig's rules and should indicate less resonance stabilization than normally expected for the system. In

CXV   CXVIa   CXVIb

general, if a molecule has *any* Kekulé structures that lead to an odd $f + g$ sum, it is abnormal from the VB point of view.[151] Note, however, that for any given structure, proper application of the rules gives an unambiguous odd or even sum.

There is the further possibility that a pseudoaromatic compound may exist as a derivative of a normal aromatic compound;[151] for example, in

CXVII

the present case the two seven-membered rings may not be equivalent, as in an azulene derivative, CXVII.

10.9 Polycyclic Compounds

Pentalene, CXIII, has been of interest since it was first postulated as a possible aromatic system by Armit and Robinson in 1922.[152] A number of attempted syntheses of pentalene and benzopentalene have failed.[153] Most of the approaches entailed a dehydrogenation of a partly hydrogenated structure. The approach of Dauben et al.[154] is particularly

---

[151] D. P. Craig, personal communication.

[152] J. W. Armit and R. Robinson, *J. Chem. Soc.*, **121**, 827 (1922).

[153] Reviewed by (*a*) E. D. Bergmann in D. Ginsburg's *Non-Benzenoid Aromatic Compounds*, p. 141, and (*b*) H. Paul, *Chem. Tech.* (*Berlin*), **8**, 189 (1956).

[154] H. J. Dauben, Jr., V. R. Ben, and S. H. K. Jiang, Abstract. 123 Meeting American Chemical Society, March 15–19 (1953), p. 9M; H. J. Dauben, Jr., and S. H. K. Jiang, *Hsua Hsüeh Pao*, **23**, 498 (1957); H. J. Dauben, Jr., S. H. K. Jiang, and V. R. Ben, *Hsua Hsüeh Pao*, **23**, 411 (1957).

important. Enolization of the diketone, CXVIII, would produce a dihydroxy-pentalene; enolization could not be detected or forced in a variety of attempts. Consequently, the pentalene structure cannot have any significant amount of resonance energy despite the HMO prediction of 2.46β

CXVIII

DE.[155]   The periphery is a $4n$ cycle and the molecule does not obey Craig's rules; according to these criteria (Sec. 10.8), the molecule should be pseudo-aromatic.

Actually a dibenzopentalene has been known for five decades. In 1912 Brand synthesized the diphenyl derivative, CXIX, as stable brown crystals.[156] The parent dibenzopentalene was synthesized much later by

CXIX

Blood and Linstead.[157,158] The compound behaves as a diphenylbutadiene and again indicates that the pentalene structure has little aromaticity.

[155] A number of MO treatments has been reported: C. A. Coulson and G. S. Rushbrooke, *Proc. Cambridge Phil. Soc.*, **36**, 193 (1940); I. Estellés and J. I. F. Alonso, *Anales real soc. españ. fis. y quim. (Madrid)*, **49B**, 267 (1953); Y. K. Syrkin and M. Dyatkina, *Acta Physiochim. U.R.S.S.*, **21**, 641 (1946); R. D. Brown, *Trans. Faraday Soc.*, **45**, 296 (1949); **46**, 146 (1950).

[156] K. Brand, *Ber.*, **45**, 3071 (1912). This is the first of twenty papers over a 36-year period on the chemistry of dibenzopentalene derivatives, the so-called diphensuccindenes.

[157] C. T. Blood and R. P. Linstead, *J. Chem. Soc.*, **1952**, 2263.

[158] Cf. also C. C. Chuen and S. W. Fenton, *J. Org. Chem.*, **23**, 1538 (1958).

A further argument relates to the dyestuff acedianthrone,[159] CXX, which has a pentalene nucleus. The crystal structure[160] reveals bond distances that indicate double and single bonds, as indicated in the pentalene moiety, although the precision is low; the a-, b-, and c-bonds are 1.41, 1.59, and 1.48 A, respectively, with errors of ±0.06 A.

CXX

Finally, a dimethyl derivative of the hydrocarbon, CXXI,[161] has been prepared and is reported to behave as a polyolefin.[162] This compound shows the same ambiguity in the application of Craig's rules as the heptalene derivative, CXVI (Sec. 10.8).

CXXI

Derivatives of heptalene, CXIV, have also elicited some theoretical[163] and synthetic[164] interest, although all attempted syntheses have been unsuccessful with the single exception of Hafner and Schneider's synthesis of dimethylaceheptylene, CXV, previously mentioned. The relatively

[159] E. Clar, *Ber.*, **72**, 2134 (1939).

[160] A. Bennett and A. W. Hanson, *Acta Cryst.*, **6**, 736 (1953).

[161] For HMO treatments of this and other unusual polycyclic systems cf. A. Rosowsky, H. Fleischer, S. T. Young, R. Partch, W. H. Saunders, Jr., and V. Bockelheide, *Tetrahedron*, **11**, 121 (1960).

[162] K. Hafner, *Angew. Chem.*, **71**, 378 (1959); K. Hafner and J. Schneider, *Ann.*, **624**, 37 (1959).

[163] G. Berthier and B. Pullman, *Bull. soc. chim. France*, **1949**, D90; G. Berthier, B. Pullman, and J. Baudet, *J. chim. phys.*, **49**, 641 (1952).

[164] Summarized by E. D. Bergmann.[153a]

large calculated DE for heptalene, 3.618β (0.301 per π-electron), as given by HMO theory, should not be taken seriously, since all supplemental criteria indicate instability. The highest occupied MO is a NBMO and oxidation should be facile; the perimeter is a $4n$ cycle and the molecule does not fit Craig's rules.

The mixed hydrocarbon, azulene, CXXII, is a stable compound with aromatic properties.[165] Azulene has a $4n + 2$ perimeter and fits Craig's rules. The HMO DE of 3.364β[166] is less than that calculated for naphthalene, 3.683β. However, the corresponding RE, 55.5 kcal./mole is considerably greater than the experimental figure, 31 kcal./mole (Table 9.5).

CXXII

In part, the discrepancy is caused by the use of too high a value of β for the 9—10 bond, which has a length of 1.49 A;[167] $k = 0.9$[168] and $k = 0.8$[169] have been used, but even the latter value reduces the DE to only 3.22β. For the most part, the discrepancy is the result of limitations inherent in the HMO procedure, especially the neglect of electronic repulsions. The calculated dipole moment is much higher than the experimental value (Sec. 6.1). These limitations are largely overcome by advanced MO calculations; for example, the Pariser-Parr method gives a calculated charge distribution[170] in good agreement with the dipole moment (cf. Sec. 16.3) and with the NMR spectrum.[171]

Among these bicyclic homologues of naphthalene, mention should be

[165] A large literature has already accumulated on the chemistry of azulene. For recent reviews see (a) K. Hafner, *Angew. Chem.*, **70**, 419 (1958), and (b) W. Keller-Schierlein and E. Heilbronner, Chaps. 5 and 6 in D. Ginsburg's *Non-Benzenoid Aromatic Compounds*, Interscience Publishers, New York (1959).

[166] Several simple LCAO treatments of azulene derivatives have been published: ref. 165b; C. A. Coulson and H. C. Longuet-Higgins, *Rev. Sci. Instr.*, **85**, 929 (1947); R. D. Brown, *Trans. Faraday Soc.*, **44**, 984 (1948); G. Berthier and A. Pullman, *Compt. rend.*, **229**, 761 (1949); M. J. S. Dewar and R. Pettit, *J. Chem. Soc.*, **1954**, 1617; G. Berthier, B. Pullman, and J. Baudet, *J. chim. phys.*, **50**, 209 (1953); ref. 169.

[167] H. M. M. Shearer and J. M. Robertson, *Acta Cryst.*, **10**, 805 (1957).

[168] A. Streitwieser, Jr., *J. Am. Chem. Soc.*, **82**, 4123 (1960).

[169] A. Pullman and G. Berthier, *Compt. rend.*, **227**, 677 (1948); B. Pullman, M. Mayot, and G. Berthier, *J. Chem. Phys.*, **18**, 257 (1950).

[170] R. Pariser, *J. Chem. Phys.*, **25**, 1112 (1956).

[171] W. G. Schneider, H. J. Bernstein, and J. A. Pople, *J. Am. Chem. Soc.*, **80**, 3497 (1958).

made of the hypothetical compound, octalene, CXXIII. This molecule has a large calculated DE, 4.189β (0.30 per π-electron), a $4n + 2$ periphery, and obeys Craig's rules. It should not be overly reactive towards polymerization or to other reagents. If the strain of the planar eight-membered rings is not too great, this compound should be a relatively stable aromatic compound. One attempted synthesis involving dehydrogenation, however, has not succeeded.[172]

CXXIII

Additional compounds of interest may be regarded as derivatives of fulvene, CXXIV, and heptafulvene, CXXV. Both are known hydro-

CXXIV   CXXV

carbons, and fulvenes have been known for many years.[173] The simple fulvenes have the olefinic properties of polyenes but are characterized by the tendency of the five-membered ring to bear negative charge. The following reaction[174] is typical:

Heptafulvene has been prepared only recently by Doering and Wiley.[175,176,177] It is an extremely sensitive red compound that must be

[172] A. Streitwieser, Jr., and S. Andreades, unpublished results.

[173] See the reviews by J. H. Day, *Chem. Rev.*, **53**, 167 (1953) and by E. D. Bergmann in J. W. Cook's *Progress in Organic Chemistry*, Vol. 3, Academic Press, New York (1955), p. 81.

[174] K. Ziegler and W. Schaefer, *Ann.*, **511**, 101 (1934).

[175] W. E. Doering in *Theoretical Organic Chemistry*, The Kekulé Symposium, Butterworths Scientific Publications, London (1959), p. 35.

[176] W. E. Doering and D. W. Wiley, *Tetrahedron*, **11**, 183 (1960).

[177] W. E. Doering, Abstracts of 16th National Organic Chemistry Symposium, Seattle, June 15–17, 1959, p. 22.

kept in dilute solution in the cold; attempted isolation yields only polymer. The dibenzo derivative, CXXVI, has also been prepared.[178] Its properties are the classical ones anticipated for such a structure.

CXXVI

The highly unsaturated character of fulvene and heptafulvene belies their HMO DE values, 1.464β and 1.997β, respectively, which are substantially higher than those of the corresponding open-chain systems, hexatriene, 0.988β, and octatetraene, 1.518β. Again, this situation is undoubtedly an artifact of the HMO method that introduces the stabilizing β of an additional bond in forming a cycle without treating the additional electronic repulsion.

The derived systems, fulvalene, CXXVII, sesquifulvalene, CXXVIII, and heptafulvalene, CXXIX, have aroused considerable interest. The HMO DE's are high: CXXVI, 3.00β; CXXVII, 3.93β; CXXVIII, 4.00β.

CXXVII   CXXVIII   CXXIX

Fulvalene has an unoccupied bonding MO and on this basis may be expected to be relatively unstable, despite the high DE-value. The hydrocarbon has been synthesized[175,177,179] but is known only in solution. Polymerization occurs at an appreciable rate in concentrations above $0.001M$. The immediate precursor in this synthesis is the dianion, CXXX,

CXXXI   CXXX

which is readily formed on treatment of dihydrofulvalene, CXXXI, with butyl lithium. The stability of the dianion is doubtless due to the vacant bonding orbital of fulvalene; that is, the dianion has a closed shell configuration.

[178] E. D. Bergmann, E. Fischer, D. Ginsburg, Y. Hirshberg, D. Lavie, M. Mayot, A. Pullman, and B. Pullman, *Bull. soc. chim. France*, **1951**, 684.

[179] Cf. also W. B. DeMore, H. O. Pritchard, and N. Davidson, *J. Am. Chem. Soc.*, **81**, 5874 (1959).

1,2,3,4-Tetraphenylfulvalene has been synthesized and is more stable;[180] perchlorofulvalene is also reported.[181] Numerous benzo derivatives have been known for some time.[173]

Sesquifulvalene has so far defied synthesis,[175] although the molecule should be reasonably stable[182] according to several criteria; it has a high DE-value, 0.33β per π-electron,[183] and a closed shell configuration. The calculated electron densities predict a substantial dipole moment; the structure, CXXXII, should be relatively important. The tetrabenzo derivative, CXXXIII, has been prepared.[183] It has a dipole moment, $0.83D$,[183] which, as expected, is smaller than that calculated by the HMO method.

CXXXII   CXXXIII

Heptafulvalene has two electrons in a low-lying antibonding MO; in consequence, we expect the compound to be rather unstable and, in particular, to oxidize readily. The hydrocarbon has been synthesized as permanganate-colored crystals which are relatively stable toward acid, oxidizing agents, and light.[175,177] The heat of hydrogenation gives an empirical RE of only 28 kcal./mole,[184] much less than the 66 kcal./mole calculated from the DE-value. The chemistry of this material so far revealed does not correspond to our qualitative MO expectations; information about its further chemistry will be awaited with anticipation. The tetrabenzo compound, CXXXIV, is a stable, colorless hydrocarbon that may not be coplanar.[185]

Some condensed benzo derivatives of these systems which are of interest are s-indacene, CXXXV, as-indacene, CXXXVI, and the hydrocarbon, CXXXVII. These compounds are unknown so far, although some

[180] E. C. Schreiber and E. I. Becker, *J. Am. Chem. Soc.*, **76**, 6125 (1954).

[181] E. T. McBee, C. W. Roberts, and J. D. Idol, *J. Am. Chem. Soc.*, **77**, 4943 (1955).

[182] J. F. Tinker, *J. Chem. Phys.*, **19**, 981 (1951).

[183] B. Pullman, A. Pullman, E. D. Bergmann, H. Berthod, E. Fischer, Y. Hirshberg, D. Lavie, and M. Mayot, *Bull. soc. chim. France*, **1952**, 73.

[184] R. B. Turner, W. R. Meador, W. E. Doering, J. R. Mayer, and D. W. Wiley, *J. Am. Chem. Soc.*, **79**, 4127 (1957).

[185] E. D. Bergmann, D. Ginsburg, Y. Hirshberg, M. Mayot, A. Pullman, and B. Pullman, *Bull. soc. chim. France*, **18**, 697 (1951).

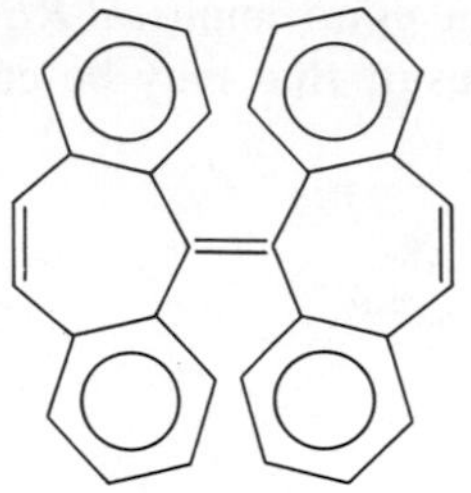

CXXXIV

attempted syntheses have been reported[186] and MO calculations have been published.[187] The synthesis of violet-black crystalline diphenyl dibenzo-*as*-indacene, CXXXVIII, has been reported.[188] The indacenes, including the dibenzo derivative, CXXXVIII, have $4n$ perimeters. CXXXV does not obey Craig's rules but CXXXVII does.

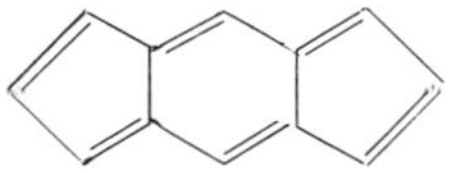

CXXXV 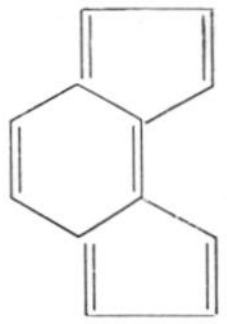   CXXXVI 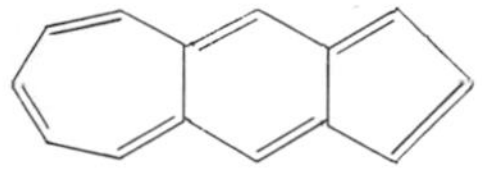   CXXXVII

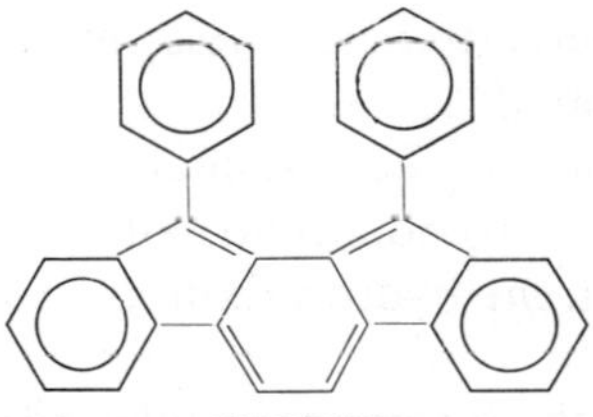

CXXXVIII

The further derivatives pyracylene, CX, and acepleiadylene, CXXXIX, should be aromatic according to Craig's rules. Pyracylene, however, has a $4n$ perimeter and is unknown to date (*vide supra*).

Acepleiadylene is a red crystalline stable material[189] which seems to have aromatic properties in agreement with its $4n + 2$ perimeter.[190]

[186] S. Dev., *J. Indian Chem. Soc.*, **30**, 729 (1953); A. D. Campbell and S. N. Slater, *J. Chem. Soc.*, **1952,** 4353; W. Baker and W. G. Leeds, *J. Chem. Soc.*, **1948,** 974.

[187] R. D. Brown, *J. Chem. Soc.*, **1951,** 2391; J. I. F. Alonso and J. L. S. Lucas, *Anales real soc. españ. fis. y quim.* (*Madrid*), **50B,** 143 (1954).

[188] A. Etienne and A. le Berre, *Compt. rend.*, **242,** 1453 (1956); for MO calculations, cf. O. Chalvet and J. Peltier, *Bull. soc. chim. France*, **1956,** 1667.

[189] V. Boekelheide and G. K. Vick, *J. Am. Chem. Soc.*, **78,** 653 (1956).

[190] For MO calculations cf. B. Pullman, A. Pullman, G. Berthier, and J. Pontis, *J. chim. phys.*, **49,** 20 (1952); B. Pullman and G. Berthier, *Compt. rend.*, **242,** 2563 (1956).

The heat of hydrogenation gives an empirical RE 21 kcal./mole greater than that of naphthalene,[191] a result that may be compared to the calculated

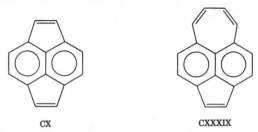

CX   CXXXIX

HMO DE difference, 2.57β. Acepleiadylene does not react with maleic anhydride under conditions that give Diels-Alder adducts with pleiadiene, CXL, and acepleiadiene, CXLI.[189]

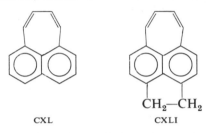

CXL   CXLI

Several heterocyclic derivatives of these interesting polycyclic hydrocarbons have been prepared. In some of the examples the heteroatom replaces a carbon in the π-system similar to the change: benzene $\rightarrow$ pyridine. In others, the heteroatom contributes two electrons and replaces a double bond of the parent hydrocarbon as in the change: benzene $\rightarrow$ pyrrole.

In the first group are the several aza-azulenes[192] and the azabenzopentalenes.[193]

Among the examples in the second group are the pseudosesquifulvalene heterocycle, CXLII,[194] and the pseudoazulene, CXLIII.[152,195] The compounds CXLIV and CXLV are also isoelectronic with azulenes and are

[191] R. B. Turner in *Theoretical Organic Chemistry*, The Kekulé Symposium, Butterworths Scientific Publications, London (1959), p. 67.

[192] W. Treibs, R. Steinert, and W. Kirchhoff, *Ann.*, **581**, 54 (1953); A. G. Anderson and J. Tazuma, *J. Am. Chem. Soc.*, **74**, 3455 (1952); T. Nozoe, T. Mukai, and I. Murata, *J. Am. Chem. Soc.*, **76**, 3352 (1954); D. Lloyd, *Chem. and Ind.*, **1953**, 921; D. Hunter, D. Lloyd, D. Marshall, D. Price, and F. Rowe, *Chem. and Ind.*, **1954**, 1068; see review by ref. 165*b*.

[193] W. Treibs, *Naturwissenschaften*, **46**, 170 (1959).

[194] J. A. Berson and E. M. Evleth, *Chem. and Ind.*, **1959**, 901.

[195] J. W. Armit and R. Robinson, *J. Chem. Soc.*, **127**, 1604 (1925); G. V. Boyd, *J. Chem. Soc.*, **1959**, 55.

CXLII   CXLIII

now known.[196] The behavior of CXLIV toward electrophilic substitution closely resembles that of azulene.[197]

CXLIV   CXLV

The generic term *cyclazines* has been proposed for heterocycles of the type CXLVI, CXLVII, and CXLVIII.[198] Cycl[3,2,2]azine has been prepared.[198] In agreement with its $4n + 2$ perimeter it is a stable aromatic system. The other cyclazines are still unknown; some MO calculations

CXLVI   CXLVII   CXLVIII

CXLIX

[196] A. G. Anderson, Jr., W. F. Harrison, R. G. Anderson, and A. G. Osborne, *J. Am. Chem. Soc.*, **81**, 1255 (1959).

[197] A. G. Anderson, Jr., and W. F. Harrison, *Tetrahedron Letters*, No. 2, 11 (1960).

[198] R. J. Windgassen, Jr., W. H. Saunders, Jr., and V. Boekelheide, *J. Am. Chem. Soc.*, **81**, 1459 (1959); V. Boekelheide, Abstracts of 16th National Organic Chemistry Symposium, Seattle, June 15–17, 1959, p. 26.

of these compounds have been published.[198,199] The stability of CXLVII will be of particular interest because of its $4n$ perimeter.

The compound CXLIX is stable; it does not hydrolyze with base.[200]

CL

Finally, we may mention the many known stable compounds that have the porphin, CL, and phthalocyanine, CLI, structure and corresponding $4n + 2$ macrocyclic ring skeletons.

CLI

[199] R. D. Brown and B. A. W. Coller, *Mol. Phys.*, **2**, 158 (1959).
[200] P. M. Maitlis, *Proc. Chem. Soc.*, **1957**, 354.

III   REACTIONS

11 Aromatic substitution

11.1 Transition State Theory

A long-standing problem in organic chemistry has been the effect of substituents on orientation in aromatic substitution. In polycyclic aromatic hydrocarbons with more than one nonequivalent position open to substitution one step has been the determination of the most reactive site. Resonance theory considerations, reasoning by analogy, etc., have been more or less successful, at least in narrowing down the reactive positions in a number of such hydrocarbons. Additional substituents render the solution still more difficult.

A large literature has already developed on the application of MO theory to this problem. Because of the variety of approaches attempted, these reactions are especially suited to introduce the ways MO theory can be used to correlate generally the effect of structure on reactivity, qualitatively and quantitatively.

The *ab initio* calculation of any reacting system in a condensed phase is beyond present techniques; our thinking is done in terms of a model for the reaction. Classically, the model has been a *mechanical* one, involving the dynamic movement and collisions of reagents. The modern approach is in terms of *transition state theory*.

For a system undergoing reaction we can characterize the extent of reaction in terms of a set of coordinates: bond distances, bond angles, and the like. Moreover, in a condensed phase in which the collision frequency with solvent is exceedingly high the system may be considered as always in thermal equilibrium; therefore, for each set of values of the bond distances and angles which characterize the reaction, a standard free energy can be defined, in principle, by allowing all other bond distances and angles, the position and configuration of solvent molecules, etc., to adopt the values that will make this free energy a minimum. For the example of a direct displacement reaction, $Y + RX \rightarrow YR + X$, the resulting function may be represented diagrammatically as in Fig. 11.1,[1] in which the curved lines represent the loci of points having the same free energy. The increasing thicknesses of the lines represent approximately increasing values of the free energy. For most reactions the general representation is that of two valleys meeting at a saddle point.[2] The dotted line in Fig. 11.1

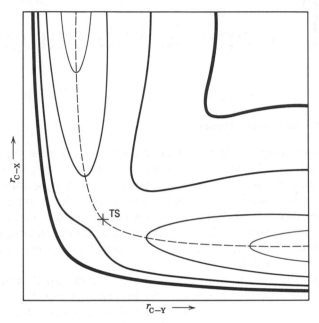

Fig. 11.1   Schematic free-energy diagram for the reaction, $X^- + RY \rightarrow RX + Y^-$. Increasing thicknesses of lines correspond to increasing values of F°.

indicates a reaction path of lowest free energy. If the free energy is plotted as a function of a distance parameter along this path, the *reaction coordinate*, the curve in Fig. 11.2 is obtained. Each point along this reaction

[1] A. Streitwieser, Jr., *Chem. Rev.*, **56**, 571 (1956).

[2] The usual treatment is given in terms of potential energy, a quantity suitable for theoretical calculations and for considerations in the gas phase. Because of the importance and complexity of solvation, the potential energy is less useful in a condensed phase. An important distinction must be acknowledged, however. A potential energy is a molecular property; a free energy is a function of a large number of molecules (cf. discussion below). For further details on transition state theory see G. E. K. Branch and M. Calvin, *The Theory of Organic Chemistry*, Prentice-Hall, Englewood Cliffs, N.J. (1941), p. 383; S. Glasstone, K. J. Laidler, and H. Eyring, *The Theory of Rate Processes*, McGraw-Hill Book Company, New York (1941); A. E. Remick, *Electronic Interpretations of Organic Chemistry*, second edition, John Wiley and Sons, New York (1949); P. D. Bartlett, "The Study of Organic Reaction Mechanisms," Chap. 1 in H. Gilman's *Organic Chemistry*, Vol. III, John Wiley and Sons, New York (1953), p. 1.

coordinate is at a minimum with respect to changes in other coordinates of the system.

In order to achieve reaction, the system must overcome a free energy maximum, $\Delta F^{\ddagger}$, the free energy of activation (Fig. 11.2), which is associated with a particular configuration, the *transition state* (TS in Fig. 11.1).

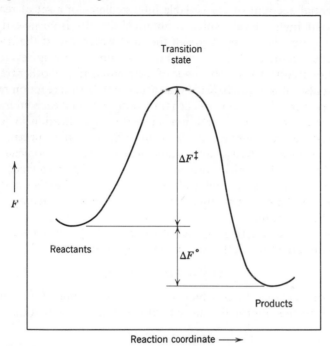

Fig. 11.2   Free-energy profile for a typical reaction.

Note that although the transition state is a maximum along the reaction coordinate it is still a minimum with respect to other coordinates.

From the postulates of the absolute rate theory,[3] the rate constant for reaction in solution is given by (1), in which κ, the *transmission coefficient*, is usually taken as unity for most reactions, $\mathbf{k}$ is Boltzmann's constant, and h is Planck's constant.

$$\text{Rate constant} = \kappa \frac{\mathbf{k}T}{h} e^{-\Delta F^{\ddagger}/RT} \qquad (1)$$

The effect of structure on rate in this theory is the effect of structure on the free energy of activation. Even a qualitative evaluation of these

[3] S. Glasstone, K. J. Laidler, and H. Eyring, *The Theory of Rate Processes*, McGraw-Hill Book Company, New York (1941).

effects requires some idea of the structure of the transition state and of the effect of changes in structure on the free energy; that is, we need to know the *reaction mechanism*.

The term reaction mechanism has meaning on several levels. On its lowest level, which is also the classical use of the term, a reaction mechanism is a sequential account of the isolable intermediates of a set of reactions. A somewhat higher level, which is obtained directly from investigations of kinetic order, is a description of the atoms involved at the transition state of the rate-determining step. For reactions involving several steps specification frequently may be made of the composition of other transition states and also of intermediates. At a rather high level a reaction mechanism is a structural description of each transition state and each intermediate involved in the reaction or reaction sequence. Specification in terms of chemical structures implies, at least in principle, the spectra, dipole moments, optical activity, and other *macroscopic* phenomena. It is the province of theoretical organic chemistry, with resonance theory, molecular-orbital theory, and the like, to evaluate the thermodynamic properties—enthalpy, entropy, and free energy—inherent in the structural description.

The free energy of activation can be dissected in the usual way into an enthalpy of activation, $\Delta H^{\ddagger}$, and an entropy of activation, $\Delta S^{\ddagger}$.

$$\Delta F^{\ddagger} = \Delta H^{\ddagger} - T \, \Delta S^{\ddagger} \tag{2}$$

These theoretical quantities, which in principle are temperature dependent, are related to the empirical constants of the Arrhenius equation,

$$k = A e^{-E^{\ddagger}/RT} \tag{3}$$

Determination of the rate constant as a function of temperature yields two parameters, A, the temperature-independent factor, and $E^{\ddagger}$, the *activation energy*, from the temperature-dependent factor. It should be emphasized that (3) holds over only a limited temperature range; however, in practice, a plot of log k versus $1/T$ over an experimentally accessible temperature range generally yields an excellent straight line from which A and $E^{\ddagger}$ are readily determined.

The relations between the two sets of quantities are given by the following equations which should be understood to hold over only a limited temperature range:

$$\Delta H^{\ddagger} = E^{\ddagger} - RT \tag{4}$$

$$A = \frac{ekT}{h} \, e^{\Delta S^{\ddagger}/R} \tag{5}$$

$$\Delta S^{\ddagger} = 4.576 \log A/T - 49.203 \tag{6}$$

In (6) the time unit is the second and the entropy unit is calories per degree (eu). As a rough first approximation, the activation energy may be associated with bonding forces and the entropy of activation with freedom of atomic motions.

When the reactivities of two systems are being compared, the question arises which experimentally determinable quantities are the most significant theoretically in the interpretation of reaction-rate data. The question is especially pertinent when we look for an experimental rate quantity for comparison with quantities calculated from MO theory. The logarithm of the ratio of the rate constants is directly proportional to the difference in the free energies of activation (7); the following argument shows that this is the quantity of fundamental significance:

$$\Delta F_a^{\ddagger} - \Delta F_b^{\ddagger} = 2.303RT \log k_a/k_b \tag{7}$$

A structural change in the transition state may change any of a number of coordinates or structural parameters. We consider the effect of a change in one such parameter, say λ, on the free energy by a Taylor's expansion:

$$\delta\Delta F^{\ddagger} = \frac{\partial\Delta F^{\ddagger}}{\partial\lambda}\,\delta\lambda + \frac{1}{2}\frac{\partial^2\Delta F^{\ddagger}}{\partial\lambda^2}\,\delta\lambda^2 + \cdots \tag{8}$$

But the free energy is a minimum with respect to λ; thus the first term in the expansion vanishes. We consider the same type of expansion for the enthalpy and entropy of activation:

$$\delta\Delta H^{\ddagger} = \frac{\partial\Delta H^{\ddagger}}{\partial\lambda}\,\delta\lambda + \frac{1}{2}\frac{\partial^2\Delta H^{\ddagger}}{\partial\lambda^2}\,\overline{\delta\lambda}^2 + \cdots \tag{9}$$

$$\delta\Delta S^{\ddagger} = \frac{\partial\Delta S^{\ddagger}}{\partial\lambda}\,\delta\lambda + \frac{1}{2}\frac{\partial^2\Delta S^{\ddagger}}{\partial\lambda^2}\,\overline{\delta\lambda}^2 + \cdots \tag{10}$$

$\Delta H^{\ddagger}$ and $\Delta S^{\ddagger}$ are not minima with respect to λ; hence the first terms on the right do not necessarily vanish in (9) and (10). What this means is that the free energy of activation varies only in the second degree with a structural change, whereas the enthalpy and entropy of activation vary to the first degree; that is, a structural change may make rather large changes in $\Delta H^{\ddagger}$ and $\Delta S^{\ddagger}$ but only relatively small changes in $\Delta F^{\ddagger}$.

Moreover, to the first degree, the changes in $\Delta H^{\ddagger}$ and $\Delta S^{\ddagger}$ are mutually compensating. By differentiating (2) with respect to λ at constant temperature, we find

$$\frac{\partial\Delta H^{\ddagger}}{\partial\lambda} = T\frac{\partial\Delta S^{\ddagger}}{\partial\lambda} \tag{11}$$

We conclude that a structural change in a reagent may make other changes in the transition state in order to keep $\Delta F^{\ddagger}$ a minimum. Such changes may make larger changes in $\Delta H^{\ddagger}$ and $\Delta S^{\ddagger}$ which are partly mutually compensating. Although there is no necessity that these effects be regular, in practice they frequently are. Thus for many series of related reactions a plot of $\Delta H^{\ddagger}$ versus $\Delta S^{\ddagger}$ yields a straight line.[4] Such linear correlations are a direct consequence of (11).

An example may help to clarify the point. In a reaction in which charged species are involved, solvation is especially important. The proximity of several solvent molecules helps to distribute the charge and lower the enthalpy. However, this proximity also limits their motions and correspondingly lowers the entropy. Moving the solvent molecules away gives them greater freedom and higher entropy, but the resulting loss of solvation energy raises the enthalpy. A balance is reached which results in the lowest free energy. If a reagent is structurally modified, the solvation balance need not be the same as before, but a different balance will affect the enthalpy and entropy more than it will affect the free energy. Ideally, we should prefer to examine a transition state with a variation of one structural feature at a time; for example, we would like to examine the effect of a substituent while keeping solvation effects unchanged. In practice, this cannot be accomplished, but of the various experimental properties the free energy of activation comes closest to this ideal.

An intermediate is frequently represented as a free energy minimum along the reaction coordinate. As such, it is flanked by two maxima which are transition states (Fig. 11.3). Although it is sometimes convenient and useful to illustrate the energetics of a reaction by a *free energy profile*, we should not forget that a total reaction can be broken into component elementary reactions which may be considered separately; for example, if Fig. 11.3 represents the conversion of A to B via an intermediate I, the total reaction is made up of four rate constants:

$$A \underset{k_{-1}}{\overset{k_1}{\rightleftharpoons}} I$$

$$I \underset{k_{-2}}{\overset{k_2}{\rightleftharpoons}} B$$

The rate constants determine the relative free-energy changes in Fig. 11.3 and, in principle, can be evaluated by experimental observations.

A useful postulate, which has become known as Hammond's postulate,[5] is that a transition state with a free energy close to that of an intermediate is likely to resemble the intermediate structurally. In Fig. 11.3, for example, the intermediate is likely to be a better model for the transition states than is either the reactant or the product.

[4] See the extensive summary and review by J. E. Leffler, *J. Org. Chem.*, **20**, 1202 (1955).

[5] G. S. Hammond, *J. Am. Chem. Soc.*, **77**, 334 (1955).

For reactions that involve changes in π-systems knowledge of the structure of the transition state enables the construction of a model of the π-system in the transition state for MO calculation. We dissect the π-energy change from the σ-energy change, solvation energies etc. To

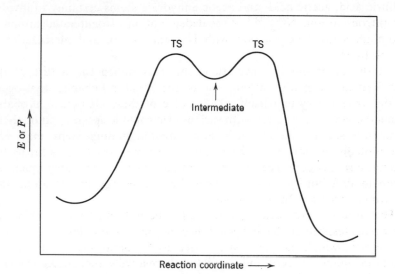

Fig. 11.3   Energy of free-energy profile for a reaction with an unstable intermediate.

the usual MO approximations, the π-energy changes for different systems can be calculated; we assume that the other changes, such as σ-bond energies and solvation energies are either constant for a related family or are proportional to the π-energy change. Hence we write

$$\log k_i = a + bW_i \tag{12}$$

in which a and b are empirical constants and W_i is some MO quantity calculated for the system i. In many cases the energy quantities are the π-bonding energies for which W_i is replaced by $(M_{TS} - M_{reactant})_i$.

11.2   Mechanism of Electrophilic Aromatic Substitution[6]

By definition, electrophilic aromatic substitution involves reaction with electron-seeking or positive species. Many individual aromatic substitution

[6] For summaries and reviews see (a) C. K. Ingold, *Structure and Mechanism in Organic Chemistry*, Cornell University Press, Ithaca (1953), Chap. 6; (b) J. Hine, *Physical Organic Chemistry*, McGraw-Hill Book Company, New York (1956), Chap. 16; (c) G. W. Wheland, *Resonance in Organic Chemistry*, John Wiley and Sons, New York

reactions fall into this category. The studies of the mechanisms of these reactions may be divided into two parts. On the one hand, numerous studies have delved into the nature of the attacking electrophilic species; this aspect is individual for different reactions. As examples, nitration in sulfuric acid, acetic acid, and acetic anhydride seems certainly to involve the nitronium ion, NO_2^+,[6a,b,d,g] the dedeuteration of deuterated aromatic compounds involves reaction with H^+ (vide infra), and mercuration is attack by Hg^{++}.[6a]

On the other hand, other studies have examined the nature of the attack at the aromatic nucleus, that is, the structural changes that occur in the aromatic system itself. The course of these changes is probably much the same for reaction with various electrophilic species, although the transition states, intermediates and rate-determining steps may and undoubtedly do vary with the reagents. We may say that as far as the aromatic ring is concerned the reaction coordinate is probably much the same in different systems, although the free energy profile along this coordinate varies with the system.

This aspect of aromatic substitution has been discussed in terms of two intermediates, which have been called π- and σ-complexes.[6e,6f,7] The π-complex, which has been symbolized as I, contains a bond between the electrophilic reagent and a structurally relatively unchanged aromatic ring. The π-complex is considered to involve a perturbation only of the aromatic π-electrons and may be equivalent to a charge-transfer complex.

I

There is clear evidence that such complexes do exist. Hydrogen chloride forms loose complexes with aromatic hydrocarbons; the solutions are not conducting and no hydrogen exchange occurs when DCl is used.[8] The complexes of halogens, tetracyanoethylene, quinones, etc., with aromatics were discussed earlier (Sec. 7.5). Silver ions form complexes with aromatic hydrocarbons; equilibrium constants and a crystal structure

(1955), p. 476; (d) E. S. Gould, Mechanism and Structure in Organic Chemistry, Henry Holt and Company, New York (1959), Chap. 11; (e) H. C. Brown and K. L. Nelson, in B. T. Brooks, S. S. Kurtz, Jr., C. E. Boord, and L. Schmerling, The Chemistry of Petroleum Hydrocarbons, Vol. 3, Reinhold Publishing Corp., New York (1955), Chap. 56; (f) K. L. Nelson, J. Org. Chem., 21, 145 (1956); (g) P. B. D. De La Mare and J. H. Ridd, Aromatic Substitution, Butterworths Scientific Publications, London (1959).

[7] M. J. S. Dewar, Electronic Theory of Organic Chemistry, Oxford University Press, New York (1949).

[8] H. C. Brown and J. D. Brady, J. Am. Chem. Soc., 71, 3575 (1949); 74, 3570 (1952).

have been determined.[9] However, the mere existence of such complexes does not implicate them necessarily as reaction intermediates.

The effect of acid strength on the rate of protodedeuteration[10] reactions, $ArD + H^+ \rightarrow ArH + D^+$, has been considered as evidence for π-complexes in this reaction. The logarithms of the observed rate constants in these reactions are proportional to Hammett's acidity function H_0 of the medium.[11,12,13] According to the Hammett-Zucker postulate,[14] this indicates a rapid equilibrium before the rate-determining step which has been formulated as[11]

This evidence for π-complexes depends solely on a postulated behavior of a particular ratio of activity coefficients which includes the activity coefficient of the transition state. However, recent work indicates that the postulate may be invalid for some cases;[15] in particular, the effect of acidity on the rate of protodedeuteration may be entirely consistent with a mechanism involving a single intermediate:

$$ArD + H^+ \rightarrow [ArHD^+] \rightarrow ArH + D^+$$

Furthermore, the protodetritiation of 1,3,5-trimethoxybenzene-2-t[16] and the protodedeuteration of azulene-1,3-d_2[17] have recently been reported

[9] S. Winstein and H. J. Lucas, *J. Am. Chem. Soc.*, **60**, 836 (1938); R. M. Keefer and L. J. Andrews, *J. Am. Chem. Soc.*, **74**, 640 (1952); N. Ogimachi, L. J. Andrews, and R. M. Keefer, *J. Am. Chem. Soc.*, **78**, 2210 (1956); R. E. Rundle and J. H. Goring, *J. Am. Chem. Soc.*, **72**, 5337 (1950); H. G. Smith and R. E. Rundle, *J. Am. Chem. Soc.*, **80**, 5075 (1958).

[10] This convenient nomenclature has been suggested by J. F. Bunnett, *Chem. Eng. News*, **30**, 4019 (1954), *J. Chem. Soc.*, **1954**, 4715; J. F. Bunnett, M. Morath, and S. Okamoto, *J. Am. Chem. Soc.*, **77**, 5055 (1955).

[11] V. Gold and D. P. N. Satchell, *J. Chem. Soc.*, **1955**, 3609, 3619, 3622; V. Gold, R. W. Lambert, and D. P. N. Satchell, *Chem. and Ind.*, **1959**, 1312.

[12] E. L. Mackor, P. J. Smit, and J. H. van der Waals, *Trans. Faraday Soc.*, **53**, 1309 (1957).

[13] G. Dallinga, A. A. V. Stuart, P. J. Smit, and E. L. Mackor, *Z. Elektrochem.*, **61**, 1019 (1957).

[14] L. Zucker and L. P. Hammett, *J. Am. Chem. Soc.*, **61**, 2791 (1939). A discussion of H_0 and its significance in mechanism is beyond the scope of this book. See review by F. A. Long and M. A. Paul, *Chem. Rev.*, **57**, 935 (1957).

[15] C. Eaborn and R. Taylor, *J. Chem. Soc.*, **1960**, 3310; also, D. S. Noyce, private communication.

[16] A. J. Kresge and Y. Chiang, *J. Am. Chem. Soc.*, **81**, 5509 (1959).

[17] J. Colapietro and F. A. Long, *Chem. and Ind.*, **1960**, 1056.

to be general acid catalyzed; such a result means that a base is required to remove the tritium or deuterium. This result is completely compatible with the single intermediate mechanism, but it is hard to reconcile with rapid equilibria involving π-complexes.

In some alkylation reactions evidence has been forthcoming for a loose complex before reaction; for example, the alkylation of benzene with boron fluoride and 2- or 3-pentanol apparently involves a complex with benzene within which a rapid carbonium ion rearrangement, $\text{Ar} \cdots (\text{C}-\overset{+}{\text{C}}-\text{C}-\text{C}-\text{C}) \rightleftharpoons \text{Ar} \cdots (\text{C}-\text{C}-\overset{+}{\text{C}}-\text{C}-\text{C})$, occurs, leading ultimately to the same mixture of 2- and 3-phenylpentane.[18]

The σ-complex, or "Wheland intermediate,"[19] has the postulated structure, II, in which the carbon atom attacked in the aromatic ring has an

II

approximately tetrahedral configuration with the entering group Z and leaving hydrogen on opposite sides of the molecular plane. A variety of evidence supports the formulation of such a structure as a key intermediate in aromatic substitution.

The most thoroughly studied cases are those for which Z = H or D. Aluminum chloride is insoluble in toluene but dissolves in the presence of hydrogen chloride to form a green complex which analyzes for

III

$(\text{ArH})_n \cdot \text{HCl} \cdot \text{AlCl}_3$.[20] Similar complexes are known for aluminum bromide and hydrogen bromide.[21] The complexes are good conductors; the

[18] A. Streitwieser, Jr., W. D. Schaeffer, and S. Andreades, *J. Am. Chem. Soc.*, **81**, 1113 (1959).

[19] G. W. Wheland, *J. Am. Chem. Soc.*, **64**, 900 (1942).

[20] H. C. Brown and A. W. Pearsall, *J. Am. Chem. Soc.*, **74**, 191 (1952).

[21] G. Baddeley, G. Holt, and D. Voss, *J. Chem. Soc.*, **1952**, 100; H. C. Brown and W. J. Wallace, *J. Am. Chem. Soc.*, **75**, 6268 (1953); D. D. Eley and P. J. King, *J. Chem. Soc.*, **1952**, 2517, 4972; J. F. Norris and J. N. Ingraham, *J. Am. Chem. Soc.*, **62**, 1298 (1940).

use of DBr leads to rapid exchange with ring hydrogens.[22] These data are in excellent agreement with the formulation of these complexes as salts of the type III.

Similar complexes are known with hydrofluoric acid. Methylated benzenes yield colored solutions in hydrofluoric acid which are excellent conductors.[23] The concentration of complex is increased by the addition of boron fluoride, and equilibrium constants have been measured for the reaction:[24]

$$ArH + HF + BF_3 \rightleftharpoons ArH_2^+ BF_4^-$$

The crystalline complexes have been isolated; at the melting point they are conducting.[25] With DF, a deuterated complex is formed, which, on heating, generates about 40% original hydrocarbon and 60% deuterated hydrocarbon, a result consistent with structure IV for the complex.

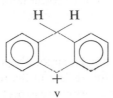

IV

Anthracene dissolves in sulfuric acid to form a solution whose spectrum is that of benzhydryl cation.[26] The formation of V seems clearly indicated. The spectra of a number of aromatic hydrocarbons have been determined in liquid hydrofluoric acid with and without boron fluoride. The spectra are consistent with the formulation of π-systems as carbonium ions with one of the original π-centers removed.[27]

V

A number of aromatic hydrocarbons in trifluoroacetic acid and boron fluoride have NMR spectra consistent with the formulation of σ-complexes; in particular, the aliphatic hydrogens of the CH_2 group are identified.[28]

[22] A. Klit and A. Langseth, Z. physik. Chem., 176, 65 (1936).

[23] M. Kilpatrick and F. E. Luborsky, J. Am. Chem. Soc., 75, 577 (1953).

[24] D. A. McCaulay and A. P. Lien, J. Am. Chem. Soc., 73, 2013 (1951).

[25] G. Olah and I. Kuhn, J. Am. Chem. Soc., 80, 6535 (1958).

[26] V. Gold and F. L. Tye, J. Chem. Soc., 1952, 2172.

[27] A. V. Stuart and E. L. Mackor, J. Chem. Phys., 27, 826 (1957); G. Dallinga, E. L. Mackor, and A. V. Stuart, Mol. Phys., 1, 123 (1958).

[28] C. MacLean, J. H. van der Waals, and E. L. Mackor, Mol. Phys., 1, 247 (1958).

In short, there is no question but that σ-complexes exist and that they have essentially the structures assigned. We next ask, "Are they necessarily reaction intermediates in aromatic substitution?" A plot of log k for protodedeuteration of some methylbenzenes versus the logarithms of relative equilibrium constants in hydrofluoric acid gives an excellent linear

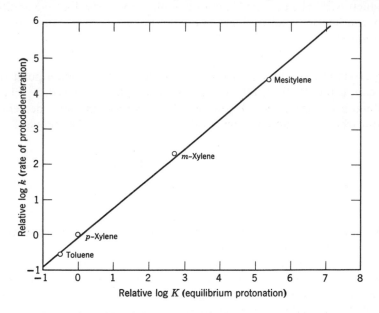

Fig. 11.4 Correlation of log k for protodedeuteration in CF_3COOH—H_2SO_4 versus log K for equilibrium protonation in HF (corrected for statistical factors). (Adapted from ref. 12.)

correlation (Fig. 11.4). From Hammond's postulate (*vide supra*), if the σ-complex is an unstable intermediate in the reaction, the transition state should closely resemble it in energy; we expect a close parallelism between rate and equilibrium constant such as that actually found. We might even expect the parallelism to extend to other substitution reactions. Even relative chlorination rates give a good correlation when plotted in this way (Fig. 11.5). The effect of substitution on rate so closely parallels the effect on equilibrium constant that the two processes undoubtedly have a basic similarity.

In an elegant and important investigation Olah and Kuhn[29] have isolated the boron tetrafluoride salts of a variety of σ-complexes at low temperatures. On warming, HBF_4 is liberated and the substituted aromatic is left.

[29] G. A. Olah and S. J. Kuhn, *J. Am. Chem. Soc.*, **80**, 6541 (1958).

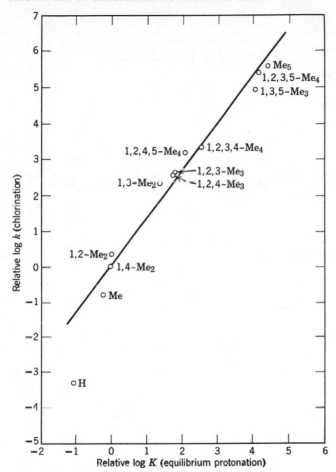

Fig. 11.5 Correlation of log k for chlorination versus log K for equilibrium protonation in HF (data summarized in ref. 6e).

There seems little question but that the intermediates in the reactions were isolated. Some typical examples from this work follow on page 320.

Further evidence on the structure of the rate-determining transition state comes from isotope effects. The general case is

$$ArH + Z^+ \underset{k_{-1}}{\overset{k_1}{\rightleftharpoons}} ArHZ^+ \underset{k_{-2}}{\overset{k_2}{\rightleftharpoons}} ArZ + H^+$$

For many reactions k_{-1} and k_{-2} may be ignored. The rate-determining step is that of k_1. The C—H bond is not changed much in this process, and substitution of the H by D should not result in a large isotope effect.

Indeed, the two most important secondary isotope effects expected are of comparable magnitude and of opposite direction.[30] In many examples of nitration and halogenation no isotope effect is observed.[31] The several

[30] A. Streitwieser, Jr., R. H. Jagow, R. C. Fahey, and S. Suzuki, *J. Am. Chem. Soc.*, **80**, 2327 (1958).

[31] L. Melander, *Acta Chem. Scand.*, **3**, 95 (1949), *Nature*, **163**, 599 (1949), *Arkiv. Kemi*, **2**, 211 (1950); W. M. Lauer and W. E. Noland, *J. Am. Chem. Soc.*, **75**, 3689 (1953); T. G. Bonner, F. Bowyer, and G. Williams, *J. Chem. Soc.*, **1953**, 2650; K. Halverson and L. Melander, *Arkiv. Kemi*, **11**, 77 (1957); P. C. Myhre, *Acta Chem. Scand.*, **14**, 219 (1960).

cases in which such isotope effects are observed are plausibly associated with k_2 being rate determining.[32] In such cases the C—H or C—D bond is being broken and an isotope effect is anticipated.

In summary, the present situation is that π-complexes or charge-transfer complexes may be intermediates in some aromatic substitutions, but direct

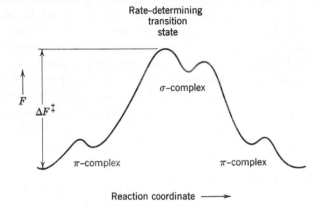

Fig. 11.6   Free-energy profile for aromatic substitution with both π- and σ-complexes.

evidence is scanty. This point is important because such intermediates have played a prominent role in some proposed mechanisms for aromatic substitution.[33,34] Furthermore, charge-transfer character has been associated with the rate-determining transition state in models for comparison with calculations (*vide infra*). The presence of such intermediates converts the free-energy profile of the reaction from that in Fig. 11.3 to something like Fig. 11.6. However, the rate of the reaction is still dependent on $\Delta F^{\ddagger}$, as indicated in Fig. 11.6; that is, the existence of low-energy π-complexes does not affect the rate and need not be considered in correlations with theory.

The σ-complex serves as a good model for the transition state. Such a

[32] U. Berglund-Larssen and L. Melander, *Arkiv. Kemi*, **6**, 219 (1953); H. Zollinger, *Helv. Chim. Acta*, **38**, 1597, 1617 (1955); E. Grovenstein and D. C. Kilby, *J. Am. Chem. Soc.*, **79**, 2972 (1957); E. Shilov and F. Weinstein, *Nature*, **182**, 1300 (1958); *Doklady Akad. Nauk. S.S.S.R.*, **123**, 93 (1958); H. Zollinger, *Experientia*, **12**, 165 (1958); *Angew. Chem.*, **70**, 204 (1958); D. B. Denney and P. P. Klemchuk, *J. Am. Chem. Soc.*, **80**, 6014 (1958); J. C. D. Brand, A . W. P. Jarvie, and W. C. Horning, *J. Chem. Soc.*, **1959**, 3844; A. Grimson and J. H. Ridd, *J. Chem. Soc.*, **1959**, 3019; S. Olsson, *Arkiv. Kemi*, **14**, 85 (1959); V. Gold, R. W. Lambert, and D. P. N. Satchell, *Chem. and Ind.*, **1959**, 1312; E. Berliner, *Chem. and Ind.*, **1960**, 177; P. C. Myhre, *Acta Chem. Scand.*, **14**, 219 (1960).
[33] S. Nagakura and J. Tanaka, *Bull. Chem. Soc. Japan*, **32**, 734 (1959).
[34] R. D. Brown, *J. Chem. Soc.*, **1959**, 2224.

model has long been useful in explaining orientation effects in aromatic substitution by using resonance theory[35] or suitable analogies such as the relative stabilities of corresponding quinones.[36]

Various models have been used for attempted correlations with MO quantities. The variety of these proposed applications form an instructive introduction to the use of MO theory in relating structure to reactivity. The proposals may be divided into two classes: those based on MO quantities calculated from the original π-system, or *isolated molecule approximations*, and those based on an altered π-system corresponding to a model of the transition state of which the most popular have been the *localization energy approximations*. The various proposals in these two categories are first reviewed for AH and compared with experiment (Secs. 11.3 and 11.4); they and their limitations are then critically examined (Sec. 11.5), and application to non-AH is considered (Sec. 11.6).

Although qualitative data on orientation in substitution, for example, can be useful for evaluating various theories, quantitative data on the relative reactivities of different positions in various compounds are of far greater value. Fortunately, a fair amount of such data has become available in recent years. One set consists of the actual basicities of aromatic hydrocarbons determined from the protonation equilibria in anhydrous hydrofluoric acid.[37] In this work the equilibrium constants were measured for the reaction:

$$ArH + HF \rightleftharpoons ArH_2^+ + F^-$$

The structure of the conjugate acid is undoubtedly that of a σ-complex; the position of the added proton was taken to be the most reactive position determined by analogous reactions. The experimental basicities cover a range of eleven powers of ten and are summarized in Table 11.1.

The related rate of hydrogen exchange has also been studied; aromatic hydrocarbons are deuterated at convenient rates in homogeneous solution in mixtures of trifluoroacetic acid-d and carbon tetrachloride containing sulfuric acid.[38] The system is complicated because more than one deuterium usually becomes introduced into several equivalent active positions as well as into others less reactive. However, the data were ingeniously analyzed to yield rate constants for each of several of the reactive positions. The identity of these positions was determined by analogy with other substitution reactions.

[35] G. W. Wheland, *Resonance in Organic Chemistry*, John Wiley and Sons, New York (1955), p. 476.

[36] W. A. Waters, *J. Chem. Soc.*, **1948**, 727.

[37] E. L. Mackor, A. Hofstra, and J. H. van der Waals, *Trans. Faraday Soc.*, **54**, 66 (1958).

[38] G. Dallinga, A. A. V. Stuart, P. J. Smit, and E. L. Mackor, *Z. Elektrochem.*, **61**, 1019 (1957).

In a typical electrophilic substitution, such as nitration, the products in principle can be isolated and identified structurally. In a series of recent investigations Dewar et al.[39] have used quantitative product determinations to evaluate relative rates of reaction of different positions within a hydrocarbon toward nitric acid in acetic anhydride. Coupling this data with results of competitive nitrations,[40] Dewar's group derived for nitration a set of *partial rate factors*, the relative reactivities of individual positions compared to a single benzene position.[41] These data are summarized in Table 11.1 as logarithms of partial rate factors relative to a 1-naphthalene position and modified to accord with more recent measurements.[42]

The available data on chlorination of aromatic hydrocarbons in acetic acid[43,44] are also summarized in Table 11.1. Relative bromination rates are available for only a few compounds.[45]

Most theoretical treatments of aromatic substitution do not give specific concern to the detailed nature of the attacking group but treat only the π-energy change within the hydrocarbon. This neglect is justified on the grounds that the attacking group is the same for the different aromatics and that the nature of the bonding to the aromatic ring at the transition state probably does not change enough to invalidate the theoretical results. The important thing, then, is the change in π-bond energy common to all electrophilic substitutions. In this event we might expect linear correlations of reactivities toward different reagents. Such linear correlations were noted for some reactions of methylbenzenes (*vide supra*). We now find linear correlations for relative reactivities in deuterodeprotonation (Fig. 11.7), nitration (Fig. 11.8), and chlorination (Fig. 11.9), plotted as logarithms versus the logarithms of relative equilibrium constants for protonation in hydrofluoric acid. The correlation lines are of varying slopes. The magnitude of the slope may be taken as a measure of the degree to which the transition state resembles the σ-complex intermediate; this point is discussed in greater detail in Sec. 11.5. For the present, we may observe that this situation resembles that found in side-chain reactions of substituted benzene derivatives which have been extensively and

[39] P. M. G. Bavin and M. J. S. Dewar, *J. Chem. Soc.*, **1956**, 164; M. J. S. Dewar and T. Mole, *J. Chem. Soc.*, **1956**, 1441; M. J. S. Dewar and E. W. T. Warford, *J. Chem. Soc.*, **1956**, 3570; M. J. S. Dewar, T. Mole, D. S. Urch, and E. W. T. Warford, *J. Chem. Soc.*, **1956**, 3572.

[40] M. J. S. Dewar, T. Mole, and E. W. T. Warford, *J. Chem. Soc.*, **1956**, 3576.

[41] M. J. S. Dewar, T. Mole, and E. W. T. Warford, *J. Chem. Soc.*, **1956**, 3581.

[42] A. Streitwieser, Jr., and R. C. Fahey, unpublished results.

[43] M. J. S. Dewar and T. Mole, *J. Chem. Soc.*, **1957**, 342 (1957).

[44] S. F. Mason, *J. Chem. Soc.*, **1959**, 1233.

[45] *Ibid.*, **1958**, 4329.

TABLE 11.1

REACTIVITIES OF AROMATIC HYDROCARBONS

Logarithms of Relative Reactivities*

Hydrocarbon	Position of Reaction	Basicity in HF, 0° Ref. 37	Deuterode-Protonation in CF₃COOD—H₂SO₄—CCl₄, 30°. Ref. 38	Nitration in Acetic Anhydride, 0°. Refs. 39, 40, 41, 42	Chlorination in Acetic Acid, 25°. Refs. 43, 44
Benzene	1			−2.10	−5.0
Naphthalene, VI	1	0	0	0	0
Anthracene, VII	2		−0.88	−0.97(−0.92)	
	1		0.70		
	2		0		
	9		3.86		
Biphenyl, VIII	2	8.1	−0.68	−1.19	} −2.1
	4	−1.7	−0.68	−1.41	
Pyrene, IX	1	6.1	2.68	1.56	3.8
1,2-Benzanthracene, X	7,12	6.6	3.35		
1,2,5,6-Dibenzanthracene, XI	7	6.5			
Perylene, XII	3	8.4	2.78	2.22	
Triphenylene, XIII	1	−0.8	−0.76	0.11	−0.7
	2		−1.25	0.11	
Chrysene, XIV	6	2.6	1.02	0.87	
Phenanthrene, XV	1	0.5		−0.11	
	2			−0.71	
	3			−0.19	
	4			−0.77	
	9			0.02	
1,2-Benzopyrene, XVI	6	0.5		2.36	1.2
Tetracene, XVII	5	11.1			
Coronene, XVIII	1	9.8		0.39	
Anthanthrene, XIX	6			2.52	
Fluoranthene, XX	3			0.46	0.4†
	8			0.25	
	7			0.08	

* Per reactive position.
† Unpublished results cited in ref. 6g.

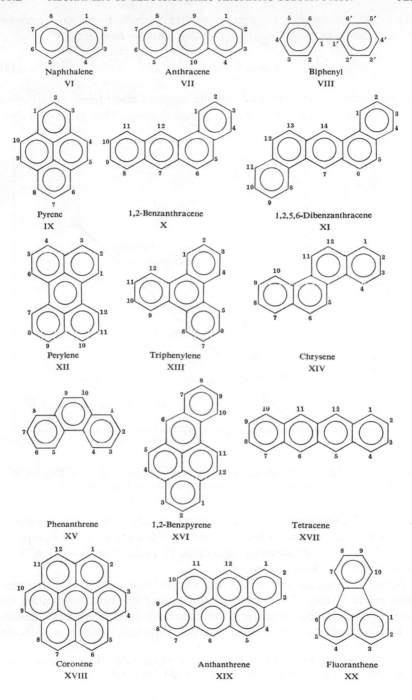

Naphthalene
VI

Anthracene
VII

Biphenyl
VIII

Pyrene
IX

1,2-Benzanthracene
X

1,2,5,6-Dibenzanthracene
XI

Perylene
XII

Triphenylene
XIII

Chrysene
XIV

Phenanthrene
XV

1,2-Benzpyrene
XVI

Tetracene
XVII

Coronene
XVIII

Anthanthrene
XIX

Fluoranthene
XX

successfully handled in Hammett's $\sigma\rho$ treatment.[46] In the same manner we may define reactivity constants σ_r and reaction constants, ρ^* by

$$\log k_r/k_{\alpha\text{-naph}} = \sigma_r\rho^* \qquad (13)$$

It would seem appropriate to define the standard electrophilic substitution rho as $\rho^* = 1$ for the protonation equilibrium; σ_r then gives directly a

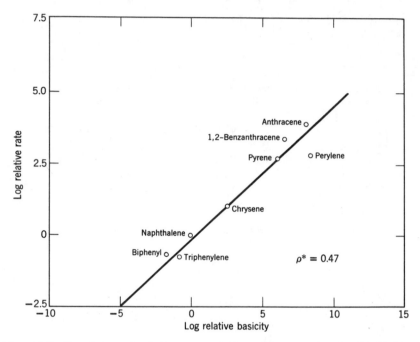

Fig. 11.7   Correlation of relative rates of deuterodeprotonation in CF_3COOD—H_2SO_4—CCl_4 versus protonation equilibrium constants in liquid HF.

measure of the relative basicity of position r in any aromatic hydrocarbon. In this way the ρ^*-values for our four reactions are simply the slopes in Figs. 11.7 to 11.9: protonation equilibria, 1; deuterodeprotonation, 0.47; nitration, 0.26; chlorination, 0.64.

Although it would seem better in principle to refer all σ_r-values to benzene, relative rates of a benzene position have not been determined for some of the reactions, and the values that have been determined are subject to substantial error. Rather, the 1-naphthalene position was chosen

[46] L. P. Hammett, *Physical Organic Chemistry*, McGraw-Hill Book Company, New York (1940); Chap. 7; H. H. Jaffé, *Chem. Rev.*, **53**, 191 (1953).

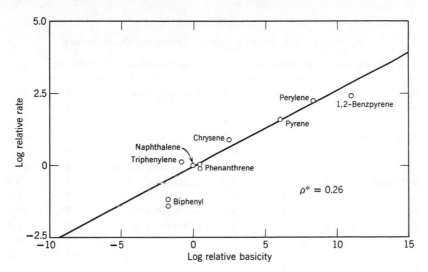

Fig. 11.8   Correlation of relative rates of nitration in acetic anhydride versus protonation equilibrium constants in liquid HF.

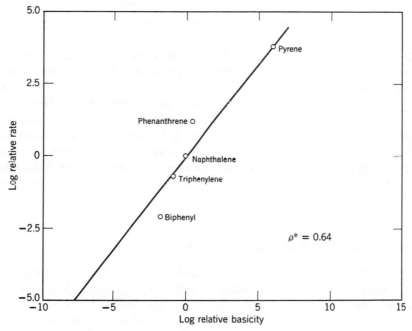

Fig. 11.9   Correlation of relative rates of chlorination in acetic acid versus protonation equilibrium constants in liquid HF.

TABLE 11.2

σ_r Values for Electrophilic Aromatic Substitution

Hydrocarbon	Position r	σ_r	Hydrocarbon	Position r	σ_r
Benzene	1	−7.8	Phenanthrene	1	−0.2
Naphthalene	1	0		2	−2.5
	2	−3.4		3	−0.5
Anthracene	1	1.1		4	−2.7
	2	0		9	0.5
	9	8.1	1,2-Benzopyrene	6	11.1
Biphenyl	2	−1.7	Tetracene	5	9.8
	4	−1.7	Coronene	1	1.7
Pyrene	1	6.1	Anthanthrene	6	10.0
1,2-Benzanthracene	7, 12	6.6	Fluoranthene	3	1.8
1,2,5,6-Dibenz-				7	0.3
anthracene	7	6.5		8	1.0
Perylene	3	8.4			
Triphenylene	1	−0.8			
	2	−2.3			
Chrysene	6	2.6			

for convenience and accuracy. Other σ_r-values may be determined directly from the protonation equilibrium at the position r; however, these values are known only for the most reactive positions. From the established ρ^*-values, the σ_r-values of other positions may be derived from the electrophilic substitutions. From the data in Table 11.1, a set of σ_r-values has been derived (Table 11.2). Such values are clearly a first approximation because of the neglect of possible varying steric effects in different reactions and because of the experimental errors in the original determinations, but they are useful for evaluating the quantitative effectiveness of various theoretical approaches.

11.3   Isolated Molecule Approximations[47]

In isolated molecule approximations the model of the transition state is generally one in which the aromatic π-system is perturbed to a relatively small degree by the attacking reagent. The tendency for reaction at different centers is determined by one of several indices defined in terms of the MO's of the original hydrocarbon.

[47] R. D. Brown, *Quart. Revs.*, **6**, 63 (1952).

The free valence concept (Sec. 2.9) is a modern version of Thiele's residual affinity. It gives numerical form to the abstraction of left-over π-bonding power. From the definition of free valence at position r, F_r (14), we see that the greater the π-bonding of position r to others, the

$$F_r = \sqrt{3} - \sum_s p_{rs} \tag{14}$$

lower the F_r-value, that is, the less π-bonding "left-over" for bonding to an attacking reagent. F_r-values for a number of aromatic hydrocarbons are summarized in Table 11.3. The quantitative relationship between these values and the experimental σ_r-values is illustrated in Fig. 11.10. We observe a rough parallelism between high F_r-values and greater reactivities, but quantitatively the correlation is only fair.[48] The average deviation of points from the line is 8% of the total range.

Qualitative theories of aromatic substitution have long been based on the idea of Coulombic attraction between the attacking positive reagent and the electron density of different positions of the aromatic ring. In HMO theory this idea becomes a correlation with q_r.[49,50] For AH, however, all $q_r = 1$, and all positions should be equally reactive to a first approximation. We may expect that the incoming charge may change the electron density at a position; the ease of this change would be reflected by the π_{rr}-value. This notion is given a quantitative derivation in Sec. 11.5; π_{rr}-values are summarized in Table 11.3. We may note the curious fact that these values resemble the F_r-values numerically to an amazing degree. Accordingly, the attempted correlation shown in Fig. 11.11 is little better than that for free valence.

Additional indices of reactivity have been proposed. In their *frontier electron theory* Fukui et al.[51] proposed the importance of the electron density of the highest bound electrons for electrophilic substitution; the most reactive position in each of a number of AH corresponds to the position of highest c_{mr}^2. The frontier electron density does not permit

---

[48] The use of F_r-values in aromatic substitution is discussed by R. Daudel, R. Jacques, M. Jean, C. Sandorfy, and C. Vroelant, *J. chim. phys.*, **46**, 249 (1949); R. Daudel, C. Sandorfy, C. Vroelant, P. Yvan, and O. Chalvet, *Bull. soc. chim. France*, **1950**, 66; F. H. Burkitt, C. A. Coulson, and H. C. Longuet-Higgins, *Trans. Faraday Soc.*, **47**, 553 (1951); B. Pullman, *Cahiers phys.*, **48**, 42 (1954); R. Daudel and O. Chalvet, *J. chim. phys.*, **53**, 943 (1956); B. Pullman and A. Pullman, in J. W. Cook's *Progress in Organic Chemistry*, Vol. 4, Butterworths Scientific Publications, London (1958) Chap. 2; N. P. Buu-Hoi, D. Lavit, and O. Chalvet, *Tetrahedron*, **8**, 7 (1960).

[49] E. Hückel, *Z. Physik.*, **72**, 310 (1931).

[50] G. W. Wheland and L. Pauling, *J. Am. Chem. Soc.*, **57**, 2086 (1935).

[51] K. Fukui, T. Yonezawa, and H. Shingu, *J. Chem. Phys.*, **20**, 722 (1952); K. Fukui, T. Yonezawa, C. Nagata, and H. Shingu, *J. Chem. Phys.*, **22**, 1433 (1954).

TABLE 11.3

QUANTITIES IN ISOLATED MOLECULE APPROXIMATIONS*

Hydrocarbon	Position	F_r	π_{rr}	S_r	Z_r
Benzene	1	0.399	0.398	0.833	1.207
Naphthalene	1	0.453	0.443	0.994	1.474
	2	0.404	0.405	0.873	1.419
Anthracene	1	0.459	0.454	1.072	1.632
	2	0.409	0.411	0.922	1.605
	9	0.520	0.526	1.312	1.677
Phenanthrene	1	0.450	0.439	0.977	1.455
	2	0.403	0.403	0.860	1.396
	3	0.408	0.409	0.893	1.447
	4	0.441	0.429	0.939	1.424
	9	0.452	0.442	0.997	1.482
Biphenyl	2	0.436	0.424	0.910	1.345
	3	0.395	0.396		
	4	0.412	0.411	0.894	1.381
Pyrene	1	0.468	0.466	1.115	1.618
	2	0.393	0.395		
	4	0.452	0.445		
Triphenylene	1	0.439	0.427	0.888	1.377
	2	0.405	0.406	0.928	1.377
Chrysene	6	0.457	0.451	1.044	1.552
Perylene	3	0.474	0.476	1.195	1.700
1,2-Benzopyrene	6	0.530	0.547	1.408	1.712
Coronene	1	0.449	0.440	0.991	1.517
Anthanthrene	6	0.531	0.557	1.504	1.769
1,2-Benzanthracene	7	0.503	0.496	1.325	1.623
	12	0.514	0.514	1.250	1.644
1,2,5,6-Dibenzanthracene	7	0.498	0.487	1.145	1.592
Tetracene	5	0.530		1.484	1.770

* F_r- and π_{rr}-values from C. A. Coulson and R. Daudel, *Dictionary of Values of Molecular Constants*, Centre de Chemie Theoriqué de France, Paris; S_r-values from refs. 51 and 52; Z_r-values from ref. 54. Additional values are available in A. Streitwieser, Jr., and J. I. Brauman, *Tables of Molecular Orbital Calculations*, Pergamon Press, New York, in press.

comparison between different hydrocarbons, but Fukui and his co-workers later defined an index called the *superdelocalizability*, S_r, defined by (15):[52]

$$S_r = 2 \sum_{j=1}^{m} \frac{c_{jr}^2}{m_j} \tag{15}$$

[52] K. Fukui, T. Yonezawa, and C. Nagata, *Bull. Chem. Soc. Japan*, **27**, 423 (1954); K. Fukui, T. Yonezawa, and C. Nagata, *J. Chem. Phys.*, **27**, 1247 (1957).

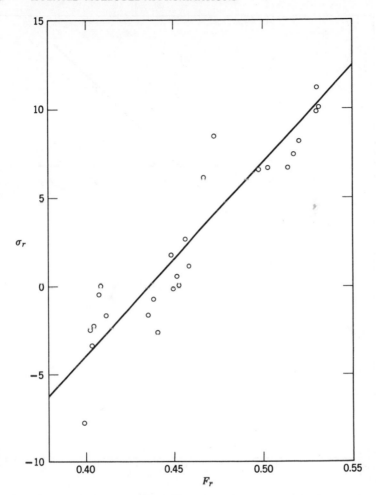

Fig. 11.10   Relation between MO F_r-values and experimental σ_r-values.

This index was derived by application of perturbation theory to a model in which the incoming group forms a weak π-bond to atom r of an otherwise unmodified π-system. S_r-values are summarized in Table 11.3. As portrayed in Fig. 11.12, the correlation with reactivities is about as good as the other reactivity indices considered so far.

The Fukui group also suggested the use of an approximate superdelocalizability, S_r', based on the use of the highest occupied MO only, since this term is usually the dominant term in (15). This proposal clearly relates the model used to the frontier electron density; however, the other terms

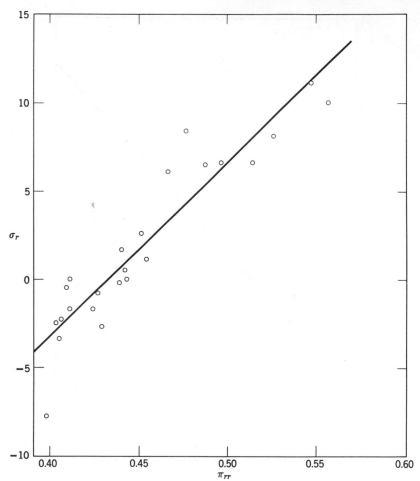

Fig. 11.11   Correlation of experimental σ_r-values with π_{rr}.

in the summation are frequently almost as important, and the S_r' approximation is a crude one.[53]

Also related closely to the frontier electron theory are the several theories based on a charge transfer model of the transition state.[33,34,54]

[53] The frontier electron theory has stimulated a substantial amount of criticism and controversy; for example, see the recent commentary of K. Fukui, T. Yonezawa, and C. Nagata, *J. Chem. Phys.*, **31**, 550 (1959); B. Pullman, *J. Chem. Phys.*, **31**, 551 (1959); H. H. Greenwood, *J. Chem. Phys.*, **31**, 552 (1959); S. S. Sung, O. Chalvet, and R. Daudel, *J. Chem. Phys.*, **31**, 553 (1959).

[54] R. D. Brown, *J. Chem. Soc.*, **1959**, 2232.

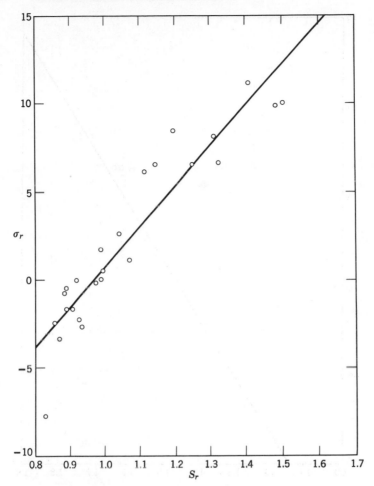

Fig. 11.12   Correlation of S_r-values with experimental reactivities.

In this model the transition state is stabilized by transfer of electron density from the aromatic molecule to the electrophilic reagent without other modification of the π-systems. Brown's perturbation treatment of such a model leads to (16) for electron transfer from the highest occupied or frontier orbital only; Z is the corresponding reactivity index.[54]

$$Z_r = Y_e - m_m + \frac{2g_e^{\ddagger 2} c_{mr}^2}{Y_e - m_m} \tag{16}$$

The two additional parameters required are Y_e, the HMO energy coefficient of the lowest vacant MO of the electrophilic reagent (m_m is

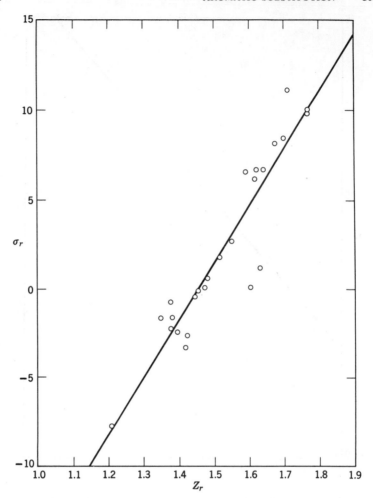

Fig. 11.13   Correlation of σ_r with Brown's Z_r-values.

the energy coefficient of the highest occupied MO of the aromatic), and $g_e^{\ddagger}$, a measure of the extent of reaction at the transition state. The similarity of the functional form to S_r' is apparent. With assignment of reasonable values to Y_e and $g_e^{\ddagger}$ (2 and 0.61, respectively) for the best fit, Brown demonstrated a good correlation with Dewar's nitration data. These Z-values are reproduced in Table 11.3. Figure 11.13 shows a good correlation with σ_r-numbers. However, the average deviation of points from the line of 6% of the total range is still rather high considering that two additional disposable parameters have been employed.

11.4   Localization Approximations

In 1942 Wheland[55] proposed the use of a reactivity index which has since become known as a *localization energy*.[47] Throughout the remainder of this book we discuss various localization energies; the present quantity may be called a *cation localization energy*, L_r^+, and is defined as the π-bonding energy required to isolate two electrons at position r from the remainder of the π-network. Such isolation results in a new π-system of one less atom and two less electrons. If the π-energy of the original system is $n\alpha + M\beta$, that of the localized system is $(n - 2)\alpha + M_r^+ \beta$ and the cation localization energy is given as

$$L_r^+ = M - M_r^+ \tag{17}$$

Since the π-bonding energy in the smaller system is often less than that of the original system, L_r^+ is usually a positive quantity.[56]

Example. Benzene has HMO π-energy of $6\alpha + 8.000\beta$. Localization of an electron pair at position 1 leaves a pentadienyl cation whose HMO π-energy is $4\alpha + 5.464\beta$. L_r^+ for benzene is $8.000 - 5.464 = 2.536$.

The localization energy is a measure of the π-energy change between the aromatic hydrocarbon and the σ-complex intermediate in aromatic substitution. If all σ-bond changes are treated as being effectively constant in this transformation, we anticipate a correlation between the calculated localization energies and the experimental reactivities. L_r^+-values are summarized in Table 11.4; the correlation with reactivities is portrayed in Fig. 11.14. The average deviation of points from the line is about the same as with Brown's Z_r-values.

Dewar's approximation method for the conjugation energy between two odd-AH units (p. 109) applies with particular effectiveness to the facile calculation of approximate localization energies,[41,57] for the localization energy is equivalent to the conjugation energy between a single atom attached to two positions of the residual molecule which is an odd-AH. The approximate localization energy is called the *reactivity number*, N_r. We consider that the position r to be localized is attached to positions s and t in the residual cation. Extension of (4.23) gives (18).

$$N_r = 2(a_{0s} + a_{0t})\beta \tag{18}$$

[55] G. W. Wheland, *J. Am. Chem. Soc.*, **64**, 900 (1942).

[56] As defined above, L_r^+ is a dimensionless number. The usual definition in the literature retains the units of β and is the negative of the number used here.

[57] M. J. S. Dewar, *J. Am. Chem. Soc.*, **74**, 3357 (1952); M. J. S. Dewar, *Record Chem. Prog.*, **19**, 1 (1958).

TABLE 11.4

CATION LOCALIZATION ENERGIES

Hydrocarbon	Position	L_r^+	N_r	L_r^ω
Benzene	1	2.536	2.31	1.603
Naphthalene	1	2.299	1.81	1.305
	2	2.480	2.12	1.519
Anthracene	1	2.25	1.57	1.23
	2	2.40	1.89	1.44
	9	2.013	1.26	0.91
Phenanthrene	1	2.318	1.86	1.247
	2	2.498	2.18	1.469
	3	2.454	2.04	1.375
	4	2.366	1.96	1.343
	9	2.299	1.80	1.347
Biphenyl	2	2.400	2.07	1.319
	3	2.544	2.31	
	4	2.447	2.07	1.366
Pyrene	1	2.190	1.51	0.990
	2	2.55	2.31	
	4	2.28	1.68	
Triphenylene	1	2.378	2.00	1.311
	2	2.477	2.12	1.398
Chrysene	6	2.251	1.67	1.166
Perylene	3	2.140	1.33	0.935
1,2-Benzopyrene	6	1.961	1.15	0.751
Coronene	1	2.306	1.80	1.190
Anthanthrene	6	1.928	1.03	0.733
1,2-Benzanthracene	7	2.101	1.43	1.02
	12	2.049	1.35	0.91
1,2,5,6-Dibenzanthracene	7	2.131	1.51	0.97
Tetracene	5	1.930	1.02	0.82

The coefficients of positions s and t are a_{0s} and a_{0t}, respectively, in the NBMO of the residual cation produced when position r is isolated or localized in the original hydrocarbon. These coefficients are evaluated by the simple procedure on p. 54.

Examples. Benzene

$$N_r = 2\left(\frac{1}{\sqrt{3}} + \frac{1}{\sqrt{3}}\right) = 2.31$$

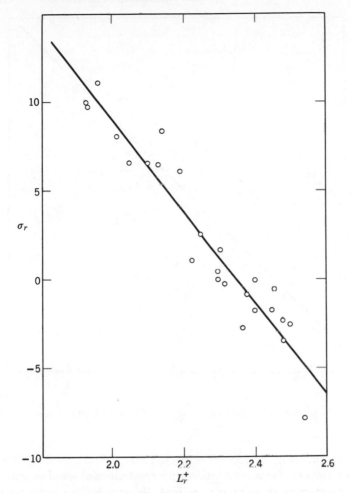

Fig. 11.14   Cation localization energies and experimental reactivities.

Naphthalene-1

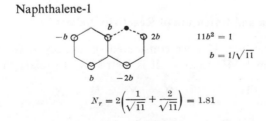

A list of N_r-values is given in Table 11.4. N_r-values are smaller in magnitude than the L_r^+-values, but there is an excellent linear correlation

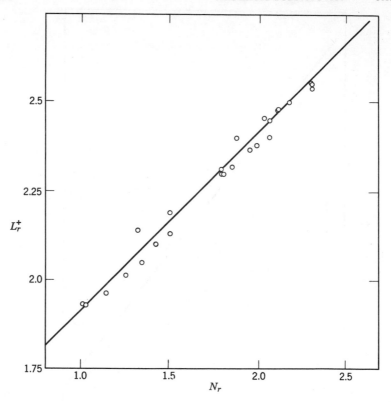

Fig. 11.15   Comparison of localization energies, L_r^+, with Dewar's reactivity numbers, N_r.

between the two (Fig. 11.15). As expected from this correlation, Dewar's reactivity numbers give a correlation with reactivities as good as that with localization energies (Fig. 11.16); indeed, this correlation is even a little better.

11.5   Comparison and Extension of Reactivity Indices

In Sec. 11.3 and Sec. 11.4 we considered six indices of reactivity that were derived from HMO theory. It is instructive to relate these indices

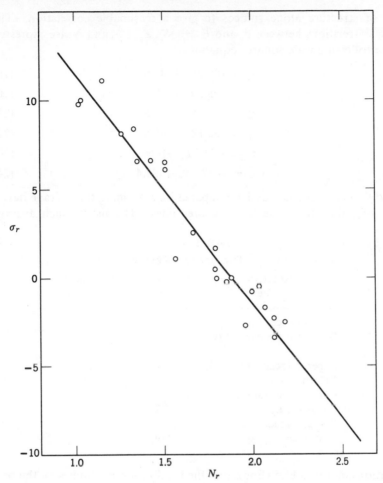

Fig. 11.16   Dewar's reactivity numbers, N_r, and experimental reactivities.

to resonance theory. The structures **XXI** contribute to the resonance hybrid of the substitution transition state.

In **XXI**a the π-electron system is little perturbed. Tendency for reaction according to this structure alone is determined by the free valence F_r or by atom polarizability π_{rr}. **XXI**b is the charge-transfer structure whose importance has been considered only in recent years. Tendency for reaction according to this structure alone is measured by the super-delocalizability S_r or by Brown's index Z_r. **XXI**c is the localization structure whose relative energy is given by the cation localization energy L_r^+ or by Dewar's reactivity number N_r. For AH consideration of any one type of structure alone suffices to give a reasonable correlation. The linear correlations between σ_r and F_r, π_{rr}, S_r, Z_r, L_r^+, and N_r are expressed in the following least squares equations:

$$\sigma_r = 108.9F_r - 47.6 \tag{19}$$

$$\sigma_r = 98.5\pi_{rr} - 42.7 \tag{20}$$

$$\sigma_r = 23.1S_r - 22.5 \tag{21}$$

$$\sigma_r = 32.1Z_r - 46.8 \tag{22}$$

$$\sigma_r = -25.7L_r^+ + 60.4 \tag{23}$$

$$\sigma_r = -12.9N_r + 24.4 \tag{24}$$

The different methods may be compared by examining the average deviations of points from the least squares lines. The list of such average

TABLE 11.5

COMPARISON OF THEORETICAL REACTIVITY INDICES

Reactivity Index	Correlation Figure	Average Deviation
Free valence, F_r	8	1.5
Self-atom polarizability, π_{rr}	9	1.4
Super-delocalizability, S_r	10	1.3
Brown's index, Z_r	11	1.2
Cation localization energy, L_r^+	12	1.1
Dewar's reactivity number, N_r	14	1.0

deviations in Table 11.5 shows that the localization methods give the best correlations but not by any overwhelming margins.[58]

In truth, the various theoretical indices of reaction are not completely independent. We know already that Dewar's reactivity numbers are a perturbation approximation to localization energies. Other interrelationships have been demonstrated. Coulson and Longuet-Higgins[59] consider that the approaching reagent alters α of the attacked atom and the β's of the bonds to the attacked atom. The π-energy change is then given as

$$\delta E_\pi = 2 \sum_s p_{rs}\delta\beta_{rs} + q_r\delta\alpha_r + \tfrac{1}{2}\pi_{rr}\overline{\delta\alpha_r^2} + \cdots \tag{25}$$

[58] A similar comparison has been made by S. Sung, O. Chalvet, and R. Daudel, *J. chim. phys.*, **57**, 31 (1960).

[59] C. A. Coulson and H. C. Longuet-Higgins, *Proc. Roy. Soc.*, **A191**, 39 (1947).

If β_{rs} goes to zero, the δE_{π} is the localization energy; for such a large $\delta\beta$ change a number of terms should be included. If $\delta\beta_{rs}$ is small and if $\delta\alpha_r$ is neglected, the first term alone suffices; from the definition of free valence, (25) becomes

$$\delta E_{\pi} \cong 2(\sqrt{3} - F_r)\delta\beta_{rs} \tag{26}$$

The change in α_r produces the second term in (25). Since all $q_r = 1$ in AH, this term is a constant, and any variation among such compounds must result from the third term. Hence we may write for reaction with AH

$$\delta E_{\pi} = c + bF_r + a\pi_{rr} \tag{27}$$

The near equality of the actual values of F_r and π_{rr} results in a correlation with either one; such correlation results from the first and dominant terms of an expansion formula for localization energies.[60]

For AH the further interdependence of the charge-transfer quantities, super-delocalizability and frontier electron density, with localization energies has also been derived.[61] For such cases the question which reactivity index is the fundamental loses a good deal of its meaning, although a good case can probably be made for localization energy. It is significant, too, that the localization methods provide the best correlations. However, even the localized model is only an approximation of the π-system of the transition state or σ-complex intermediate, for the reacting center is transformed at least partially into a tetrahedral center that can hyperconjugate with the remaining cationic π-system. Several studies in which such hyperconjugation has been included have shown that no significant change results; that is, localization energies without hyperconjugation correlate well with localization energies including hyperconjugation.[62,63]

In the calculation of the cation localization energy the difference is taken of the π-energies of two separate π-systems, one of which is neutral and the other cationic. In some previous examples the comparison of cations with neutral systems gave limited success; *for example* ionization potentials (cf. Chap. 12). We may enquire whether the limitations are due in

[60] For further discussion of the interrelations of these quantities see M. J. S. Dewar, *J. Am. Chem. Soc.*, **74**, 3355 (1952); H. H. Greenwood, *Trans. Faraday Soc.*, **48**, 585 (1952); O. Chalvet, *Ann. Chim.*, **9**, 97 (1954); H. Baba, *Bull. Chem. Soc. Japan*, **30**, 147 (1957); refs. 47 and 48.

[61] K. Fukui, T. Yonezawa, and C. Nagata, *J. Chem. Phys.*, **26**, 831 (1957); I. Samuel, *Comp. rend.*, **249**, 1893 (1959). The relationship between S_r- and Z_r-values has also been demonstrated: K. Fukui, K. Morokuma, T. Yonezawa, and C. Nagata, *J. Chem. Phys.*, **32**, 1743 (1960).

[62] E. L. Mackor, A. Hofstra, and J. H. van der Waals, *Trans. Faraday Soc.*, **54**, 66 (1958).

[63] R. Daudel and O. Chalvet, *J. chim. phys.*, **1956**, 943.

part to the same reasons, namely, neglect of electronic repulsion terms. Dallinga et al.[38] have reported the results of SCF calculations of the localized cations; an excellent correlation is found with experimental rates of deuterodeprotonation.

The applicability of the ω-technique has also been suggested.[62] Ideally, this method should be applied to the σ-complexes with inclusion of hyperconjugation and with iterations taken to self-consistency. No such calculations have been reported. Alternatively, a simple appoximation method is readily applicable. According to the ω-technique (Sec. 4.5), a charge, ζ_r, on atom r results in a change in α of $\omega\zeta_r\beta$. From the first approximation of the change in energy from change in α (4.29), we have

$$\Delta E_\omega^+ = \sum_r q_r\,\delta\alpha_r = \sum_r \omega(1 - \zeta_r)\zeta_r\beta \qquad (28)$$

Since the total charge is unity,

$$\Delta E_\omega^+ = \omega\beta - \omega\sum_r \zeta_r^2\beta \qquad (29)$$

The total π-energy change may be decomposed to a localization energy as before, plus a correction term to the carbonium ion energy, ΔE_ω^+, because of the effect of charge on Coulombic attraction:

$$L_r^\omega = M - (M^+ + \Delta E_\omega^+) = L_r^+ + \omega\sum_s \zeta_s^2 - \omega \qquad (30)$$

This expression constitutes the definition of what may be called an omega localization energy, L_r^ω, and may be readily evaluated with $\omega = 1.4$ because ζ_s-values are rapidly obtained for odd-AH systems. Values calculated with (30) are listed in Table 11.4; the correlation with experimental reactivities is shown in Fig. 11.17. The results are encouraging; the average deviation is now less than 0.8 σ_r-units, a value less than the probable error of some of the experimental quantities.

The added term has the effect of giving greater stability to the conjugate acids that possess a perinaphthenyl cation nucleus, XXII (cf. Sec. 12.1). Examples are the pyrene-1 and perylene-3 positions.

XXII

The slope of the correlation line between σ_r and L_r^+ corresponds to an effective β for the basicity equilibrium of -35 kcal./mole. Clearly, if partial rate factors for other reactions are plotted against L_r^+, straight lines would develop with slopes corresponding to smaller values of β; the effective values of β are in fact given by $-35\rho^*$ kcal./mole. Such variable slopes have been recognized[41,43,44,45] and were interpreted in terms of the

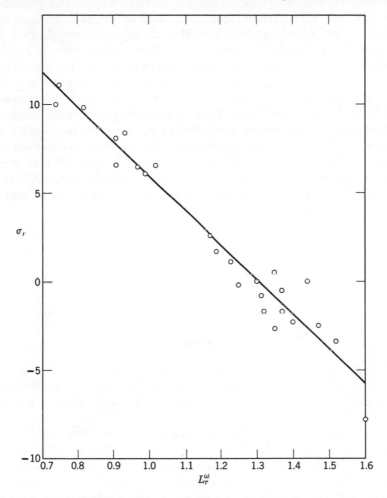

Fig. 11.17   Correlation of omega localization energies with experimental reactivities.

relative structures of transition states much as in Sec. 11.2. Dewar[41] has added an approximate MO derivation that justifies this interpretation. The value of -35 kcal./mole (-1.5 ev) is less than the value of β_0 indicated by other correlations; however, even the conjugate acid or σ-complex intermediate is not completely localized; hyperconjugation with the methylene group is doubtless important and will raise this magnitude.

11.6   Applications to Other Hydrocarbons

Although quantitative data on relative reactivities are not available for other alternant hydrocarbons, qualitative data on orientation effects are plentiful. The actual prediction of this chemistry is complicated by potential effects of steric hindrance, which we have so far not touched upon. Two types of steric hindrance are of primary importance. The first results from steric strains within a planar molecule that assist the bending of a bond during reaction. In perylene, for example, the deuterodeprotonation rates[38] and the basicities[27] of the 1- and 3-positions are identical, although reactivity indices predict the 3-position to be the more reactive ($N_r =$ values: 1, 1.45; 3, 1.33). The 1-position has the same steric environment of the 4,5-positions of phenanthrene. It has been suggested[64] that steric crowding between these hydrogens in the reacting hydrocarbon XXIII is relieved when one of the hydrogens is displaced from the plane during reaction, thus producing an additional driving force for reaction.

XXIII

A similar argument applies to the comparison of the 7- and 12-positions in 1,2-benzanthracene, X. In this system the effect has been amplified by appropriate methyl substituents.[64] This type of steric effect is a *steric acceleration*.

The second important steric hindrance effect results from the bulk of the attacking reagent. Attack by a bulky reagent may be impeded at a carbon too close to other positions. This effect of *steric retardation* is frequently exemplified in many examples of Friedel-Crafts acylations. The frequent tendency for acylation of the 2-naphthalene position rather than the 1-naphthalene position is well known and is usually attributed to the bulk of the complex electrophilic reagent.

These problems will be evidenced by the failure of an attempted correlation between reactivity and basicity of different positions. In the absence of such quantitative comparisons the organic chemist must still, as he

[64] G. Dallinga, P. J. Smit, and E. L. Mackor, in G. W. Gray's *Steric Effects in Conjugated Systems*, Butterworths Scientific Publications, London (1958), Chap. 13.

has in the past, temper his calculated predictions with intuition and analogy. These considerations apply especially to various kinds of *ortho*-positions.[65]

Even with these complications, the theory is still rather useful. Dewar's reactivity numbers are especially convenient for AH because of their ease of calculation. For example, only a few minutes are required to determine the N_r-values for all of the positions in 1,2-benzpyrene, XXIV. The

XXIV

6-position is clearly the most basic position by far, but this position is a hindered position of the 9-anthracene type, which does not generally react with succinic anhydride and aluminum chloride. The next most active positions are 1- and 3-positions. Both are similar positions sterically and the 1-position is predicted to be the more reactive. When 1,2-benz-pyrene is allowed to react with succinic anhydride and aluminum chloride, the product of acylation at the 1-position is isolated.[66] Gore[67] has published reactivity numbers for several large polycyclic hydrocarbons to supplement the variety of hydrocarbons treated by Dewar.[57]

Biphenylene, XXV, constitutes an even more impressive example, for no

XXV

other cyclobutane compounds serve as suitable analogies for application of resonance theory, and the 1–8 positions are relatively far apart so that steric hindrance problems should be absent. The Dewar reactivity numbers in XXV show that the 2-position is without question the more reactive. This preference is confirmed by the atom polarizabilities ($\pi_{1,1} = 0.419$; $\pi_{2,2} = 0.443$)[68] and the cation localization energies ($L_1^+ = 2.403$; $L_2^+ = 2.352$).[68,69] The conclusion is unaltered when the ω-technique

[65] For a further discussion cf. R. D. Brown, *J. Am. Chem. Soc.*, **75**, 4077 (1953).

[66] N. P. Buu-Hoi and D. Lavit, *Tetrahedron*, **8**, 1 (1960).

[67] P. H. Gore, *J. Chem. Soc.*, **1954**, 3166.

[68] R. D. Brown, *Trans. Faraday Soc.*, **46**, 146 (1950).

[69] F. H. Burkitt, C. A. Coulson, and H. C. Longuet-Higgins, *Trans. Faraday Soc.*, **47**, 553 (1951).

approximation is used.[70] Significantly, the free valence index makes the opposite prediction: $F_1 = 0.428$, $F_2 = 0.420$.[71]

In fact, all substitution reactions—nitration, halogenation, acylation, mercuration, etc.—occur exclusively in the 2-position.[72] Moreover, the localization methods predict a reactivity similar to naphthalene; the experimental conditions are not inconsistent with such a prediction. This example shows that the free valence may be the least trustworthy of the reactivity indices.

Unfortunately, the theory is not as general for non-AH. The nonuniform charge distribution in these hydrocarbons produces variations in the $q_r \delta \alpha_r$ term in (25), and the interdependence of the correlations may be expected

XXVI

to break down. Furthermore, qualitative orientation data are available for only relatively few such hydrocarbons. Two compounds for which some information is available are fluoranthene, XX, and azulene, XXVI. The various reactivity indices for these hydrocarbons are summarized in Table 11.6.

The various reactivity indices for fluoranthene show a notable disagreement in predictions. The charge-transfer indices predict that the 7-position is at least as reactive as the 3-position. The electron densities point to the 8-position as the most reactive with the 3-position far behind. On the other hand, the localization energies give the 3-position as the most reactive and the 8-position as having an intermediate value. Brown[47] has considered that the electron densities are a primary index if the transition state occurs early in the reaction, that is, if the transition state resembles the initial reactants. On the basis of (25), $\delta \beta_{rs}$ would then be small and the $q_r \delta \alpha_r$ term would be dominating. However, localization energy would be a primary index if the transition state occurred late in the reaction; that is, if the transition state resembled the intermediate. According to this

[70] $L_1^\omega = 1.358$; $L_2^\omega = 1.315$.

[71] R. D. Brown, *Trans. Faraday Soc.*, **45**, 296 (1949); J. D. Roberts, A. Streitwieser, Jr, and C. M. Regan, *J. Am. Chem. Soc.*, **74**, 4579 (1952); J. I. F. Alonso and F. Peradejordi, *Anal. real soc. españ. fis. y quim. (Madrid)*, **50B**, 253 (1954).

[72] W. Baker, M. P. V. Boarland, and J. F. W. McOmie, *J. Chem. Soc.*, **1954**, 1476; W. Baker, J. W. Barton, and J. F. W. McOmie, *J. Chem. Soc.*, **1958**, 2658, 2666; W. Baker, J. F. W. McOmie, D. R. Preston, and V. Rogers, *J. Chem. Soc.*, **1960**, 414.

view, the proper index for non-AH depends on the position of the transition state along the reaction coordinate (cf. Fig. 11.18). A given position might be more reactive than another toward reagents that give "early" transition states and less reactive in reactions with more localized or "late" transition states. The relative reactivities then *cross over* as a function of the reaction.

TABLE 11.6

REACTIVITY INDICES FOR FLUORANTHENE AND AZULENE

Position	F_r	q_r	π_{rr}	L_r^+	S_r	Z_r	L_r^ω
			Fluoranthene*				
1	0.453	0.947	0.440	2.466	0.818	1.407	1.238
2	0.398	1.005	0.400	2.503	0.860	1.407	1.389
3	0.470	0.959	0.462	2.341	0.930	1.447	1.142
7	0.438	0.997	0.427	2.371	0.936	1.447	1.167
8	0.409	1.008	0.410	2.435	0.872	1.407	1.218
			Azulene*†				
1	0.480	1.173	0.425	1.924	1.634	1.667	0.598
2	0.420	1.047	0.419	2.362	0.962	1.523	1.072
4	0.482	0.855	0.438	2.551	0.674	1.536	1.425
5	0.429	0.986	0.429	2.341	0.980	1.578	1.097
6	0.454	0.870	0.424	2.930	0.652	1.523	1.525

* Calculations by the author.
† Ref. 75.

Non-AH cases for which q_r and L_r^+ make similar predictions are said to be *noncrossing*. The predictions in this case do not depend on the detailed free-energy reaction profile and are likely to be relatively valid. On the other hand, when different predictions are made, the noncrossing rule is said to be violated and correlations are not reliable. Fluoranthene clearly violates the noncrossing rule. Experimentally, the 3-position is the most reactive and the 8-position is second.[73] Nitration in acetic anhydride gives, in addition, smaller amounts of the 1- and 7-nitrofluoranthenes.[74]

Both q_r and L_r^+ predict the azulene 1-position to be by far the most reactive. The noncrossing rule is obeyed and the prediction should be valid. Note that free valence and polarizability predict otherwise. In fact, the 1-position is by far the most reactive. The low L_r^+- and L_r^ω-values, 1.924 and 0.598, respectively, indicate that this position should be rather basic.

[73] J. von Braun and G. Manz, *Ann.*, **488,** 111 (1931); **496,** 170 (1932); S. H. Tucker and M. Whalley, *Chem. Rev.*, **50,** 483 (1952).
[74] A. Streitwieser, Jr., and R. C. Fahey, unpublished results.

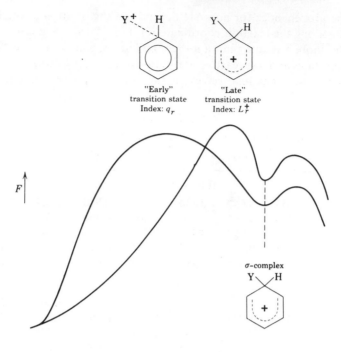

Fig. 11.18   Idealized free energy reaction profiles that exemplify the application of different MO reactivity indices.

And it is; azulene can be extracted into moderately strong sulfuric acid as the cation, XXVIII, and reprecipitated on dilution.[75] Note that XXVII is a tropylium cation derivative (Sec. 10.6).

XXVII

From this discussion it would seem that considerations of electron density and localization energy should provide at least qualitative indications of relative reactivities. Unfortunately, Dewar's simple approximation

[75] For a recent review of azulene chemistry see E. Heilbronner in D. Ginsburg's *Non-Benzenoid Aromatic Compounds*, Interscience Publishers, New York (1959), Chap. 5.

method cannot be applied to non-AH compounds and the localization
energies are available only by individual calculation. Relatively few such
values have been published. Furthermore, relatively few additional experimental data are available with which to test the calculations. For example,
pleiadiene, XXVIII, and acepleiadylene, XXIX, give mono-nitro products
on nitration but the positions of nitration are unknown.[76] The electron

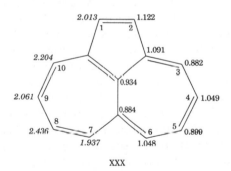

densities shown in XXVIII and XXIX indicate that the reactive positions
should be the 5- and 7-positions, respectively,[77] although Brown's Z_r-index
gives the 7-position as the active position in XXVIII.[54] Unfortunately, the
localization energies are not reported.

Fig. 11.19   Electron densities and cation localization energies (italics) for aceheptylene,
XXX.

The 3,5-dimethyl derivative of aceheptylene, XXX, has been synthesized.[78] The hydrocarbon reacts with dimethylformamide and phos-
phorous oxychloride to yield an aldehyde (Vilsmeier reaction) whose
structure has not been established. The electron densities and localization
energies of the parent hydrocarbon as indicated in Fig. 11.19 point to the

[76] V. Boekelheide and G. K. Vick, *J. Am. Chem. Soc.*, **78**, 653, (1956).
[77] B. Pullman, A. Pullman, G. Berthier and J. Pontis, *J. chim. phys.*, **49**, 20 (1952).
[78] K. Hafner and J. Schneider, *Ann.*, **624**, 37 (1959).

1- and 6-positions as the most reactive; analogy with fluoranthene suggests, that the latter should be the more reactive.[79]

Addition reactions of olefins commonly proceed in two steps, of which the first is an electrophilic reaction similar to the first step in aromatic substitution. The intermediate carbonium ion generally does not liberate a proton because the system does not gain the resonance stabilization characteristic of the corresponding aromatic reaction but rather reacts with a nucleophilic reagent to give net addition:

$$R_2C{=}CR_2 + Y^{\oplus} \longrightarrow R_2\overset{\displaystyle Y}{\underset{\displaystyle |}{C}}{-}\overset{\oplus}{C}R_2 \overset{X^{\ominus}}{\longrightarrow} R_2\overset{\displaystyle Y}{\underset{\displaystyle |}{C}}{-}\overset{\displaystyle X}{\underset{\displaystyle |}{C}}R_2$$

The various MO reactivity indices may be expected to apply in an obvious manner to such reactions of families of olefins. Relatively few correlations have been attempted, but these have given promising results.[80]

11.7   Heteroatom Compounds

The reactions of heterosubstituted and heterocyclic compounds provide an important and formidable problem for even qualitative molecular orbital interpretation. These compounds are nonalternant and the problem even arises as to which reactivity index to use. The introduction of heteroatoms requires additional parameters. Two types of approach have been treated in the literature. On the one hand, we can determine whether MO theory can account for the principal aspects of heteroatom orientations. Thus, in the first successful attack on this problem, Wheland and Pauling[81] demonstrated that in substituted benzene derivatives there are parameter values of not unreasonable magnitude which give electron density distributions consistent with observed orientation phenomena. Later, Wheland[82] showed that suitable parameter values give rise to localization energies also consistent with qualitative orientation behavior.

In modern organic chemical theory the effects of substituents have been interpreted on the basis of conjugative and inductive effects. A variation in the conjugative power of a substituent has an obvious counterpart in the relative value of k_{C-X} for its bond to the aromatic ring. Coupled with a suitable value for h_X, the two MO parameters reproduce all trends

[79] A. Streitwieser, Jr., and J. B. Bush, unpublished calculations.
[80] For example, cf. S. Sato and R. J. Cvetanovic, *J. Am. Chem. Soc.*, **81**, 3223 (1959).
[81] G. W. Wheland and L. Pauling, *J. Am. Chem. Soc.*, **57**, 2086 (1935).
[82] G. W. Wheland, *J. Am. Chem. Soc.*, **64**, 900 (1942).

commonly associated with electromeric effects of electron-attracting or electron-donating substituents.[83,84,85]

Attempts have been made to reproduce within an MO framework the traditional consequences of an inductive effect, a polarization induced along bonds by an electronegative or electropositive substituent:

$$C \rightarrow C \rightarrow C \rightarrow X \qquad \text{or} \qquad C \leftarrow C \leftarrow C \leftarrow X$$

The assumed MO counterpart has been the auxiliary inductive parameter, which in different applications assigns an h-value to the carbon attached to the substituent and perhaps also to successive carbons down a chain (Sec. 5.6). These techniques can reproduce most of the qualitative orientational effects of substituents, but their quantitative or even semiquantitative performance is not at all impressive. The procedure necessarily neglects completely the electrostatic charge-dipole and dipole-dipole interactions across space—the field effect—which alone can account at least qualitatively for all of the salient resultants usually associated with the inductive effect. The inability of the theory to handle quantitatively an electrostatic interaction foreign to its framework constitutes a severe limitation in application to reactivities of heteroatom systems. The argument is exemplified in the discussions that follow.

Sandorfy et al.[86] have considered the use of F_r and π_{rr} as reactivity indices for heteroatom systems, but on the basis of other results with non-AH (Sec. 11.5) it would seem that attention could be restricted to electron densities and localization energies. Most treatments in the literature use one or the other or both of these indices.

The qualitative electromeric consequences of the various parameters are given in simple fashion by approximation techniques. Replacement of the carbon of an aromatic hydrocarbon by a heteroatom such as nitrogen to give a heterocycle or the attachment to the carbon of a purely inductive substituent such as an ammonium group is treated simply by altering the α of the affected carbon. The effect that this alteration has on electron densities elsewhere in the molecule is determined by the atom polarizabilities. In an even-AH a positive h_r, if r is a starred position, produces in the main a decrease in electron density of every unstarred position.[83] Thus the type of structural change mentioned decreases the reactivity of *ortho-* and *para*-positions to electrophilic attack. To a first approximation, the other starred positions are little altered; if an auxiliary inductive parameter is

[83] M. J. S. Dewar, *J. Am. Chem. Soc.*, **74**, 3350 (1952).

[84] H. H. Jaffe, *J. Am. Chem. Soc.*, **77**, 274 (1955).

[85] C. Sandorfy, *Can. J. Chem.*, **36**, 1739 (1958).

[86] C. Sandorfy, C. Vroelant, P. Yvan, O. Chalvet, and R. Daudel, *Bull. soc. chim. France*, **1950**, 304.

used to produce a $\Delta\alpha$ at the adjacent carbon, all positions suffer diminished electron density, the other starred positions less than the unstarred. Alternatively, electrostatic interaction with the dipole or charge of such a substituent may superimpose enhanced or diminished reactivity at all positions in the ring.

The effect of an inductive change on localization energy is also approximated simply, at least for alternant compounds. We need only the first approximation (30), since electron densities in the starting AH are unity everywhere and are readily evaluated for the localized odd-AH cation.

$$\Delta E = q_r \, \Delta\alpha_r = q_r h_r \tag{30}$$

To this approximation, an inductive change at a starred position of the starting AH causes no change in localization energy of any other starred position but alters the localization energy of an unstarred position by an amount dependent on the charge density at that position in the localized cation.

An example provided by Longuet-Higgins[87] is instructive. We consider the effect of methyl substitution by using the inductive model (Sec. 5.7) on the localization energy for reaction at the 9-position of anthracene. The charge densities of the localized cation are indicated in XXXI.

XXXI

To the first approximation, methyl substitution at the 1- and 3-positions should have no effect; methyl groups should have an equal rate-enhancing effect at the 2- and 4-positions and a much more powerful effect at the 10-position. Mackor et al.[88] have shown how this simple technique quantitatively accounts for the basicities of various methyl-substituted aromatic hydrocarbons.

The conjugation of many substituents with an aromatic ring must be explicitly considered. Some, such as halogen, hydroxy, and amino, have electron pairs to donate to the ring π-system. A simple model is the —CH_2^- group;[89] that is, $h_X = 0$, $h_{CX} = 1$ in which X is an appropriate united atom. Similarly, electron-attracting conjugative groups, such as nitro, cyano, and ester, may be simply modeled by a —CH_2^+ group.[89]

[87] H. C. Longuet-Higgins, *J. Chem. Phys.*, **18**, 283 (1950).

[88] E. L. Mackor, G. Dallinga, J. H. Kruizinga, and A. Hofstra, *Rec. trav. chim.*, **75**, 836 (1956); G. Dallinga, P. J. Smit, and E. L. Mackor, *Mol. Phys.*, **3**, 130 (1960).

[89] J. D. Roberts and A. Streitwieser, Jr., *J. Am. Chem. Soc.*, **74**, 3357 (1952).

Hence AH's with substituents become odd-AH cations or anions, in which the substituent united atom is a starred position. Clearly, a cationic substituent will lower the electron density of all other starred positions; in this approximation it has no effect on unstarred positions. Direct calculations of typical examples show a similar pattern of changes in localization energies.[89] In the next approximation the united atom models are given an appropriate change in α. An increase in the electronegativity (positive h), together with a corresponding inductive alteration in the α of the attached carbon, results in lowering the electron density at all positions, the effect being greater at the other starred positions.[83] Such is the gross effect actually observed with nitro groups.

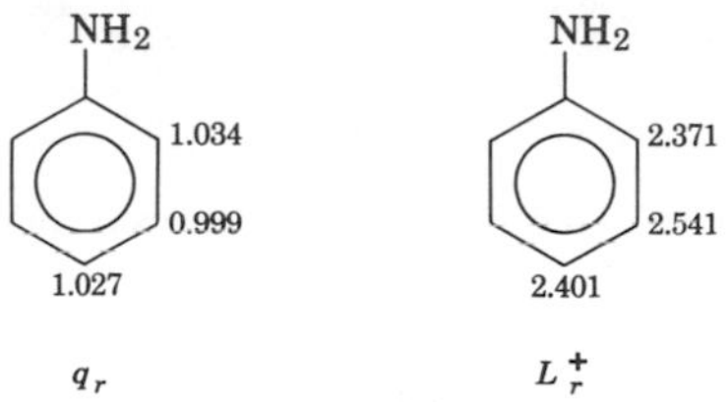

Fig. 11.20   Electron densities and localization energies in aniline with $h_N = 1.5$, $h' = 0.1$, $k_{C-N} = 0.8$ (ref. 93).

These qualitative results of approximation methods are borne out generally by the several complete calculations of electron densities and localization energies of substituted benzenes with systematic changes in the heteroatom parameters.[85,90,91] Additional calculations have been reported of specific compounds by using appropriate explicit values for the additional parameters.[92]

Some examples will show the limits of applicability of this technique. Aniline may be approximated with the parameter values $k_{CN} = 0.8$, $h_N = 1.5$, $h' = 0.1$, in which the last value is assigned to the carbon attached to the nitrogen. This model gives the electron densities and localization energies summarized in Fig. 11.20.[93] Compared with benzene having $q = 1.000$, $L^+ = 2.536$, we find the orientation of aniline to be o-p with activation, in qualitative agreement with experiment.

[90] C. Sandorfy, *Bull. soc. chim. France*, **1949**, 615.

[91] M. J. S. Dewar, *J. Chem. Soc.*, **1949**, 463.

[92] E. Heilbronner and M. Simonetta, *Helv. Chim. Acta*, **35**, 1049 (1952); M. Simonetta and A. Vaciago, *Nuovo cimento*, **11**, 596 (1954); S. L. Matlow and G. W. Wheland, *J. Am. Chem. Soc.*, **77**, 3653 (1955); S. S. Perez, *Anales real soc. españ. fís. y quim.* (*Madrid*), **53B**, 479 (1957); S. Basu and J. N. Chaudhuri, *Proc. Natl. Inst. Sci. India*, Pt. A, **24**, 130 (1958).

[93] Unpublished calculations of the author.

If a similar model is applied to chlorobenzene with $k_{C-Cl} = 0.4$, $h_{Cl} = 2, h' = 0.2$, the results in Fig. 11.21 are obtained.[93] The calculations say that chlorobenzene should give *meta*-substitution with a rate about that of benzene! Only if an inductive parameter is extended at least to the *ortho*-position will the calculations agree qualitatively with experiment. Again, the neglected electrostatic interactions with substituent dipoles may play a role. If the chlorobenzene model is repeated with $h'' = 0.1$

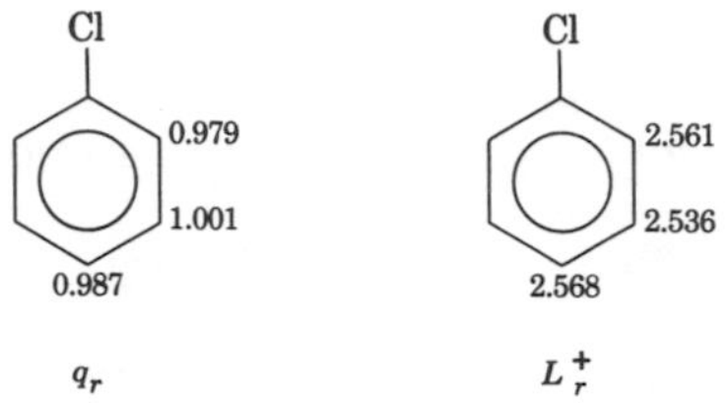

Fig. 11.21 Electron densities and localization energies in chlorobenzene with $h_{Cl} = 2$, $h' = 0.2$, $k_{C-Cl} = 0.4$ (ref. 93).

at the *ortho*-positions, we find for L_r^+ : *m*-, 2.60; *p*-, 2.57. We now obtain *para*-orientation with deactivation. Unfortunately, the handling of the *ortho*-position is now no longer straightforward, since the α-value different from the other positions implies a changed σ-bond energy difference.[82,91] This characteristic is the most serious limitation of the application of localization theory to heterosubstituted compounds.

On the other hand, if the auxiliary inductive parameter is not used at all—if $h' = 0$—chlorobenzene becomes *o-p* directing with activation. This is the π-energy result. We could then say that the electrostatic effect of the C—Cl dipole inhibits production of a positive charge in the ring and results in over-all deactivation. The same electrostatic effect in different magnitude must also exist for aniline and anisole but must be less than the activating effect of the conjugating substituent. No quantitative attempts have yet been reported to compare experimental reactivities with calculated π-energy changes plus calculated field effects.

A number of investigations has been concerned with the orientation in electrophilic substitution of heterocyclic ring systems, particularly the nitrogen heterocycles of the pyridine type. Two additional problems are encountered in such systems. In the first it must be recognized that reaction at the carbon attached to the heteroatom possibly involves a σ-bond change different from that of the other positions. For example, reaction at the β- and γ-positions of pyridine involves conversion of two C_{sp^2}—C_{sp^2} bonds to C_{sp^2}—C_{sp^3} bonds. Reaction at the α-position, however, involves in place of one of these changes the conversion of an N_{sp^2}—C_{sp^2} bond to an N_{sp^2}—C_{sp^3} bond. These energy changes may differ substantially; hence, comparison of electron densities or localization energies of such α-positions with other positions may not be a valid criterion of reactivity. Unfortunately, this problem is almost wholly ignored in the literature.

Another problem results from the basicity of many of these heterocycles. Since many electrophilic substitutions take place in acid solution, the actual reactant may be the conjugate acid in which substantially different parameter values would be required. Nevertheless, calculations of pyridine put the highest electron density and the lowest magnitude of the localization energy at the β-position in agreement with the observed chemistry.[86,94,95] Similarly, in quinoline the 5- and 8-positions are calculated to be the most reactive,[86,94,96,97] in agreement with experiment. Agreement between theory and experiment has been observed in other analogous nitrogen heterocycles.[97,98]

The five-membered heterocycles such as furan, and pyrrole are isoelectronic with cyclopentadienyl anion. The conjugation of an unshared electron pair with the remaining carbon π-network is expected to provide relatively high electron densities at some of these positions. This expectation is borne out in all simple MO calculations of these compounds.[99,100] Brown,[100,101] in particular, has studied a variety of mono- and di-hetero systems with systematic variation of parameters and has shown that the principal aspects of the chemistry of these compounds can be successfully interpreted. The parameter values derived from these investigations are in satisfactory agreement with those summarized in Sec. 5.8, except that variable and sometimes rather large auxiliary inductive parameters are

[94] C. Sandorfy and P. Yvan, *Bull. soc. chim. France*, **1950**, 131.

[95] R. D. Brown and M. L. Heffernan, *Australian J. Chem.*, **9**, 83, (1956); **10**, 211 (1957).

[96] C. Sandorfy and P. Yvan, *Compt. rend.*, **229**, 715 (1949); R. D. Brown and R. D. Harcourt, *J. Chem. Soc.*, **1959**, 3451.

[97] M. J. S. Dewar and P. M. Maitlis, *J. Chem. Soc.*, **1957**, 2521; R. D. Brown and R. D. Harcourt, *Tetrahedron*, **8**, 23 (1960).

[98] R. D. Brown and R. D. Harcourt, *Tetrahedron*, **8**, 23 (1960).

[99] H. C. Longuet-Higgins and C. A. Coulson, *Trans. Faraday Soc.*, **43**, 87 (1947); S. Nagakura and T. Hosoya, *Bull. Chem. Soc. Japan*, **25**, 179 (1952); J. De Heer, *J. Am. Chem. Soc.*, **76**, 4802 (1954); L. Melander, *Arkiv Kemi*, **8**, 361 (1956); **11**, 397 (1957); S. Carra and S. Polezzo, *Gazz. chim. ital.*, **88**, 1103 (1958); J. I. F. Alonso, R. Domingo, and L. C. Vila, *Rec. trav. chim.*, **78**, 215 (1959).

[100] R. D. Brown, *Australian J. Chem.*, **8**, 100 (1955); R. D. Brown and B. A. W. Coller, *Australian J. Chem.*, **12**, 152 (1959).

[101] I. M. Bassett and R. D. Brown, *J. Chem. Soc.*, **1954**, 2701; I. M. Bassett, R. D. Brown, and A. Penfold, *Chem. and Ind.*, **1956**, 892; R. D. Brown and M. L. Heffernan, *J. Chem. Soc.*, **1956**, 3683, 4288.

apparently required.[102] In view of the rather severe limitations of the MO models used, the significance of the observed qualitative agreements is still in question.

[102] For additional calculations of such heterocyclic ring systems see S. Basu, *Proc. Natl. Inst. Sci. India*, **21A**, 173 (1955); L. Melander, *Acta Chem. Scand.*, **9**, 1400 (1955); J. I. F. Alonso, L. C. Vila, and R. Domingo, *J. Am. Chem. Soc.*, **79**, 5839 (1957); M. M. Kreevoy, *J. Am. Chem. Soc.*, **80**, 5543 (1958).

12 Carbonium ions

12.1 Properties of Carbonium Ions

A carbonium ion may be defined as a compound whose structure on application of the usual rules of valency contains a trivalent carbon with a formal positive charge. Since such structures can be written for all organic compounds, the definition normally refers to the most stable structures in VB terminology. Which structures to consider becomes a relative matter; for example, the question of whether crystal violet is merely an ammonium salt, as implied in I, or a carbonium ion, as in II, is a moot point.

In practice, "carbonium ion" frequently relates to a property of a π-system, at least to the degree that σ- and π-bonds may be considered separately. A π-system carbonium ion is simply one in which the π-nuclear potential exceeds by one the supply of π-electrons. According to this definition, crystal violet is unambiguously a carbonium ion.

Not only are such carbonium ions important intermediates in many reactions, but some are sufficiently stable to be isolated. In all such cases the positive charge is actually distributed among several atoms so that no single carbon is highly cationic; that is, the available electrons are spread out as far as possible to reduce electronic repulsion. Early examples were provided by the triarylmethyl cations. These examples include not only the triarylmethane dyes such as crystal violet but also hydrocarbon cases such as numerous salts of triphenylmethyl cation itself.[1]

Other classes are of more recent vintage and include the tropylium cations (Sec. 10.6) and perinaphthenyl cation, III.[2] In HMO theory the

III   IV

positive charge in III is distributed equally among the 1-, 3-, 4-, 6-, 7-, and 9-positions. It is amusing that this theory places none of the cationic charge at the 13-position in the center; the corresponding resonance structure, IV, contains a cyclododecahexaene structure which violates the $4n + 2$ rule! Although delocalization energies are not reliable indicia of stability, the high DE-value of perinaphthenyl cation 5.827β (for comparison, phenanthrene with one more carbon has $DE = 5.448$) is certainly suggestive. The recognition of this structure as a stable unit augurs some interesting chemistry. Aromatic hydrocarbons whose conjugate acids contain this unit are highly reactive in electrophilic substitutions; examples are the pyrene-1 position, V, and the perylene-3 position, VI. Reid[3] has synthesized some interesting hydrocarbons in which the perinaphthenyl system is fused to a stable anionic structural unit; an example is VII. This hydrocarbon is about as basic as the azulenes because of the stability of the conjugate acid, VIII.

Perinaphthenone, IX, is a yellow ketone that dissolves in acids and may be recovered unchanged on dilution; its basicity is undoubtedly due to the stability of the hydroxyperinaphthenyl cation structure of the conjugate acid.

[1] These include not only the highly colored stable complexes of triarylmethyl chlorides and Lewis acids but also the crystalline triarylmethyl perchlorates. For examples see M. Gomberg, *Ber.*, **35**, 1822 (1902); A. Baeyer, *Ber.*, **38**, 1162 (1905); A. Tschitschibabin, *Ber.*, **40**, 1817 (1907); M. Gomberg and L. Cone, *Ann.*, **370**, 193 (1909); K. A. Hofmann and H. Kirmreuther, *Ber.*, **42**, 4861 (1909); K. A. Hofmann, H. Kirmreuther, and A. Thal, *Ber.*, **43**, 185 (1910); K. Brand and O. Stallman, *J. prakt. Chem.* [2], **107**, 358, 373 (1924).

[2] R. Pettit, *Chem. and Ind.*, **1956**, 1306, *J. Am. Chem. Soc.*, **82**, 1972 (1960).

[3] D. H. Reid, *Tetrahedron*, **3**, 339 (1958).

V   VI

H   H
VIII   VII

It is significant that the ketone has an abnormally high dipole moment.[4]

In simple resonance theory the positive charge in benzyl cation is distributed among the o-, p-, and α-positions (Fig. 12.1a). The same result

IX

is given by HMO theory with a charge distribution as in Fig. 12.1b. The ω-technique gives the slightly modified charge distribution as in Fig. 12.1c.[5] SCF results are not greatly dissimilar. Pople's treatment[6] (Sec. 16.3) gives the charge distribution in Fig. 12.1d, in which the principal difference is the relatively high electron density at the 1-position. It seems characteristic of this method in which Coulombic energies are explicitly incorporated that charge tends to pile up at tertiary positions. Nevertheless, all of the theories have the positive charge at the *meta*-position considerably less than at the *ortho*- and *para*-positions.

[4] Y. A. Zhdanav, O. A. Osipov, O. E. Shelepin, and V. A. Kogan, *Proc. Acad. Sci. U.S.S.R.* (*English Translation*), **128,** 809 (1959).

[5] A. Streitwieser, Jr., unpublished calculations; cf. A. Streitwieser, Jr., and P. M. Nair, *Tetrahedron*, **5,** 149 (1959).

[6] A. Brickstock and J. A. Pople, *Trans. Faraday Soc.*, **50,** 901 (1954); J. A. Pople, *J. Phys. Chem.*, **61,** 6 (1957).

These distributions are in qualitative accord with organic chemical experience. For example, the effect of methyl substitution has been considered to provide an indication of charge distribution.[7,8] In solvolysis of benzyl p-toluenesulfonates in acetic acid at 40°, a p-methyl substituent results in a rate increase of 56.5-fold, whereas a m-methyl group provides

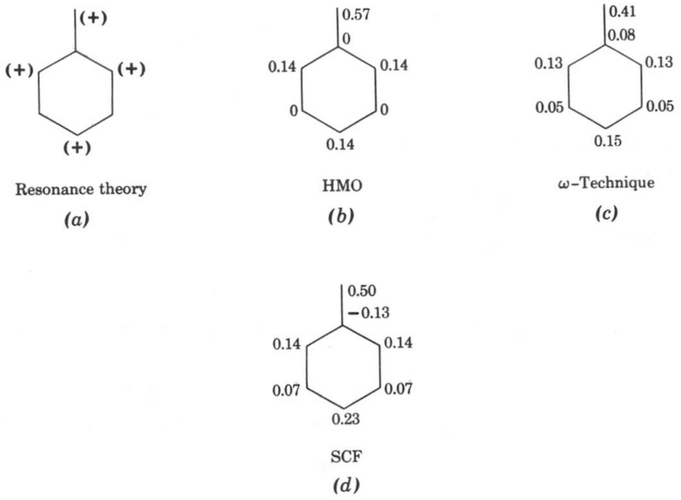

Fig. 12.1   Charge distribution in benzyl cation.

only a 2.6-fold rate enhancement.[9] The stabilization of carbonium ions by methyl groups is more important the greater the cationic character at the substituent site. The greater influence of a *para*-methyl group compared to a *meta*-methyl group implies a higher positive charge at the *para*-position of benzyl cation. This use of the effect of methyl groups as a measure of charge distribution also follows directly from the inductive model of a methyl group (Sec. 5.7) and (4.29). [Compare the effect of methyl groups on electrophilic aromatic substitution (Sec. 11.7).]

Unfortunately, this harmonious tableau is muddled by similar comparisons for triphenylmethyl cation (Fig. 12.2). Again, simple resonance theory and HMO theory (a) put positive charge only on the *ortho*- and *para*-carbons; the ω-technique (b) puts a small amount of positive charge in the *meta*-positions, but the SCF method (c) makes the *meta*-positions more positive than the *ortho*-positions. The NMR spectra of triphenylmethyl cation and deuterated derivatives have been interpreted in just this

[7] C. G. Swain and W. P. Langsdorf, *J. Am. Chem. Soc.*, **73**, 2813 (1951).

[8] A. Streitwieser, Jr., *Chem. Rev.*, **56**, 571 (1956).

[9] A. Streitwieser, Jr., and R. H. Jagow, unpublished results; quoted in ref. 8, Table 20.

way, namely, that the positive charge density decreases in the order $p > m > o$.[10]

Recent comparisons for β-phenylallyl cation cast suspicion on the SCF calculations. The charge densities are given in Fig. 12.3 for this cation.

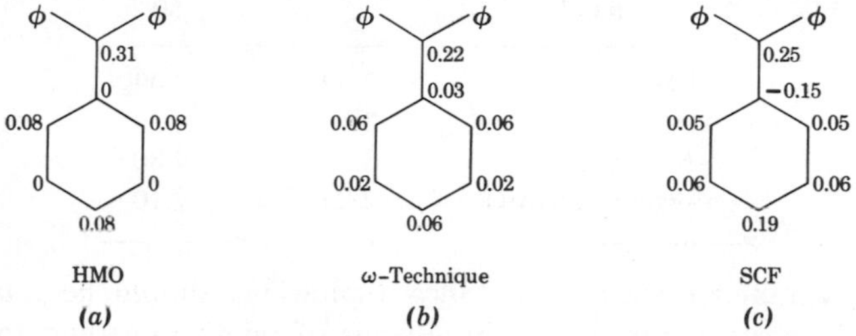

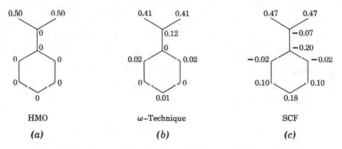

Fig. 12.2   Charge distribution in triphenylmethyl cation.

The simple LCAO methods, (a) and (b), give little or no positive charge density in the phenyl ring, whereas the SCF method (c) puts about as much positive charge in the *para*-position as in triphenylmethyl cation. This type of cross conjugation is contrary to organic chemical experience. For

Fig. 12.3   Charge distribution in β-phenylallyl cation.

example, solvolysis rates for several β-phenylallyl *p*-toluenesulfonates are summarized in Table 12.1.[11] The noteworthy point in these data is the relatively small rate-enhancing effect of *p*-methyl groups. To the extent that the effects of such substitution reflect charge density distribution, the simple LCAO methods clearly give a more satisfactory portrait of the charge distribution than the SCF method.

On the other hand, SCF energies for cations are unquestionably better than the HMO values; the limitations of the HMO method in handling cations was seen particularly in the treatment of the ionization potentials

[10] R. Dehl, W. R. Vaughan, and R. S. Berry, *J. Org. Chem.*, **24**, 1616 (1959).
[11] A. Streitwieser, Jr., and J. B. Bush, unpublished results.

TABLE 12.1

SoLVOLYSIS RATES OF β-ARYLALLYL (ATROPYL) TOSYLATES

$$10^5 k(\sec^{-1})$$

ROTs	AcOH 75°	HCOOH 60°
Allyl	2.09	7.49
Atropyl	2.12	6.59
p-Methylatropyl	3.15	9.80
m-Methylatropyl	2.25	7.10

of organic radicals (Sec. 7.3). Such limitations should be retained in proper perspective during the discussions of relative energies to follow.

12.2   Carbonium Ion Equilibria

The protonations of aromatic hydrocarbons treated in Chap. 11 are representative of equilibria of the carbonium ion type. Other examples, such as the ionization equilibria of triarylchloromethanes and the pseudo-basicity of arylcarbinols, may be handled in the same general manner: the total energy change for each compound is divided into σ-bond changes and solvation energies which are assumed to be constant for members of a related series and π-bond changes which are calculated by an MO method. Suitable π-models of reactant and product lead to the corresponding bonding energy coefficients, M_{reactant} and M_{product}. The existence of a linear correlation implies

$$\Delta F_i = a + b(M_{\text{product}} - M_{\text{reactant}})_i \tag{1}$$

The effective value of β for the equilibrium is b.

Triarylmethyl chlorides are known to be partially ionized in liquid sulfur dioxide solution; conductance measurements have provided equilibrium constants for several cases.[12] The elegant work of Lichtin has not only given a number of precise values for a number of compounds but has also shown that the equilibrium constants obtained in the usual manner are complicated by ion-pair equilibria:[13]

$$Ar_3CCl \rightleftharpoons Ar_3C^+Cl^- \rightleftharpoons Ar_3C^+ + Cl^-$$

    [12] K. Ziegler and H. Wollschitt, *Ann.*, **479**, 90 (1930); K. Ziegler and W. Mathes, *Ann.*, **479**, 111 (1930).
    [13] N. N. Lichtin and P. D. Bartlett, *J. Am. Chem. Soc.*, **73**, 5530 (1951); N. N. Lichtin and H. Glazer, *J. Am. Chem. Soc.*, **73**, 5537 (1951); N. N. Lichtin and H. P. Leftin. *J. Phys. Chem.*, **60**, 160, 164 (1956).

Nevertheless, a correlation has been demonstrated between the free energies for the composite equilibria of several triarylchloromethanes and HMO energy differences between the cations and the covalent halides, the π-energies of which were taken as the sum of the π-energies of the substituent aromatic hydrocarbons.[14] The results in Table 12.2 are shown plotted

TABLE 12.2

IONIZATION EQUILIBRIA OF TRIARYLCHLOROMETHANES IN
LIQUID SULFUR DIOXIDE AT 0.00–0.12°

No. in Fig. 12.4	Ar_1	Ar_2	Ar_3	ΔM_i	ΔF_i kcal./mole
1	phenyl	phenyl	phenyl	1.8004	5.48
2	m-biphenylyl	phenyl	phenyl	1.7990	5.64
3	m-biphenylyl	m-biphenylyl	m-biphenylyl	1.7960	6.09
4	p-biphenylyl	phenyl	phenyl	1.8184	4.52
5	p-biphenylyl	p-biphenylyl	phenyl	1.8396	3.72
6	p-biphenylyl	p-biphenylyl	p-biphenylyl	1.8578	3.13
7	β-naphthyl	phenyl	phenyl	1.8134	4.80
8	α-naphthyl	phenyl	phenyl	1.8544	3.96

in Fig. 12.4. The m-biphenylyl compounds are less reactive than expected, perhaps because of limitations of the assumption of normal β for the bond between the phenyl rings or because of the electron attracting inductive effect associated with unsaturated groups. Because the bond between phenyls undoubtedly has $k < 1$ (Sec. 5.1), these compounds actually ionize less than is indicated by the calculations. Similarly, the nonplanarity of triarylmethyl cations means that the bonds to the central carbon should also have $k < 1$; that is, phenyl substituents do not stabilize cations as much as indicated in the calculations. The slope of the correlation line is about $\beta = -42$ kcal./mole, or -1.8 ev. The point for α-naphthyl-diphenylmethyl chloride is off this line, undoubtedly because of increased steric hindrance to coplanarity in the corresponding cation.

In strong acids carbinols behave as pseudo-bases because of the equilibrium:

$$ROH + H^+ \rightleftharpoons R^+ + H_2O$$

Equilibrium constants for a number of carbinols have been determined; the negative logarithm is defined as the pK_{R^+} for the carbinol by analogy with the pK_b of normal bases. Wheland[15] suggested many years ago that such equilibrium constants should correlate with π-energy differences, but

[14] A. Streitwieser, Jr., *J. Am. Chem. Soc.*, **74**, 5288 (1952).
[15] G. W. Wheland, *J. Chem. Phys.*, **2**, 474 (1934).

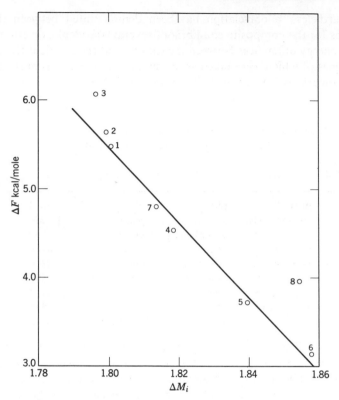

Fig. 12.4   Ionization equilibria of triarylchloromethanes in liquid sulfur dioxide at 0°.
Numbers refer to compounds in Table 12.2.

no experimental data were available at that time to test the prediction.
Gold[16] has shown recently that the pK_{R^+} data of Deno, Jaruzelski, and
Schriesheim[17] for several arylcarbinols give a linear correlation with the
π-energy difference between a model of the carbinol and that of the cation.
The technique is exactly equivalent to that previously used for the triarylchloromethanes.   The correlation has been extended to a number of
tropylium alcohols by Naville, Meuche, Strauss, and Heilbronner.[18]

[16] V. Gold, *J. Chem. Soc.*, **1956**, 3944.   In this paper the π-energy cited for fluorenyl
cation is actually that of the radical.

[17] N. C. Deno, J. J. Jaruzelski, and A. Schriesheim, *J. Org. Chem.*, **19**, 155 (1954);
J. Am. Chem. Soc., **77**, 3044 (1955).

[18] D. Meuche, H. Strauss, and E. Heilbronner, *Helv. Chim. Acta*, **41**, 57, 414 (1958);
G. Naville, H. Strauss, and E. Heilbronner, *Helv. Chim. Acta*, **43**, 1221 (1960).   The
pK-values cited by these authors are derived on the basis of an H_0 rather than a C_0
treatment.   Their values have been translated in Table 12.3 to a C_0 basis, using their
empirical equations for the interrelation of these quantities.

Additional data have been contributed by Dauben,[19] who has estimated some pK_{R^+}-values from results of hydride exchanges in several hydrocarbons:

$$R'—H + R^+ \rightleftharpoons R'^+ + R—H$$

TABLE 12.3

PSEUDO-BASICITY OF CARBINOLS

No. in Fig. 12.5	Compound	ΔM_i	pK_{R^+}*
1	Benzyl alcohol	0.721	−22−−27
2	9-Fluorenol	1.161	−14.0
3	Benzhydrol	1.301	−13.3
4	Tribenztropyl alcohol	1.455	~ −15
5	2,3,6,7-Dibenztropyl alcohol	1.544	−3.7
6	Perinaphthenyl alcohol	1.695	0–2
7	Benztropyl alcohol	1.772	1.7
8	Triphenylcarbinol	1.800	−6.63
9	(Naphtho-1,2)tropyl alcohol	1.830	2.2
10	Cyclopropenyl alcohol	2.000	0–10
11	Tropyl alcohol	2.000	4.7
12	Triphenylcyclopropenyl alcohol	2.309	3.1
13	2,3,4,5-Dibenztropyl alcohol	1.646	−5.8
14	(Naphtho-2,3)tropyl alcohol	1.724	0.3
15	Diphenylcyclopropenyl alcohol	1.903	0.3

* Compiled from refs. 17, 18, 19, 20, and 21.

The various data are summarized in Table 12.3; the correlation, pictured in Fig. 12.5, is excellent. Deviations are observed for 1,2,3,4-dibenztropylium, X, and tribenztropylium cation, XI, which undoubtedly possess substantial steric hindrance to coplanarity,[18] and for the phenyl-substituted compounds, undoubtedly because of the decreased resonance

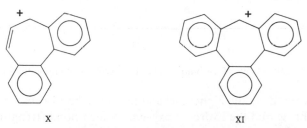

interaction across the substituent single bond. This effect of phenyl groups was observed for the triarylchloromethane ionizations and is obvious again in our discussion of the acidity of hydrocarbons in Sec. 14.1.

[19] H. J. Dauben, Jr., personal communication.

Both the triphenylcyclopropenyl[20] and diphenylcyclopropenyl[21] alcohols are less basic than expected from the correlation of Fig. 12.5. Hindrance to coplanarity is probably not an important factor here; the discrepancy is

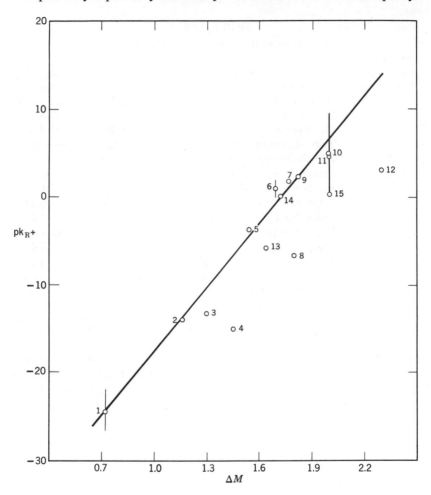

Fig. 12.5 Pseudobasicity of carbinols. Numbers refer to compounds in Table 12.3.

probably caused by the higher *s*-character in the bond between the three-membered ring and the hydroxyl oxygen. Added bond strength from this source would result in lower ease of dissociation.

We should note a further implicit assumption. In this correlation we

[20] R. Breslow and H. W. Chang, *J. Am. Chem. Soc.*, **83**, 2367 (1961).
[21] R. Breslow, J. Lockhart, and H. W. Chang, *J. Am. Chem. Soc.*, **83**, 2375 (1961).

have considered primary, secondary, and tertiary carbinols, although most are indeed secondary. In the conversion of a primary alcohol to the carbonium ion one C_{sp^3}—C_{sp^3} and two C_{sp^3}—H bonds in the carbinol are converted to a C_{sp^2}—C_{sp^3} and two C_{sp^2}—H bonds in the carbonium ion. For a secondary alcohol two C_{sp^3}—$C_{sp^3} \to C_{sp^2}$—C_{sp^3} and only one C_{sp^3}—H $\to$ C_{sp^2}—H conversions occur, whereas for a tertiary carbinol three C_{sp^3}—C_{sp^3} $\to C_{sp^2}$—C_{sp^3} conversions occur. Our implicit assumption is that the σ-energy change associated with a C_{sp^3}—$C_{sp^3} \to C_{sp^2}$—C_{sp^3} conversion is the same as that for a C_{sp^3}—H $\to C_{sp^2}$—H conversion. There is no a priori reason why this assumption should be valid, yet its significance is normally ignored in papers in the literature. At present, insufficient data exist to establish the validity of the assumption, although indications are that the errors associated with its use are likely to be small.

12.3   Carbonium Ion Reactions

The successful application of MO methods to carbonium ion equilibria prompts extension to reactions in which carbonium ions are intermediates. Among such reactions is the broad class of solvolytic displacement reactions, or S_N1 reactions.[8] Closely related are the direct displacement, or S_N2 reactions:

$$RX + Y^- \to RY + X^-$$

in which X and Y are Lewis bases which need not be anions. These reactions follow simple second-order kinetics and undoubtedly proceed through transition states of the type of structure XII.[22] This structure is

usually approximated in orbital terms as an overlapping of orbitals on X and Y with a p-orbital of the central carbon (Fig. 12.6).[23] The LCAO description is derived as follows: the orbitals, φ_X and φ_Y, are used to form two pseudo-orbitals as the linear combinations,

$$\psi^+ = \frac{1}{\sqrt{2}} (\varphi_X + \varphi_Y) \quad \text{and} \quad \psi^- = \frac{1}{\sqrt{2}} (\varphi_X - \varphi_Y)$$

[22] A thorough discussion of the pertinent evidence is beyond the scope of this book; see ref. 8 or almost any book on advanced organic chemistry or reaction mechanisms.

[23] R. J. Gillespie (*J. Chem. Soc.*, **1952**, 1002) has suggested that d-orbitals of the central carbon may contribute significantly to such a structure; cf. M. J. S. Dewar, *J. Chem. Soc.*, **1953**, 2885. A carbon $3d_{z^2}$-atomic orbital has different symmetry properties than a $2p_z$-orbital and interacts only with ψ^+ rather than ψ^- (*vide infra*).

Only ψ^- is allowed by symmetry to interact appropriately with the central p-orbital.[24] The net result is that ψ^- is treated as a heteroatom, which contributes two electrons and involves a h_ψ and a $k_{C\psi}$. Note that of the four electrons originally associated with the bases, X and Y, two can participate in the π-lattice and two are contained in the effectively isolated pseudo-orbital, ψ^+.[25]

Fig. 12.6 Orbital picture of intermediate or transition state in direct displacement reaction.

This simple picture serves to illustrate much of the known chemistry of the reaction. For example, the greater conjugation energy in C—C—C—X compared to C—X explains the greater reactivity of allyl compounds compared to methyl derivatives in displacement reactions.[26]

ψ^- is expected to be effectively more electronegative than carbon. Atom polarizabilities for butadiene and for styrene readily show that direct displacements on allyl and benzyl derivatives put positive charge density at the γ- and *para*-positions, respectively, in the transition state; this accounts for the rate acceleration associated with methyl substitution at these positions.[8] With this model, Daudel and Chalvet[24] have demonstrated an excellent correlation between calculated energy differences and experimental rates of reaction of arylmethyl chlorides with potassium iodide in acetone.

Some cases of solvolytic displacement reactions, particularly with primary alkyl halides, are undoubtedly direct displacements by solvent. In such cases structure XII also represents a rather unstable intermediate. Indeed, even for ordinary direct displacement reactions, recent evidence indicates that an unstable intermediate of such structure is involved.[27]

[24] R. Daudel and O. Chalvet, *Compt. rend.*, **242**, 2150 (1956), *J. chim. phys.*, **1956**, 943.

[25] The reader should note that we can calculate XII in the usual way in terms of three 3-center MO's. The treatments are equivalent; that given in the text simply makes use of the symmetry of the system and emphasizes certain features of the results.

[26] We can go a step further. Because of the expected effect of such conjugation energy on the rate of direct displacement and the resemblance of the MO model to corresponding hetero-compounds, we might anticipate a correspondence between direct displacement rates and reduction potentials of corresponding aldehydes. Although no such relationship seems to have been tested in the literature, the reader may verify that a plot of log k for reaction of ArCH$_2$Cl with potassium iodide in acetone (Table 12.4) versus reduction potentials for ArCHO (Table 7.3) gives an excellent straight line.

[27] S. Winstein, D. Darwish, and N. J. Holness, *J. Am. Chem. Soc.*, **78**, 2915 (1956).

As the effective electronegativity of ψ increases or as β for the C—ψ bond diminishes, the central carbon system increasingly resembles that of a carbonium ion. As the stability of the carbonium ion increases, the system is less dependent on the nucleophilicity of the incoming group. In many cases (limiting solvolyses) the rate-determining step does not involve an attack by an incoming nucleophilic reagent. Such cases are associated with leaving groups that are weak Lewis bases, relatively stable carbonium ion systems, or non-nucleophilic solvents. The first intermediate is usually described as an ion pair which may have covalent bonding between the ions:

$$R \quad X \rightarrow R^+X^- \rightarrow \text{products}$$

Even in such cases the transition states can be treated by the preceding MO method. However, the carbonium ion nature of the first intermediate suggests trying the same approach that succeeded for carbonium ion equilibria (Sec. 12.2). A number of data have become available on solvolysis rates of arylmethyl compounds in various solvents and are summarized in Table 12.4.

An interesting feature of these data is that a plot of any one set against any other gives a reasonably good linear correlation. Some examples are shown in Fig. 12.7. The noteworthy aspect of these correlations is that they have widely varying slopes. A related situation occurs in the varying magnitudes of the effects on rate associated with substitution in a benzene ring; in fact, the magnitude of the rate enhancement in various solvolytic reactions caused by *para*-methyl substitution has been used to indicate the relative carbonium ion character of the transition state.[7,8] We are clearly observing the same phenomenon in the varying slopes of Fig. 12.7. The less nucleophilic and more ionizing the solvent, the "better" the leaving group, the larger the slope. This situation is also reminiscent of the varying relative rates in electrophilic aromatic substituion (Sec. 11.2). In the same manner described for aromatic substitution we may define a "σ" and a "ρ" such that "σ" is characteristic of the arylmethyl moiety and "ρ" is characteristic of the reaction. If the solvolysis of arylmethyl chlorides in 80% aqueous ethanol at 50° is taken as our standard reaction, with "ρ" defined as 1.00, "ρ"-values for the other reactions are given directly from the slopes in Fig. 12.7. These values are tabulated in Table 12.5. If "σ" for the benzyl system is taken as the standard, 0.00, values for other systems are obtained from (2).

$$``\sigma_i" = \frac{\log \dfrac{k_i}{k_{C_6H_5CH_2X}}}{``\rho"} \tag{2}$$

TABLE 12.4

SOLVOLYSIS OF ARYLMETHYL COMPOUNDS

$-\log k$, sec^{-1}, 25°

Ar	ArCH$_2$Cl 80% Aq. Alc.* Ref. 28	ArCH$_2$Cl 79.5% Aq. Dioxane Ref. 29, 31	ArCH$_2$Cl 50.7% Aq. Dioxane Ref. 29	ArCH$_2$Cl HCOOH 0.38M H$_2$O Refs. 28, 30	ArCH$_2$Cl 6.1% H$_2$O, 54.1% HCOOH, 39.8% dioxane Refs. 28, 32, 35	ArCH$_2$OTs AcOH† Ref. 33	ArCHClCH$_3$ 80% Aq. Acetone Ref. 34	ArCCl(CH$_3$)$_2$ EtOH‡ Ref. 28	ArCH$_2$Cl KI, Acetone§ Refs. 29, 32, 35
Phenyl	5.66	8.18	6.63	6.94	7.61	4.77	6.17	4.83	3.19
2-Naphthyl	5.37			5.58	7.09	3.63	5.32	3.90	2.86
2-Phenanthryl	5.41			5.55	7.13	3.58	5.48	3.90	2.91
3-Phenanthryl	5.16	8.01	6.29	4.80	6.45	3.04	5.04	3.62	2.85
2-Anthracyl									
2-Triphenylyl	5.38			5.43		2.52	4.46		
2-Pyrenyl		7.64	7.74		7.42	4.04			
8-Fluoranthyl					5.36	2.17			
1-Naphthyl	5.03	7.73	6.00	4.36	6.26	2.78	4.93	4.17	2.61
1-Phenanthryl	5.16			4.52	6.55			4.32	2.64
9-Phenanthryl	5.10	7.87	6.26	4.31	6.38	2.70	5.23	3.92	2.58
6-Chrysyl	4.38			2.89			4.61		
4-Pyrenyl			6.06		6.10	2.59			2.58
1-Pyrenyl		5.52	2.87		3.03		2.25		2.08
3-Fluoranthyl					5.21	1.00			2.25
9-Anthracyl		4.76	1.31						1.42
1,2-Benz-10-anthracyl									1.56

* At 50°.   † At 40°.   ‡ At 0°.   § Second-order rate constants, l, mole^{-1}, sec.$^{-1}$.

28 M. J. S. Dewar and R. J. Sampson, *J. Chem. Soc.*, **1957**, 2946, 2952.
29 M. Planchen, P. J. C. Fierens, and R. H. Martin, *Helv. Chim. Acta*, **42**, 517 (1959).
30 M. J. S. Dewar and R. J. Sampson, *J. Chem. Soc.*, **1956**, 2789.
31 P. J. C. Fierens and J. Berkowitch, *Tetrahedron*, **1**, 129 (1957).
32 P. J. C. Fierens, H. Hannaert, J. V. Rysselberge, and R. H. Martin, *Helv. Chim. Acta*, **38**, 2009 (1955).
33 A. Streitwieser, Jr., R. H. Jagow, and R. M. Williams, unpublished results.
34 E. Berliner and N. Shieh, *J. Am. Chem. Soc.*, **79**, 3849 (1957).
35 G. Geuskens, G. Klopman, J. Nasielski, and R. H. Martin, *Helv. Chim. Acta*, **43**, 1927, 1934 (1960).

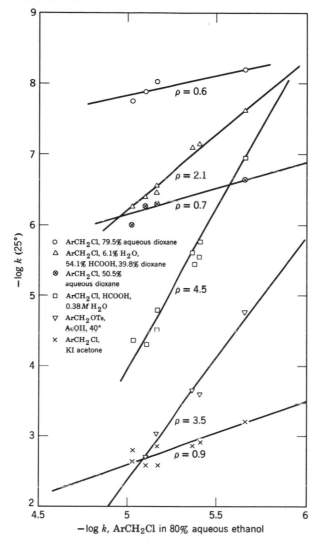

Fig. 12.7   Correlation of solvolyses in various solvents with solvolysis in 80% aqueous ethanol.

Average values for various arylmethyl systems are summarized in Table 12.6; the errors are listed as the average deviations. With a few exceptions, the "σ"-values are reasonably constant for a given system. Exceptions are found for the pyrene compounds. The solvolyses of 2-pyrenylmethyl chloride, XIII, in an aqueous formic acid-dioxane mixture and of the p-toluenesulfonate in acetic acid are a little faster than the parent benzyl

TABLE 12.5

"ρ" FOR VARIOUS SOLVOLYTIC REACTIONS

Compounds	Solvent	Temperature	"ρ"	$-\beta$, kcal.
$ArCH_2Cl$	80% aqueous ethanol	50°	(1)	17
$ArCH_2Cl$	79.5% aqueous dioxane	25°	0.6	10
$ArCH_2Cl$	50.5% aqueous dioxane	25°	0.7	12
$ArCH_2Cl$	6.1% H_2O, 54.1% HCOOH, 39.8% dioxane	25°	2.1	35
$ArCH_2Cl$	HCOOH, $0.38M$ H_2O	25°	4.5	76
$ArCH_2OTs$	CH_3COOH	40°	3.5	59
$ArCHClCH_3$	80% aqueous acetone	25°	2.5	42
$ArCCl(CH_3)_2$	EtOH	0°	2.6	44
$ArCH_2Cl$	KI, acetone	25°	0.9	15

TABLE 12.6

"σ" VALUES AND MO QUANTITIES FOR ARYLMETHYL SYSTEMS

No. in Figs. 12.8 and 12.11	$ArCH_2$ System	"σ"-value	No. of Reactions	$M_{ArCH_2^+} - M_{ArH}$	$2a_0$	ΔE_ω
1	Phenyl	(0)		0.721	1.511	1.578
2	2-Naphthyl	0.31 ± 0.03	6	0.744	1.454	1.642
3	2-Phenanthryl	0.30 ± 0.04	6	0.736	1.474	1.644
4	3-Phenanthryl	0.46 ± 0.05	8	0.754	1.429	1.712
5	2-Anthracyl	0.67 ± 0.03	2	0.769	1.372	1.734
6	3-Triphenylyl	0.31 ± 0.03	2	0.745	1.455	1.681
7	2-Pyrenyl	0.15 ± 0.06*	2	0.717	1.512	1.594
8	8-Fluoranthyl	0.73†	1	0.764	—	1.732
9	1-Naphthyl	0.68 ± 0.10	6	0.812	1.342	1.805
10	1-Phenanthryl	0.52 ± 0.02	3	0.803	1.361	1.808
11	9-Phenanthryl	0.56 ± 0.03	6	0.813	1.336	1.796
12	6-Chrysyl	1.09 ± 0.19	2	0.838	1.281	1.893
13	4-Pyrenyl	0.70 ± 0.06	3	0.828	1.287	1.838
14	1-Pyrenyl	2.2‡	1	0.868	1.206	1.999
15	3-Fluoranthyl	1.08	1	0.780	—	1.806

* Widely divergent results were obtained in aqueous dioxane: "σ" $= -1.54$ in 50.7% aqueous dioxane, $+0.93$ in 79.5% aqueous dioxane (ref. 29).

† This result is based on acetolysis of the tosylate. The KI-acetone displacement gives a similar result, 0.64; but solvolysis of the chloride in formic acid dioxane-water gives the anomalously high value of 1.07 (ref. 35).

‡ Aqueous dioxane results are much larger: 5.2 and 4.6 in 50.7 and 79.5% aqueous dioxane, respectively (ref. 29). Since 4-pyrenylmethyl chloride also gave divergent results in these solvents, the high values are discounted. See text.

compound. XIII is reported to be much faster than benzyl chloride in 79.5% aqueous dioxane but much slower in 50.7% aqueous dioxane; in fact, XIII is reported[29] to solvolyze faster in 79.5% aqueous dioxane than in 50.5% aqueous dioxane. In contrast, all other alkyl halides solvolyze much slower in the less aqueous medium. 1-Pyrenylmethyl chloride, XIV, is reported[29] to be exceedingly reactive in the aqueous dioxane media; in this case the difference in solvolysis rates in the two solvents is much greater (by a factor of almost ten) than are other benzylic halides. These results are probably in error, perhaps as a consequence of the relatively low solubility and rates of solution of pyrene compounds.

CH$_2$Cl

CH$_2$Cl

XIII   XIV

As in analogous aromatic substitution, the present "σ"-values may be compared with MO quantities in place of the relative rate data. As mentioned above, we can try the same method that proved successful in carbonium ion equilibria, namely the comparison with π-energy differences of carbonium ions and the corresponding aromatic hydrocarbons; that is, $\Delta M = M_{\mathrm{ArCH_2}^+} - M_{\mathrm{ArH}}$. The values are listed in Table 12.6 and the correlation is displayed in Fig. 12.8.

We see immediately that there are two linear correlations both of which, at best, are only fair. Compounds of the α-naphthylcarbinyl type are slower than anticipated, undoubtedly as a result of steric hindrance to coplanarity of the methylene group caused by the proximity of the peri-hydrogen[30] (Fig. 12.9). This explanation finds confirmation in the increased retardation of 1-naphthyl positions in solvolyses of 1-arylethyl[34] and 2-arylpropyl chlorides.[28]

The slopes of the linear correlations for the benzylic and 1-naphthyl positions are drawn parallel,[30] although there is little justification for this procedure. The slope multiplied by the various "ρ-"values gives the effective β-value for each reaction. Dewar and Sampson have shown by an ingenious approximation technique how this effective value of β gives a measure of bonding to incoming and outgoing bases during reaction. Their method[28,30] is slightly modified in the following.

In the absence of direct ΔM values, many of which involve laborious calculations unless a computer is available, approximate values may be obtained for AH systems from the cycle:

$$\mathrm{Ar} \xrightarrow{\Delta E_1} \mathrm{ArC{=}C} \xrightarrow{\Delta E_2} \mathrm{ArC^+}$$

$$\Delta M$$

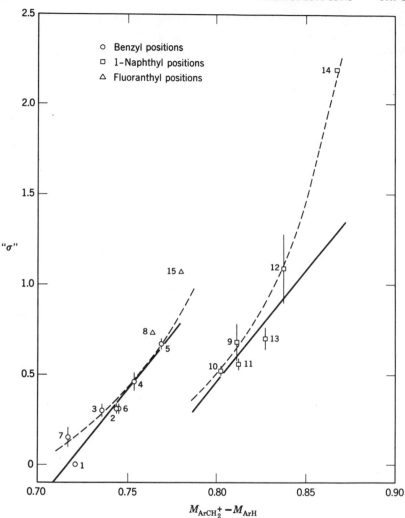

Fig. 12.8   Correlation of "σ"-values with HMO ΔM-values.

Fig. 12.9   Illustrating steric strain between coplanar methylene group in a 1-naphthyl type of position and the peri-hydrogen.

ΔE_1 is approximately constant and ΔE_2 is given approximately as $-2a_0\beta$ (4.23), in which a_0 is the NBMO coefficient of the exocyclic carbon in the carbonium ion; hence $\Delta M \simeq C - 2a_0\beta$. Values of $2a_0$ are listed in Table 12.6; the rather good correlation with actual ΔM-values is shown in

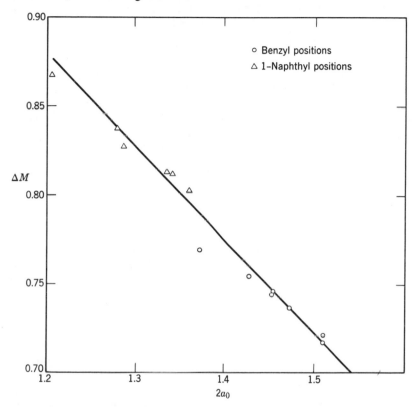

Fig. 12.10   Correlation of $2a_0$ of carbonium ions with actual ΔM-values.

Fig. 12.10. The slope is not -1, as expected from the derivation, but only -0.52, probably because ΔE_1 does reflect somewhat the changes in conjugating powers of the Ar group.

We may proceed further to the ArC—X model for the transition state discussed for $S_N 2$ reactions. Again we use a sequence:

$$\text{ArC}^+ \xrightarrow{\Delta E_3} \text{ArC}\overset{k}{\text{—}}\text{C} \xrightarrow{\Delta E_4} \text{ArC}\overset{k}{\text{—}}\text{X}^h$$

The bond integral coefficient of the leaving group (and entering group for a nonlimiting solvolysis) pseudo-orbital is k, and h is the corresponding Coulomb integral coefficient. ΔE_3 may be approximated as $2a_0 k\beta$.

Because ArC—C is an AH, ΔE_4 is given approximately as $qh\beta = h\beta$. Hence the over-all change, Ar → ArC—X, is given approximately by constant $- 2a_0(1 - k)\beta$. It follows from this that the more limiting the solvolysis, the weaker the bond to the leaving group, the smaller the magnitude of k, and the greater the magnitude of the slope of the correlation or the effective value of β for the reaction.

Instead of straight lines, the correlations in Fig. 12.8 may actually be curved upward. Such a situation would seem to be required by the relatively high reactivity of 1-pyrenylmethyl derivatives. We could account for such an effect on the basis that a solvolysis should be more limiting the more stable the carbonium ion; that is, k in ArC—X for the transition state may very well be a function of ΔM such that k is smaller the larger the value of ΔM.[28]

The solvolyses of the fluoranthylmethyl compounds are faster than expected from the other compounds. This is especially true of the 3-derivative. Fluoranthene is the only non-AH studied in this group and clearly does not follow the same quantitative correlation that represents the AH derivatives. The suggestion has been advanced that low bond order between the benzene and naphthalene rings in fluoranthene is primarily responsible;[35] perhaps also the neglected electron repulsion characteristics affect differently the non-self-consistent field of the non-AH.[36] Short of actual SCF calculations, we could investigate the ω-technique approach (Sec. 4.5).

Mason[37] has noted some remarkable linear correlations of a number of solvolysis and other carbonium ion reactions with the sums of the squares of the Hückel charge densities for odd-AH cations, $\Sigma \zeta_r^2$. Such an expression has been related to a solvation energy via a Born charging process. An exactly similar term results as a first approximation to an ω-technique. According to the ω-approximation, each α_r is altered by an amount $\omega(1 - q_r) = \omega\zeta_r$. The effect on the energy of the carbonium ion to the first approximation is then given by (3), derived from (4.29).

$$\delta E_\omega = \omega \sum_r q_r \zeta_r = \omega\left(1 - \sum_r \zeta_r^2\right) \tag{3}$$

If the transition state is approximated as the carbonium ion, the net energy change is given by

$$\Delta E_\omega = M_{\text{ArCH}_2^+} - M_{\text{ArH}} + \delta E_\omega$$
$$= \Delta M + \omega\left(1 - \sum_r \zeta_r^2\right) \tag{4}$$

[36] C. A. Coulson and M. J. S. Dewar, *Discussions Faraday Soc.*, **2**, 54 (1947).
[37] S. F. Mason, *J. Chem. Soc.*, **1958**, 808.

ΔE_ω-values from such a first approximation are recorded in Table 12.6; the correlation with "σ"-values is shown in Fig. 12.11. The correlations are improved in many respects. The 1-pyrenyl compound is still too reactive, although, if parallel correlations for benzyl and 1-naphthyl positions are no longer drawn, a linear correlation can be made to include

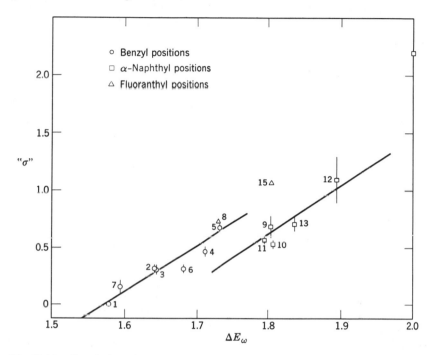

Fig. 12.11   Correlation of "σ" values with energy differences, ΔE_ω, derived from a first approximation ω-technique.

the 1-pyrene point. The 3-fluoranthyl point is also still too reactive, although it does fit the correlation for benzyl positions. A more complete ω-treatment gives no better results.

In (4) the $\Sigma \zeta_r^2$ term is generally dominant; hence a dependence of reaction rate on this term for carbonium ion reactions follows from the ω-technique as well as from the solvation energy approach. However, a further corollary of (4) is a proportionality between $\Sigma \zeta_r^2$ and the ionization potentials of odd-AH radicals (i.e., ΔM vanishes). An excellent linear correlation is actually observed[38] for this gas phase system in which solvation energies do not enter.

[38] A. G. Harrison, P. Kebarle, and F. P. Lossing, *J. Am. Chem. Soc.*, **83**, 777 (1961); cf. also A. Streitwieser, Jr., and P. M. Nair, *Tetrahedron*, **5**, 149 (1959).

Before any extensive set of ΔM-values became generally available for $ArH \rightarrow ArCH_2^+$, correlations were observed between solvolysis rates and localization energies; that is, instead of correlating with an increase in conjugation energy, rates of formation of carbonium ions actually correlated with a decrease in conjugation energy.[32,34] However, this correlation is not independent of those already discussed but may be derived from them. We recall that the localization energy for an even-AH is the π-bonding energy required to remove atom t bound to atoms r and s (Sec. 11.4) and is approximately proportional to $(a_{0r} + a_{0s})$ (Sec. 11.4) in which a_{0r} is the NBMO coefficient of atom r in the localized cation system. The additional π-bonding energy which results from adding atom u to atom t correlates with solvolysis rates and is proportional to b_{0u}, the NBMO coefficient of atom u in the resulting cation that now has two atoms more than the localized cation. Because of the properties of NBMO's of odd-AH compounds (p. 54), it may be readily shown that

$$b_{0u} = \frac{a_{0r} + a_{0s}}{\sqrt{(a_{0r} + a_{0s})^2 + 1}} \qquad (5)$$

Finally, we should mention that an S_N2 reaction, $ArCH_2Cl + KI$, has been included in this solvolysis discussion. It correlates as well as the solvolyses and has, as expected, a relatively low effective β-value. This parallel is in agreement with the prevailing thought that S_N1 and S_N2 reactions have basic similarities.

In principle, the effect of substituents on the rate of formation of carbonium ions should be determinable by the usual techniques for handling heteroatom substituents. No such treatments have been reported, but we might expect more or less the same success and limitations found for aromatic substitutions (Sec. 11.7); in particular, the type of field effect due to interactions with a substituent bond dipole may be significant as in the aromatic substitution case. The parallelism between S_N1 and S_N2 rates does not hold generally for benzylic compounds with heteroatom substituents probably for this reason.

The reaction of carbonium ions with Lewis bases is the microscopic reverse of a heterolytic dissociation such as the first step in a solvolysis. For reactive carbonium ions the transition state obviously resembles the carbonium ion. The best approach for determining the position of reaction in a conjugated carbonium ion is based on a perturbation of the carbonium ion. A carbon bearing a positive charge is relatively electronegative. Reaction with base disperses some of the charge and results in a lowering of the effective electronegativity at the carbon; that is, the reaction may be associated with a small positive $\delta\alpha$, or equivalently, a small negative h (Sec. 5.1). Because $\delta E \cong q_r\delta\alpha = q_r h\beta_0$, the greater the electron density at the position of reaction, the more bonding energy is decreased. Hence reaction should occur at the position of lowest electron density or of highest positive charge. This conclusion is identical to that resulting from a model based on simple electrostatic interaction between cation and base. For odd-AH cations the electron density is the same as the electron density of the frontier orbital which for this reaction is the lowest vacant MO; hence frontier electron theory[39] (Sec. 11.3) makes the same prediction.

Alternatively, the weak bond between the carbonium ion and the incoming base may be considered explicitly as contributing to the total π-framework. If the electronegativity difference of the incoming group is neglected, the energy change is the same as ΔE_3, which for cations of odd-AH is given simply as $2a_0k\beta$. Reaction occurs preferentially at the position of highest a_0—but this position is also that of highest positive charge. The different approaches lead to the same conclusions.

According to these predictions, benzyl cation should react with bases at the exocyclic carbon, and it does. In cinnamyl cation, XV, the α- and

XV

γ-carbons have identical a_0 and ζ-values. The theory does not differentiate between these positions. In fact, products of reaction at both positions are generally obtained; for example, reaction of cinnamyl or α-phenylallyl chlorides with silver acetate in acetic acid yields 53 to 58 % of the primary acetate.[40]

Pilar[41] has considered the problem of an allyl cation with varying α- and β-integrals. For a cation of the type of XVI, the atom polarizabilities

$$R—\overset{}{\underset{3}{C}}\cdots\overset{+}{\underset{2}{C}}\cdots\underset{1}{C}$$

XVI

of allyl cation ($\pi_{1,1} = 0.441$, $\pi_{1,2} = -0.177$, $\pi_{1,3} = -0.264$) show that an electron-donating R group such as an alkyl group which gives a negative h on the 3-carbon, decreases q_3 and increases q_1. Reaction of α-methylallyl and of α,α-dimethylallyl cations with bases occurs preferentially at the α-position as predicted.[40]

[39] K. Fukui, T. Yonezawa, and H. Shingu, J. Chem. Phys., **20**, 722 (1952).

[40] See the extensive summary of such reactions by R. H. DeWolfe and W. G. Young, Chem. Rev., **56**, 753 (1956).

[41] F. L. Pilar, J. Chem. Phys., **29**, 1119 (1958); **30**, 375 (1959).

12.4   Carbonium Ion Rearrangements

Carbon skeleton rearrangements frequently accompany carbonium ion reactions. It seems remarkable that such rearrangements are far less general in radical and carbanion reactions; however, this pattern is entirely explicable by simple LCAO theory.

A structure such as XVIII is usually written for the transition state of a carbonium ion rearrangement; such a structure apparently also occurs as an intermediate in some such reactions.

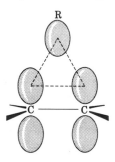

The orbital description of XVIII may undoubtedly be approximated as follows: $2p_z$-orbitals of C_α and C_β overlap π-wise with each other. An orbital of R overlaps partly σ-wise and partly π-wise with the C_α- and C_β-orbitals. Such a model is pictured in Fig. 12.12. Berson and Suzuki[42]

Fig. 12.12   Atomic orbital description of a carbonium ion rearrangement transition state or intermediate.

have recently discussed the question of the configuration of an alkyl group, R, during rearrangement. The conclusion that R is still tetrahedral or close to tetrahedral derives from the relative ineffectiveness of conjugating functions as in allyl and benzyl groups in migration aptitudes. Furthermore, these authors showed experimentally that the migrating group bears little positive charge.

A simple LCAO calculation may be carried out for XVIII by assigning appropriate values to k_{CR} and to h_R (Sec. 5.1). The calculation is simply that of an altered cyclopropenyl cation. The reader may show that for $h < 2k^2$ the secular equations have only one negative root. The C—R

[42] J. A. Berson and S. Suzuki, *J. Am. Chem. Soc.*, **81**, 4088 (1959).

bond is expected to be rather strong; hence k is relatively large. For most cases of interest XVIII has only one bonding MO which is occupied by the two electrons in XVIII originally associated with the C—R bond in the unrearranged cation. Only if h is rather large does a second MO become bonding to allow rearrangement of a carbanion. An example is the Stevens rearrangement:[43]

$$C_6H_5\overset{\overset{O}{\|}}{C}CH_2 \diagdown \diagup CH_2C_6H_5 \quad\overset{-OH}{\rightleftharpoons}\quad C_6H_5\overset{\overset{O}{\|}}{C}\overset{\ominus}{C}H \diagdown \diagup CH_2C_6H_5$$

$$\overset{+}{N} \qquad\qquad \overset{+}{N}$$

$$CH_3 \diagdown CH_3 \qquad CH_3 \diagup CH_3$$

$$C_6H_5\overset{\overset{O}{\|}}{C}CHCH_2C_6H_5$$

$$N(CH_3)_2$$

The presence of a highly electronegative ammonium nitrogen undoubtedly lowers the entire energy scale of the MO's to allow reaction via the modified cyclopropenium anion.

Rearrangement of carbonium ions is facilitated when the rearranged carbonium ion is more stable than the starting cation. The MO analogue may be exemplified by the following case:

$$\underset{XIX}{C{=}C{-}\overset{\overset{R}{|}}{C}{-}\overset{+}{C}} \;\rightarrow\; \underset{XX}{C{=}C{-}\overset{R}{\overset{+}{\triangle}}C} \;\rightarrow\; \underset{XXI}{C{=}\overset{+}{C}{=}C{-}\overset{\overset{R}{|}}{C}}$$

The product allyl cation, XXI, is more stable than the starting unconjugated cation, XIX. The increased stabilization is reflected in the electronic energy of XX. If we assume for simplicity that $k = 1$, $h = 0$, the calculation for XX is just that for vinylcyclopropenyl cation. The energy difference, $\Delta E_{XIX\to XX}$ is given as $\Delta E_\sigma + 2\alpha + 6.42\beta$, in which ΔE_σ is the term associated with loss of σ-bond energy. The corresponding energy change for XVII → XVIII is given as $\Delta E_\sigma + 2\alpha + 4.00\beta$. The vinyl-substituted case clearly involves less loss in bonding energy and should rearrange faster.[44]

---

[43] T. S. Stevens et al., *J. Chem. Soc.*, **1928**, 3193; **1930**, 2107, 2119; **1932**, 55, 1926; **1934**, 279.

[44] More extensive calculations on this theme have been presented recently by H. E. Zimmerman and A. Zweig, *J. Am. Chem. Soc.*, **83**, (1961).

Carbonium ion rearrangements are also facilitated by an increase in positive charge density at the α-carbon in the original cation; for example, such rearrangements are less common in S_N2 reactions than in related S_N1 solvolyses. This fact also follows from MO considerations. Consider the carbonium ion, XXII, which is encumbered by a nucleophilic group, Y. Letting $k_{CY} = 1$, $h_Y = 1$, we derive $\Delta E_{XXII \rightarrow XXIII} = \Delta E_\sigma + 2\alpha + 3.37\beta$. This case involves a greater loss in bonding energy than the original example, XVII → XVIII; rearrangement should be less facile.

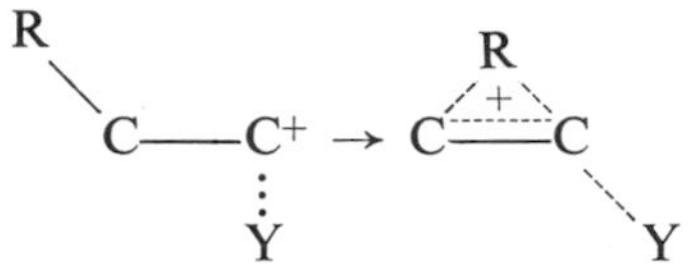

XXII   XXIII

Burr[45] has given an MO treatment of the transition state for rearrangement of an aryl group based on a model in which C_α, C_β, and the aromatic carbons are in one plane, XXIV. This model is almost certainly incorrect

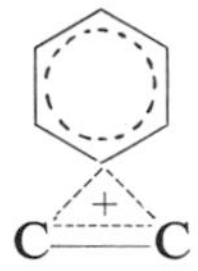

XXIV

and leads to erroneous results. One piece of evidence, for example, is the actual isolation of the intermediate in a reaction involving a rearranging p-phenolate anion.[46] The structure of this neutral compound is undoubtedly XXV, in which the plane of the phenyl group is perpendicular to the C—C bond.

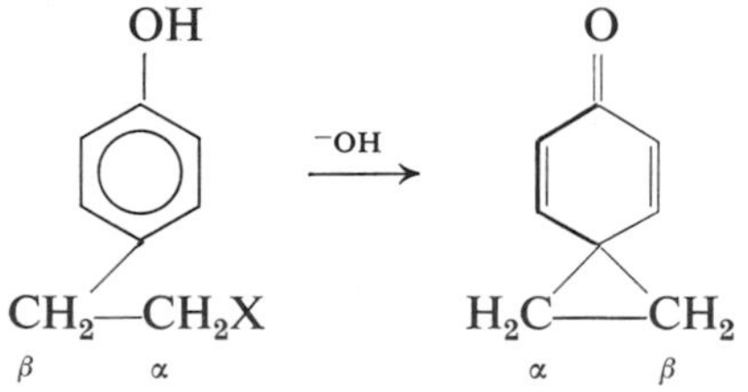

XXV

The corresponding ethylene-phenonium ion intermediates may be represented approximately by the orbital picture in Fig. 12.13a in which both a p_z orbital and an sp^2-orbital on the benzene ring overlap with the

[45] J. G. Burr, Jr., *Chem. and Ind.*, **1956**, 798.
[46] S. Winstein and R. Baird, *J. Am. Chem. Soc.*, **79**, 756 (1957).

two *p*-orbitals of the ethylene moiety. This picture is completely equivalent to a model in which the *p*- and *sp²*-orbitals in the benzene ring are rehybridized to form two hybrid *sp⁵*-orbitals to overlap with the ethylene *p*-orbitals.

All of these orbitals cannot contribute to the same MO for the reason that they have different symmetry properties. The most convenient

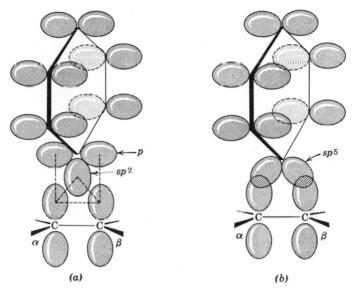

(a)   *(b)*

Fig. 12.13   These orbital pictures of a phenonium ion are equivalent.

procedure is to form pseudo-orbital linear combinations from the *p*-orbitals on C_α and C_β essentially as in the treatment of the S_N2 transition state (p. 367):

$$\varphi^+ = \frac{1}{\sqrt{2}} (\varphi_{C_\alpha} + \varphi_{C_\beta}) \qquad (6)$$

$$\varphi^- = \frac{1}{\sqrt{2}} (\varphi_{C_\alpha} - \varphi_{C_\beta}) \qquad (7)$$

The symmetry properties of φ^+ are the same as those of φ_{sp^2} on the benzene ring with respect to reflection through the plane of the benzene ring. These two orbitals can form two MO's, the bonding orbital of which will house the two electrons originally associated with the benzene-carbon σ-bond in the starting material. φ^- has the symmetry of the benzene-p_z orbitals. If we take, as usual, $H(\varphi^-, \varphi^-) = h\beta_0$ and $H(\varphi^-, \varphi_{\text{benzene-}p_z}) = k\beta_0$, the resulting secular equations are exactly those of a modified benzyl cation having six π-electrons (Fig. 12.14). This antisymmetrical part of the electronic structure will vary with the structure of the aryl group. Because of this relationship, we may anticipate a correlation between the relative ease of forming a phenonium ion and

Fig. 12.14   Antisymmetrical part of ethylenephenonium ion.

the relative ease of forming a corresponding benzyl cation.[44] Appropriate data are available in the relative migratory aptitudes of aryl groups in pinacol reactions:

$$
\underset{\substack{\displaystyle | \quad | \\ Ar_2 \quad Ar_2}}{\overset{\substack{OH \quad OH \\ \displaystyle | \quad |}}{Ar_1-C-C-Ar_1}} \rightarrow \underset{\substack{\displaystyle | \\ Ar_2}}{\overset{\substack{O \ Ar_1 \\ \displaystyle \| \ |}}{Ar_1C \ C-Ar_2}} + \underset{\substack{\displaystyle | \\ Ar_2}}{\overset{\substack{O \ Ar_1 \\ \displaystyle \| \ |}}{Ar_2C \ C-Ar_1}}
$$

Some values are listed in Table 12.7.[47] A comparison with solvolysis of substituted benzyl tosylates (Table 12.7, Fig. 12.15) shows a good correlation.

TABLE 12.7

MIGRATORY APTITUDES AND SOLVOLYSIS RATES

Aryl group	Relative Migratory Aptitude. Pinacol Rearrangement*	Relative Rate $ArCH_2OTs$, 76.6 mole % H_2O in Acetone, 25.3°†
Phenyl	1.00	1.00
m-Anisyl	1.6	0.61
p-Anisyl	500	25,000
m-Tolyl	1.95	1.8
p-Tolyl	15.7	30
p-Biphenylyl	11.5	17‡
p-Bromophenyl	0.7	0.41

* See ref. 47.

† J. K. Kochi and G. S. Hammond, *J. Am. Chem. Soc.*, **75**, 3445, 3452 (1953).

‡ Estimated from mutual correlations among the data of †, refs. 33 and 34.

[47] Such migratory aptitudes are the result largely of the work of W. E. Bachman. For further discussion and references cf. G. W. Wheland, *Advanced Organic Chemistry*, second edition, John Wiley and Sons, New York (1949), p. 513.

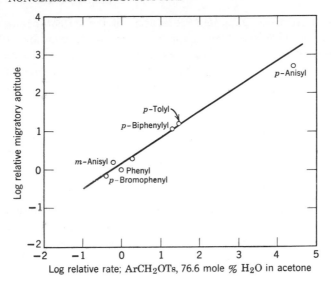

Fig. 12.15   Comparison of relative migratory aptitudes with solvolysis rates of benzyl tosylates.

12.5   Nonclassical Carbonium Ions

The so-called *nonclassical carbonium ions* are best represented by a bridged structure. Although isolated examples have been postulated at various times in the chemical literature, the first comprehensive proposal that such cations are important intermediates in organic reactions was developed by Dewar.[48]   At present, the relative stabilities of many such

XXVII   XXVI   XXVIII

cations, particularly of the type of XVIII in open chain systems, are still subject to considerable controversy. Nevertheless, in some cases, particularly with bicyclic structures, the evidence is substantial that nonclassical carbonium ions are important *intermediates* in many reaction sequences. One early example is the cation, XXVI, which seems clearly to be involved in solvolytic reactions of camphene hydrochloride, XXVII, and of

[48] M. J. S. Dewar, *The Electronic Theory of Organic Chemistry*, Oxford University Press, London (1949).

isobornyl derivatives, XXVIII.[8] Reactions of the parent *exo*-norbornyl compounds, XXIX, also form a corresponding bridged cation, XXX.[49,50] Such cations may be considered as the intermediate stage in a 1,2-rearrangement. Their structures are probably similar to those discussed in Sec. 12.4 (cf. Fig. 12.12); for example, the structure of XXX may probably

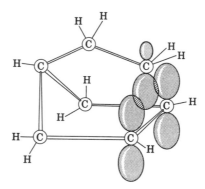

XXIX   XXX

be represented approximately as in Fig. 12.16. Such a structure derives from delocalization of the positive charge density of a carbonium ion center by overlap with a suitably oriented carbon-carbon single bond. The relative importance of such delocalization in the bicyclo[2:2:1]heptane

Fig. 12.16   Possible orbital structure of the nonclassical norbornyl cation.

compounds is undoubtedly due to the rigid orientation and strain within this compact ring system. As the electrons in the single bond become less firmly held, the importance of delocalization should become magnified. The single bond within a cyclopropane ring seems clearly to have such enhanced conjugating power. Some reactions of cyclopropylmethyl

[49] J. D. Roberts, W. Bennett, and R. Armstrong, *J. Am. Chem. Soc.*, **72**, 3329 (1950); J. D. Roberts and W. Bennett, *J. Am. Chem. Soc.*, **76**, 4623 (1954); J. D. Roberts, C. C. Lee, and W. H. Saunders, Jr., *J. Am. Chem. Soc.*, **76**, 4501 (1954); **77**, 3034 (1955).

[50] S. Winstein, B. K. Morse, E. Grunwald, H. W. Jones, J. Corse, D. Trifan, and H. Marshall, *J. Am. Chem. Soc.*, **74**, 1127 (1952); S. Winstein and D. Trifan, *J. Am. Chem. Soc.*, **74**, 1147, 1154 (1952); S. Winstein and K. C. Schreiber, *J. Am. Chem. Soc.*, **74**, 2156 (1952).

derivatives seem to involve carbonium ion intermediates in which de-localization is important;[8,51] for example, XXXI → XXXII. A particularly

interesting example has recently been provided by Winstein, Sonnenberg, and De Vries[52] who showed that solvolysis of *cis*-3-bicyclo[3.1.0]hexyl tosylate, XXXIII, involves the unusual cation, XXXIV.

The extension to the looser π-electrons of a double bond is obvious; suitably oriented double bonds clearly provide substantial stabilization of carbonium ions. The examples include such cases as the homoallylic cations, XXXV[53] and XXXVI,[54] derived from cholesteryl and norbornenyl systems, respectively, and the unusually stable cations, XXXVII,[55] XXXVIII,[56] and XXXIX.[57]

[51] C. G. Bergstrom and S. Siegel, *J. Am. Chem. Soc.*, **74**, 145, 254 (1952); J. D. Roberts and V. C. Chambers, *J. Am. Chem. Soc.*, **73**, 5034 (1951); J. D. Roberts and R. H. Mazur, *J. Am. Chem. Soc.*, **73**, 2509, 3542 (1951); J. D. Roberts and W. Bennett, *J. Am. Chem. Soc.*, **76**, 4623 (1954); S. Winstein, H. M. Walborsky, and K. C. Schreiber, *J. Am. Chem. Soc.*, **72**, 5795 (1950); R. H. Mazur, W. N. White, D. A. Semenow, C. C. Lee, M. S. Silver, and J. D. Roberts, *J. Am. Chem. Soc.*, **81**, 4390 (1959); S. Winstein and E. M. Kosower, *J. Am. Chem. Soc.*, **78**, 4347, 4354 (1956); **81**, 4399 (1959).

[52] S. Winstein, J. Sonnenberg, and L. De Vries, *J. Am. Chem. Soc.*, **81**, 6523 (1959).

[53] C. W. Shoppee, *J. Chem. Soc.*, **1946**, 1138; R. H. Davies, S. Meecham, and C. W. Shoppee, *J. Chem. Soc.*, **1955**, 679; C. W. Shoppee and D. F. Williams, *J. Chem. Soc.*, **1955**, 686; S. Winstein and R. Adams, *J. Am. Chem. Soc.*, **70**, 838 (1948); L. C. King and M. J. Bigelow, *J. Am. Chem. Soc.*, **74**, 6238 (1952); S. Winstein, E. Clippinger, A. H. Fainberg, R. Heck, and G. C. Robinson, *J. Am. Chem. Soc.*, **78**, 328 (1956); S. Winstein and E. Clippinger, *J. Am. Chem. Soc.*, **78**, 2784 (1956).

[54] J. D. Roberts and W. Bennett, *J. Am. Chem. Soc.*, **76**, 4623 (1954); J. D. Roberts, W. Bennett, and R. Armstrong, *J. Am. Chem. Soc.*, **72**, 3329 (1950); J. D. Roberts, C. C. Lee, and W. H. Saunders, Jr., *J. Am. Chem. Soc.*, **77**, 3034 (1955); S. Winstein, H. M. Walborsky, and K. Schreiber, *J. Am. Chem. Soc.*, **72**, 5795 (1950); S. Winstein and M. Shatavsky, *Chem. and Ind.*, **1956**, 56.

[55] S. Winstein, M. Shatavsky, C. Norton, and R. B. Woodward, *J. Am. Chem. Soc.*, **77**, 4183 (1955); S. Winstein and M. Shatavsky, *J. Am. Chem. Soc.*, **78**, 592 (1956); W. G. Woods, R. A. Carboni, and J. D. Roberts, *J. Am. Chem. Soc.*, **78**, 5653 (1956).

[56] S. Winstein and C. Ordronneau, *J. Am. Chem. Soc.*, **82**, 2084 (1960).

[57] G. Leal and R. Pettit, *J. Am. Chem. Soc.*, **81**, 3160 (1959).

XXXV

XXXVI

XXXVII

XXXVIII

XXXIX

The incorporation into many of the particularly stable nonclassical cations of a cyclopropenyl type of unit is clearly not without significance. A general MO treatment of homoallylic cations has been introduced by Winstein and Simonetta.[58] Within a cation such as that in Fig. 12.17

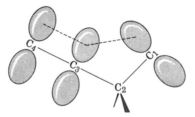

Fig. 12.17   Overlap of *p*-orbitals in a homoallylic cation.

both σ- and π-overlap are important between the *p*-orbitals on atoms 1 and 3. In the absence of reasonable alternatives the simple assumption may be made that β for the homo-interaction is proportional to the total overlap integral.[59] Both σ- and π-contributions to the total overlap integral must be determined from an assumed model of the geometry of the system (p. 15). The calculation of the homoallylic cation reduces to that of a modified allyl cation in which one of the β's is replaced by some suitable

---

[58] M. Simonetta and S. Winstein, *J. Am. Chem. Soc.*, **76**, 18 (1954).

[59] Winstein and Simonetta[58] actually assumed that β is proportional to $S/(1 + S)$. Values derived by this method are not significantly different from the simple proportionality to S.

value, $k\beta_0$. The extension of the procedure to cations of the type of **XXXVII** (Fig. 12.18) has been shown by Woods, Carboni, and Roberts.[60] The MO calculation reduces to that of a cyclopropenyl cation in which two of the β's are replaced by appropriate $k\beta$-values. A reasonable value for k for the cation, **XXXVII**, for example, seems to be about 0.3.[60,61]

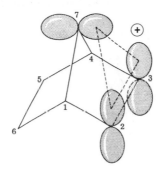

Fig. 12.18 p-Orbital overlap in the 7-norbornenyl nonclassical cation.

An interesting application of even qualitative aspects of MO theory may be made to the chemistry of some nonclassical cations. Wiley[62] has recently determined the rate of acetolysis of **XL** (OBs $= p$-bromo-benzenesulfonate); comparison with the data of Bartlett and Giddings[63]

OCH$_3$

OBs

OCH$_3$

XL

OBs

XLI

OCH$_3$ OBs

OCH$_3$

XLII

OBs

XLIII

[60] W. G. Woods, R. A. Carboni, and J. D. Roberts, *J. Am. Chem. Soc.*, **78**, 5653 (1956).
[61] A. Streitwieser, Jr., and J. B. Bush, unpublished calculations.
[62] G. A. Wiley, unpublished results.
[63] P. D. Bartlett and W. P. Giddings, *J. Am. Chem. Soc.*, **82**, 1240 (1960).

for the unsubstituted compound, XLI, showed that the two methoxy groups provided a rate enhancement of about 27-fold. In the related 7-benzonorbornenyl compounds, XLII and XLIII, the methoxy groups gave a rate enhancement of only 4-fold. It seems difficult to account for such a difference with simple resonance theory. Application of elementary perturbation concepts to MO theory provides a satisfactory interpretation.

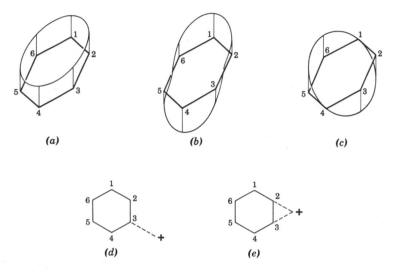

Fig. 12.19   Bonding MO's in benzene and interaction with an external carbonium ion center.

We consider first the three bonding MO's of benzene (Fig. 12.19). Addition of electron-donating methoxy groups at positions 1 and 4 will substantially lower the energies of (a) and (b) but will leave (c) unchanged because in (c) the methoxy groups are joined at a nodal position. The cation from XLI may be approximated as a benzyl cation, in which the bond to the exocyclic carbon represents a rather weak interaction ($k \simeq 0.3$) (Fig. 12.19d). This interaction will alter the energy of MO (a) but little because of the relatively large energy difference between this orbital and that of the perturbing cation. The benzene MO's (b) and (c) have substantial magnitude at position 3, and both will be affected by the perturbing cation in (d). Because of the effect of the methoxy groups on (b), the 1,4-dimethoxy groups will substantially stabilize the cation (d). In (e), however, symmetry precludes any interaction with (b). Interaction with (a) is again relatively small because of the energy difference; only (c) is substantially affected by the perturbing cation in (e). Yet this is the orbital that cannot interact with the two methoxy groups. In sum, because (a) is completely bonding, it is not greatly affected by interaction with the external cation in (e); the methoxy groups but not the external cation can interact with (b); the reverse is true for (c). The net result is that the methoxy groups in LXII should produce substantially less stabilization

XLIV

of the nonclassical cation than ordinarily expected. Similar considerations for the isomer, XLIV, predict that substantial stabilization of the carbonium ion by the the methoxy groups should ensue. These qualitative considerations that are based only on perturbation theory and symmetry properties are substantiated by more complete MO calculations.[61]

13 Radicals

13.1 Stability of Radicals

The average bond energy used in Chap. 9 for the evaluation of resonance energies is a hypothetical property of a bond which, summed over all of the bonds in a molecule, yields a close approximation to the total atomization energy or its equivalent for a large number of compounds. As such, this property differs fundamentally from the *bond dissociation energy*, a quantity that can be measured directly, at least in principle. The bond dissociation energy D is defined as the energy required to dissociate a molecule into two radicals:

$$\text{A} - \text{B} \rightleftharpoons \text{A} \cdot + \text{B} \cdot; \qquad \Delta H = D$$

Several techniques have been developed for the determination of bond dissociation energies; values are now available for a number of compounds.[1,2] As defined, the bond dissociation energy is an equilibrium heat. Some of the methods actually yield the energy of activation for the dissociation; however, since the combination of two radicals generally involves but little activation energy, for many cases the activation energy and the actual dissociation energy are close in value.

If one of the radicals is kept constant, the effect of structure in the second radical on D can be divided, in principle, into a σ- and a π-energy term. We assume that the σ-energy term is identical to D for an analogous system which involves no π-conjugation. Hence the difference in D for the system of interest from that of the standard compound is associated with the π-energy change which we may compare with MO calculations. For example, D-values for dissociation of allyl-Y and benzyl-Y compounds into Y radical and allyl and benzyl radicals, respectively, are known for several compounds (Table 13.1). The values are lower than those for the

[1] M. Szwarc, *Quart. Revs.*, **5**, 22 (1951).
[2] T. L. Cottrell, *The Strengths of Chemical Bonds*, second edition, Butterworths Scientific Publications, London (1958). A summary of many bond dissociation energies is given in an appendix.

corresponding methyl halides, and, in fact, the differences are rather constant for different leaving groups. These differences may be associated with the increase in π-conjugation energy attending the formation of allyl and benzyl radicals. The average values for these differences are: allyl-Y,

TABLE 13.1

BOND DISSOCIATION ENERGIES FOR ALLYL AND BENZYL COMPOUNDS[*]

Bond, R—Y	D Me-Y kcal.	D Allyl-Y kcal.	D Benzyl-Y kcal.	$D_{allyl\text{-}Y} - D_{MeY}$ kcal.	$D_{benzyl\text{-}Y} - D_{MeY}$ kcal.
R—H	102	80	83	−22	−19
R—Me	83	61.5	63	−21.5	−20
R—NH$_2$	80		59		−21
R—Cl	∼80	60	68?	∼−20	∼−12
R—SH	70?		53		−17?
R—SMe	73?		51		−22?
R—Br	67	46	51	−21	−16
R—I	53	36	39	−17	−14
			Average:	−20	−18

[*] Taken from ref. 2.

-20 ± 1 kcal., benzyl-Y, -18 ± 3 kcal., and may be related to the corresponding HMO ΔM-values, 0.828 and 0.721, respectively:

$$\Delta M(\text{allyl}) = M(\text{allyl}) - M(\text{ethylene}) = 2.828 - 2.000 = 0.828$$
$$\Delta M(\text{benzyl}) = M(\text{benzyl}) - M(\text{benzene}) = 8.721 - 8.000 = 0.721$$

The corresponding effective values of β_0 are -24 ± 1 kcal. and -25 ± 4 kcal., respectively. This agreement justifies use of this method and allows application to other systems.

The calculated difference in D between the central bonds in ethane and in bibenzyl is $2(0.828)\beta_0$, or 40 kcal.; the experimental values of $D_{Me\text{-}Me} = 83$ kcal., and $D_{benzyl\text{-}benzyl} = 47$ kcal. lead to an experimental difference of 36 kcal. However, a similar approach with the central bond in hexaphenylethane leads to a calculated D of -3 kcal., a value much lower than the experimental value of 11 kcal.[2,3] The discrepancy is undoubtedly due to the noncoplanarity of triphenylmethyl radical. The twisting of the benzene rings from a plane gives decreased conjugation interaction and increases D. The repulsion energy between the ortho-hydrogens has been estimated by Pritchard and Sumner[4] who calculate the angle of twist of

[3] Compare the earlier calculation of E. Hückel, Trans. Faraday Soc., 30, 40 (1934).
[4] H. O. Pritchard and F. H. Sumner, J. Chem. Soc., 1955, 1041.

the phenyl rings as 57°. Even with such a large angle of twist, the *DE* of triphenylmethyl radical is still considerable and undoubtedly is largely responsible for the stability of this radical.[5] Tris-*p*-biphenylylmethyl radical in which the odd electron is more extensively delocalized shows less of a tendency to dimerize.[6] Nevertheless, the dissociation of hexaarylethanes clearly is not determined by radical π-energies alone. The dimerization of 9-phenylfluorenyl radical is accompanied by a greater loss in π-energy than

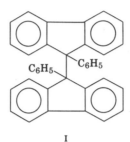

I

is triphenylmethyl radical (3.67β and 3.60β, respectively); yet, the dimer, I, is more stable than hexaphenylethane. The planarity of the fluorene ring system probably makes the actual difference even greater. Certainly a substantial portion of the tendency of hexaarylethanes to dissociate comes from steric repulsions between the aryl rings so crowded together.[7] The relief of such strain by dissociation is, of course, not considered in the MO calculations. The tying together of several rings as in I should reduce the strain and stabilize the ethane.

The energy required to form an aryloxy radical, ArO·, from a phenol, ArOH, should also be related to the corresponding π-energy difference. Suitable approximations to this quantity also provide satisfactory approaches. Hush,[8] for example, finds a linear correlation for the energy difference between the isoconjugate ArH and ArCH$_2$ systems. In this approximation, the conjugation of the phenolic oxygen with the ring is neglected. The correlation of the energies of highest occupied orbitals

[5] For calculation of some other radicals see E. Hückel, *Z. Elektrochem.*, **43**, 752 (1937); M. M. Kreevoy, *Tetrahedron*, **2**, 354 (1958).

[6] E. Müller, I. Müller-Rodloff, and W. Bunge, *Ann.*, **520**, 235 (1935); P. W. Selwood and R. M. Dobres, *J. Am. Chem. Soc.*, **72**, 3860 (1950); T. L. Chu and S. I. Weissman, *J. Am. Chem. Soc.*, **73**, 4462 (1951).

[7] C. S. Marvel, M. B. Mueller, C. M. Himel, and J. F. Kaplan, *J. Am. Chem. Soc.*, **61**, 2771 (1939); C. S. Marvel, J. F. Kaplan, and C. M. Himel, *J. Am. Chem. Soc.*, **63**, 1892 (1941); K. Ziegler, *Ann.*, **551**, 150 (1942). For further discussion see W. Hückel, *Theoretical Principles of Organic Chemistry*, Vol. I, Elsevier Publishing Co., Amsterdam (1955), p. 178.

[8] N. S. Hush, *J. Chem. Soc.*, **1953**, 2375.

with the closely related "critical oxidation potentials" of Fieser has been discussed earlier (p. 186).

We next inquire whether similar correlations hold for rates of reactions in which radicals are formed. Important in this type of reaction is the radical displacement reaction:

$$R—Y + Z· \rightarrow R· + Y—Z$$

Phenols act as inhibitors for autoxidation of organic compounds. There seems to be little question that phenols inactivate the chain-carrying radicals of such oxidation, although the details of the process have been a matter of controversy.[9] Bolland and ten Have[10] have suggested that the inhibition reaction is a hydrogen abstraction by a peroxide radical:

$$ArOH + ROO· \rightarrow ArO· + ROOH$$

In agreement with such a process, the autoxidation efficiencies correlate with the same MO functions that give correlation for the dissociation energies.[11,12]   However, Hammond and co-workers[13] conclude that inhibition involves formation of a complex between the phenol and the peroxide radical:

$$ArOH + ROO· \rightarrow [complex]$$

If this complex is of the charge-transfer type (i.e., one in which an electron transfer structure such as $ArOH^+ ROO^-$ is important; cf. Sec. 7.5), a correlation with the highest occupied orbital of the phenol is still to be expected. Fueno, Ree, and Eyring[12] have gone a step further and have considered formation of the complex to involve a structure that contains an extended π-network:

They treat the interaction of the oxygen with the radical, Z, as a perturbation and show that this model will also account for the observed inhibition efficiencies.   Its general importance will become increasingly evident in this chapter.

[9] For further details, see C. Walling, *Free Radicals in Solution*, John Wiley and Sons, Inc., New York (1957), p. 430.

[10] J. L. Bolland and P. ten Have, *Trans. Faraday Soc.*, **43**, 201 (1947).

[11] D. S. Davies, H. L. Goldsmith, A. K. Gupta, and G. R. Lester, *J. Chem. Soc.*, **1956**, 4926.

[12] T. Fueno, T. Ree, and H. Eyring, *J. Phys. Chem.*, **63**, 1940 (1959).

[13] G. S. Hammond, C. E. Boozer, C. E. Hamilton, and J. N. Sen, *J. Am. Chem. Soc.*, **77**, 3238 (1955).

A number of radical displacement reactions have been studied in which Y is hydrogen and Z· is a halogen atom or a radical such as trichloromethyl radical.[14] Relative reactivities of a number of C—H bonds towards CCl_3 are compared with the calculated increases in HMO π-energies in Table 13.2. Clearly, no correlation exists with the same type of MO

TABLE 13.2

REACTIVITIES OF C—H BONDS TOWARDS TRICHLOROMETHYL
RADICAL*

R—H R	Relative Rate	$M_{R·} - M_{RH}$[†]
Benzyl	0.42	0.72
Diphenylmethyl	3.35	1.30
Triphenylmethyl	7	1.80
9-Fluorenyl	47	1.35
α-Vinylbenzyl	12	1.38
Hept-3-en-1-yl	4.9	0.83‡

* Taken from E. C. Kooyman, *Discussions Faraday Soc.*, **10**, 163 (1951).
† Hyperconjugation is neglected.
‡ Calculated as allyl.

π-energy difference that accounts for variations in dissociation energies. Fluorene, for example, is exceptionally reactive, yet possesses a lower ΔM than either triphenylmethane or allylbenzene. The C—H bond dissociation energy in chloroform is greater than that of the C—H bonds compared in Table 13.2; the radical displacement reactions are exothermic. Consequently, the transition state for reaction probably does not resemble the hydrocarbon radical structure. This seems to be especially true for chlorine atom reactions; HCl has a bond strength of 103 kcal., which is greater than that of any hydrogen bound to a tetrahedral carbon.

$$X \cdot H - R \leftrightarrow X - H \cdot R \leftrightarrow X^- \dot{H} R^+$$
$$\text{IIa} \qquad\qquad \text{IIb} \qquad\qquad \text{IIc}$$

In the transition state the hydrogen is undoubtedly bonded in part to both incoming and leaving radicals. The relative electronegativities of Cl and CCl_3 lead to significant contributions of IIc to the resonance hybrid of the transition state.[15]

[14] For a review see ref. 9, Chap. 8.
[15] F. R. Mayo and C. Walling, *Chem. Rev.*, **46**, 191 (1950); H. C. Brown and A. B. Ash, *J. Am. Chem. Soc.*, **77**, 4019 (1955); C. Walling and B. Miller, *J. Am. Chem. Soc.*, **79**, 4181 (1957); J. Kenner, *Tetrahedron*, **3**, 78 (1958); G. A. Russell, *J. Org. Chem.*, **23**, 1407 (1958); *Tetrahedron*, **5**, 101 (1959); E. S. Huyser, *J. Am. Chem. Soc.*, **82**, 394 (1960).

The relative importance of IIc is reflected in the significance of polar factors in these radical reactions. Toward chlorine atoms cyclohexane is more reactive than toluene, despite its $\sim$17 kcal. greater C—H bond dissociation energy, probably because of the electron-attracting inductive effect of the benzene ring.[16]

The dipole character of the reaction site and the net positive charge at the carbon atom shows up in the resonance and inductive effects of substituents in reactions with substituted toluenes; in general, reaction is facilitated by electron-donating groups and is hindered by groups that are electron attracting.[15,17] The importance of resonance interactions is demonstrated for reactions of bromine atoms[18] or N-succinimidyl[18] or trichloromethyl[19] radicals with substituted toluenes by significantly better Hammett σ-ρ-correlations obtained with the use of σ^+-constants than with σ-constants; σ^+-constants tend to give better correlations when the transition state has important benzyl cation character.[20]

III

These effects are also shown by the extended π-system model for the transition state, III. As an example, we may examine a model with all β's equal, $h_H = 0$ and $h_X =$ some positive number. The change in energy between toluene and the transition state, III, is dissected into σ- and π-energy terms as usual. The latter is simply $\Delta E_\pi = E_{IV} - E_{\text{benzene}}$.

IV

If h_X were zero, IV would be cinnamyl radical, an AH having $M = 11.38$. Since all q's equal unity for odd-AH radicals, the first approximation to ΔE_π is $3.38 + h_X$ (cf. Sec. 4.4). Toluene bearing an electron-attracting

---

[16] H. C. Brown and G. A. Russell, *J. Am. Chem. Soc.*, **74**, 3995 (1952); **77**, 4578 (1955).

[17] E. C. Kooyman, R. van Helden, and A. F. Bickel, *Koninkl. Ned. Akad. Wetenschap. Proc.*, **B56**, 75 (1953); R. van Helden and E. C. Kooyman, *Rec. trav. chim.*, **73**, 269 (1954).

[18] G. A. Russell, *J. Org. Chem.*, **23**, 1407 (1958).

[19] E. S. Huyser, *J. Am. Chem. Soc.*, **82**, 394 (1960).

[20] Y. Okamoto and H. C. Brown, *J. Org. Chem.*, **22**, 485 (1957).

or electron-donating substituent in the *para*-position may be approximated by benzyl cation or anion, respectively. The corresponding model for the

X

V

transition state, V, has 9 or 11 electrons, respectively. If h_X were zero, both would have the same energy and ΔE_π would be the same for both; q at the X-position clearly is larger, however, for the 11-electron case. Hence the introduction of a positive h_X leads, in the first approximation, to a greater π-bonding energy for the electron-donating substituent. Actual values calculated for $h_X = 1$ are as follows:

X

8.000 → 12.656 $\Delta E_\pi = 4.656$

X

+ 8.721 → + 13.392 $\Delta E_\pi = 4.671$

X

− 8.721 → − 13.764 $\Delta E_\pi = 5.043$

For a given type of C—H bond, in the absence of polar effects, relative rates of hydrogen abstraction by radicals may parallel the dissociation energies and the calculated π-energy differences;[21] for example, we might anticipate a correlation between such reactions for a series of methylarenes, $ArCH_3$, and the MO quantities, $E(ArCH_2^\cdot) - E(ArH)$. Unfortunately, suitable data are not yet available for comparison.

13.2   Aromatic Substitution

In principle, we should be able to treat the reactions of multiple bonds with radicals by many of the same techniques used for comparable electrophilic substitutions. The French workers have advocated particularly the

[21] M. Szwarc and J. H. Binks in *Theoretical Organic Chemistry*, Special Publication no. 12, The Chemical Society, Butterworths Scientific Publications, London (1959), p. 262.

use of free valence for such correlations,[22] although Burkitt, Coulson, and
Longuet-Higgins[23] conclude that the localization energy provides the best
theoretical measure of the reactivity of a position towards free radicals
(cf. Chap. 11). Both are important concepts and both have limitations in
this application.[24]

Kooyman and Farenhorst[25] determined the relative reactivities of
trichloromethyl radical toward aromatic hydrocarbons and observed an

TABLE 13.3

RELATIVE REACTIVITIES OF HYDROCARBONS TOWARD $CCl_3^{\cdot}$

Hydrocarbon	Relative rate (Ref. 25)	Maximum F_r	Lowest $L_r^{\cdot}$
Benzene	$\sim 10^{-4}$	0.398	2.54
Biphenyl	7×10^{-4}	0.436	2.38
Triphenylene	$\sim 7 \times 10^{-3}$	0.439	2.38
Phenanthrene	4×10^{-3}	0.452	2.30
Naphthalene	10^{-2}	0.452	2.30
Chrysene	3.3×10^{-2}	0.457	2.24
Pyrene	0.3	0.469	2.19
Stilbene	0.5	0.481	2.16
1,2,5,6-Dibenzanthracene	1.85	0.498	2.13
Anthracene	11.0	0.520	2.01
Tetracene	26	0.530	1.93
1,2-Benzanthracene	30	0.514	2.04
1,2-Benzpyrene	70	0.529	1.96

excellent correlation with the highest free valence index of the hydrocarbon;
however, for AH the free valences and the localization energies are linearly
related to a good approximation[23] (Sec. 11.5). Kooyman and Farenhorst's
data correlate equally well with the radical localization energies, $L_r^{\cdot}$, of
the most reactive positions; the slope of the correlation corresponds to an
effective value of β of -20 kcal. (Table 13.3, Fig. 13.1).

[22] For some recent reviews see B. Pullman and A. Pullman in J. W. Cook's *Progress in Organic Chemistry*, Vol. 4, Butterworths Scientific Publications, London (1958), p. 31, and R. Daudel in I. Prigogine's, *Advances in Chemical Physics*, Vol. I, Interscience Publishers, New York (1958), p. 165.

[23] F. H. Burkitt, C. A. Coulson, and H. C. Longuet-Higgins, *Trans. Faraday Soc.*, **47**, 553 (1951).

[24] For the application of frontier electron theory see K. Fukui, T. Yonezawa, and C. Nagata, *J. Chem. Phys.*, **21**, 174 (1953).

[25] E. C. Kooyman and E. Farenhorst, *Trans. Faraday Soc.*, **49**, 58 (1953).

Szwarc and co-workers[26] in extensive research have developed a simple procedure for determining the relative reactivities of unsaturated compounds toward methyl radicals and have determined the "methyl affinities" of a

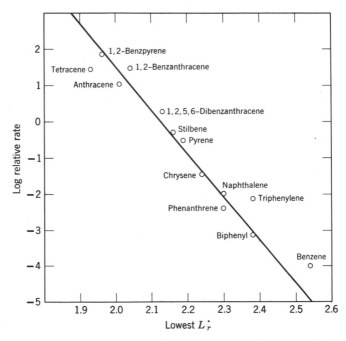

Fig. 13.1   Relative reactivities of aromatic hydrocarbons towards CCl_3^-.

large number of compounds. In effect, the procedure amounts to maintaining a dilute solution of acetyl peroxide and the hydrocarbon to be

[26] M. Levy and M. Szwarc, *J. Chem. Phys.*, **22**, 1621 (1954); *J. Am. Chem. Soc.*, **77**, 1949 (1955); M. Szwarc, *J. Polymer Sci.*, **16**, 367 (1955); M. Levy, M. S. Newman, and M. Szwarc, *J. Am. Chem. Soc.*, **77**, 4225 (1955); A. Rembaum and M. Szwarc, *J. Am. Chem. Soc.*, **77**, 4468 (1955); F. Leavitt, M. Levy, M. Szwarc, and V. Stannett, *J. Am. Chem. Soc.*, **77**, 5493 (1955); R. P. Buckley, F. Leavitt, and M. Szwarc, *J. Am. Chem. Soc.*, **78**, 5557 (1956); R. P. Buckley and M. Szwarc, *J. Am. Chem. Soc.*, **78**, 5696 (1956); M. Gazith and M. Szwarc, *J. Am. Chem. Soc.*, **79**, 3339 (1957); A. R. Bader, R. P. Buckley, F. Leavitt, and M. Szwarc, *J. Am. Chem. Soc.*, **79**, 5621 (1957); A. Rajbenbach and M. Szwarc, *J. Am. Chem. Soc.*, **79**, 6343 (1957); R. P. Buckley and M. Szwarc, *Proc. Roy. Soc.*, **A240**, 396 (1957); F. Leavitt, V. Stannett, and M. Szwarc, *Chem. and Ind.*, **1957**, 985; F. Leavitt, V. Stannett, and M. Szwarc, *J. Polymer Sci.*, **31**, 193 (1958); R. P. Buckley, A. Rembaum, and M. Szwarc, *J. Chem. Soc.*, **1958**, 3442; J. H. Binks and M. Szwarc, *J. Chem. Soc.*, **1958**, 226; F. Carrock and M. Szwarc, *J. Am. Chem. Soc.*, **81**, 4138 (1959); J. Gresser, J. H. Binks, and M. Szwarc, *J. Am. Chem. Soc.*, **81**, 5004 (1959); A. Rajbenbach and M. Szwarc, *Proc. Roy. Soc.*, **A251**, 394 (1959). For reviews see M. Szwarc, *J. Phys. Chem.*, **61**, 40 (1957), and ref. 21.

tested in isoöctane at an appropriate temperature. The acetyl peroxide decomposes to yield methyl radicals:

$$(CH_3COO)_2 \longrightarrow 2CH_3^\cdot + 2CO_2$$

The methyl radicals may abstract a hydrogen from the solvent to yield methane:

$$CH_3^\cdot + SH \longrightarrow CH_4 + S\cdot$$
$$S\cdot \longrightarrow \text{other products}$$

The methyl radicals may add to the unsaturated hydrocarbon, in which case methane is not produced:

$$CH_3^\cdot + M \longrightarrow CH_3M\cdot \longrightarrow \text{products}$$

The decrease in methane production caused by the added unsaturated compound is a measure of the reactivity of the substance relative to the solvent reaction.

TABLE 13.4

METHYL AFFINITIES OF AROMATIC HYDROCARBONS AND ATOM
LOCALIZATION ENERGIES

Hydrocarbon	Relative Methyl Affinity Per Reactive Position* $85°$	Lowest L_r'
Benzene	1.0	2.54
Biphenyl	7.5	2.38
Naphthalene	33.0	2.30
Phenanthrene	40.5	2.30
Chrysene	172	2.24
Pyrene	187	2.19
1,2,5,6-Dibenzanthracene	1110	2.13
1,2-Benzanthracene	3090	2.04
Anthracene	2460	2.01
1,2-Benzpyrene	4020	1.94
Tetracene	13900	1.93
Fluoranthene	480†	2.21

* From summary in ref. 21 and †.
† M. Szwarc, private communication.

Coulson[27] showed that the methyl affinities correlate well with localization energies. The correlation has been additionally developed by Szwarc.[21,26] (Table 13.4, Fig. 13.2). The slope of the correlation corresponds to $\beta = -11$ kcal., a rather small value, which suggests that localization has

[27] C. A. Coulson, *J. Chem. Soc.*, **1955**, 1435.

not proceeded far at the transition state. Furthermore, the correlation has its limitations. Product studies on the methylation of naphthalene give a 1-/2-ratio of 6.1 at 85°.[28] Localization energies with $\beta = -11$ kcal. give

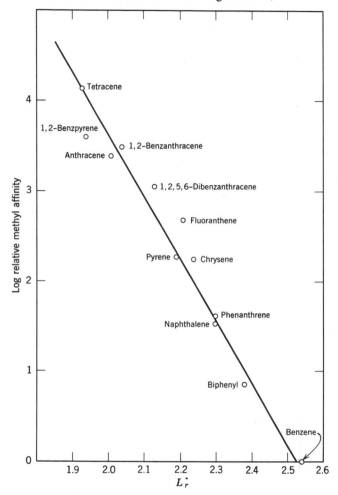

Fig. 13.2   Methyl affinities and localization energies.

a calculated ratio of 11. The 1-position may be less reactive than calculated because of steric interference by the *peri*-hydrogen. No correction for such possible steric hindrance has been made in the correlation of Fig. 13.2.

If the transition state resembles Fig. 13.3, the incoming methyl group provides an extension of the π-system by one atom. The use of such an

[28] J. A. Kent and R. O. C. Norman, *J. Chem. Soc.*, **1959**, 1724.

extended conjugation model for substitution on benzene, for example, would correspond to the use of benzyl radical as the model for the transition state instead of pentadienyl radical, the localized system. Because of the interrelationship between both types of MO quantities—localization of a position and extension of the π-system at the position by on atom—for AH systems, at least, we would expect the π-extension model to fit about as well as the localization model (compare the corresponding situation in

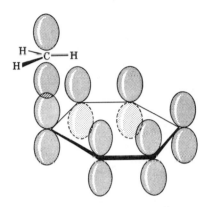

Fig. 13.3   Possible transition state for reaction of benzene with methyl radical.

cations, Sec. 12.3). The π-extension model, however, permits us to allow for the electron-donating effect of methyl groups by assigning a small negative h_X-value to the exocyclic position. Such a model would account for the increase in methyl affinity caused by electron-attracting substituents. The π-extension model would seem to be an attractive one for further study; it has not been considered in this connection in the literature.

The corresponding model for electrophilic substitution has been alluded to briefly (Sec. 11.5). In this model the transition state for reaction with benzene, for example, is handled as a benzyl cation, perhaps with different values for α of the exocyclic atom and for β of the bond to this atom. In short, the model is handled in the same manner as the explicit consideration of the methylene-type group of the localized atom using a Mulliken or conjugation model for hyperconjugation (Sec. 5.7).

Szwarc and Binks[21] have shown that the reaction of olefins with methyl radicals will follow the same localization energy correlation given by aromatic hydrocarbons if allowance is made for the variation of β with bond distance.[29]

For heterocyclic compounds derived from even-AH, it is easy to show

[29] See also B. Pullman and J. Effinger, *Bull. soc. chim. France*, **1958**, 482.

that the localization of carbon positions is unaffected by the heteroatom
to the first approximation: the π-energy of both the heterocycle and the
localized system is altered to the first approximation by an amount, $qh_X\beta$.
Both systems are AH in which q is unity at all positions; hence the correction is the same for both systems and cancels when the energy difference is

TABLE 13.5

METHYL AFFINITIES OF SOME NITROGEN HETEROCYCLES

Compound	Relative Methyl Affinity (Ref. 21)
Benzene	1
Pyridine	3
Pyrazine	~18
Naphthalene	22
Quinoline	29
Isoquinoline	36
Anthracene	820
Acridine	430
Phenazine	~250

taken. The effect of a nitrogen, for example, seems to be fairly small
(Table 13.5).[30] The argument does not hold, of course, for any special
reaction at the nitrogen position.

Relative reactivities of unsaturated compounds have been determined
toward ethyl,[31] n-propyl,[32] and isopropyl radicals.[33] The values are all
close to the methyl affinities.

Aromatic substitution by radicals is a significant synthetic process for
aryl radicals. These result from several important reactions: thermal
decomposition of diazonium chlorides or hydroxides,

$$ArN{=}NX + Ar'H \longrightarrow ArAr' + HX + N_2$$

[30] R. D. Brown, *J. Chem. Soc.*, **1956**, 272, has used the orientation effects in pyridine
toward the comparable phenyl radicals (*vide infra*) to derive the parameter values for
nitrogen: $h_N = 0.5$, $k_{CN} = 1$. Further calculations of B. Pullman and J. Effinger,
International Colloquium for Calculation of Wave Functions, CNRS, Paris, 1958, p. 351,
show that the correlation of methyl affinities of nitrogen heterocycles are indeed insensitive to the particular parameter values used. For comparable considerations of the
methyl affinities of quinones, see B. Pullman and S. Diner, *International Colloquium for
Calculation of Wave Functions*, p. 365, and T. Fueno and J. Furukawa, *Bull. Inst. Chem.
Research, Kyoto Univ.*, **36**, No. 4, 81 (1958).

[31] J. Smid and M. Szwarc, *J. Am. Chem. Soc.*, **78**, 3322 (1956).

[32] *Ibid.*, *J. Am. Chem. Soc.*, **79**, 1534 (1957).

[33] J. Smid and M. Szwarc, *J. Chem. Phys.*, **29**, 432 (1958).

decomposition of acylarylnitrosoamides,

$$ArN(NO)COR \longrightarrow [ArN{=}NOCOR] + Ar'H \longrightarrow ArAr' + N_2 + RCOOH$$

and the decomposition of diacyl peroxides,

$$(ArCOO)_2 + Ar'H \longrightarrow ArAr' + CO_2 + ArCOOH$$

The last reaction generally gives the highest yields of relatively tar-free products. Relative rate and orientation data are becoming increasingly available for such reactions, particularly for substituted benzene substrates.[34] Of the polycyclic aromatic hydrocarbons, naphthalene has been the most thoroughly studied. Several studies with phenyl radicals from different sources agree on yielding 80% reaction at the 1-position, 20% at the 2-position.[35] Together with the relative rates of reaction of naphthalene and benzene,[36] partial rate factors toward phenyl radicals are benzene, 1; 1-naphthalene, 17; 2-naphthalene, 4. A plot of these three points against atom localization energies yields $\beta \simeq -9$ kcal. for the correlation. With this value, the calculated partial rate factors are benzene, 1; 1-naphthalene, 26; 2-naphthalene, 2, in only fair agreement with the experimental results. The low β-value suggests, as for the methyl radical, that localization is but little underway at the transition state; the relatively poor correlation also suggests that steric interactions with *peri*-hydrogens may be important. Nevertheless, the qualitative order of reactivity of phenanthrene positions toward phenyl radicals, $9 > 1 > 3 > 2$, agrees completely with theory.[37]

Relative rates of arylation appear to be dependent not only on the nature of polar substituents in the aromatic substrate but also on the polar groups within the aryl radical. Reaction is facilitated when the substituents in the two rings are of opposite polarity. Examples are summarized in Table 13.6. Nitrobenzene is less reactive than toluene towards *p*-nitrophenyl radical; the order of reactivities is reversed toward *p*-tolyl radical.[34b] Such behavior is explicable in resonance language with the use of

[34] For recent reviews see (a) O. C. Dermer and M. T. Edmison, *Chem. Rev.*, **57**, 77 (1957); (b) D. R. Augood and G. H. Williams, *Chem. Rev.*, **57**, 123 (1957); (c) D. H. Hey in W. A. Water's *Vistas in Free Radical Chemistry*, Pergamon Press, New York (1959), p. 209; (d) G. H. Williams, *Homolytic Aromatic Substitution*, Pergamon Press, New York (1960).

[35] R. Huisgen and G. Sorge, *Ann.*, **566**, 162 (1950); R. Huisgen and R. Grashey, *Ann.*, **607**, 46 (1957); D. I. Davies, D. H. Hey, and G. H. Williams, *J. Chem. Soc.*, **1958**, 1878; B. A. Marshall and W. A. Waters, *J. Chem. Soc.*, **1959**, 381.

[36] R. Huisgen, F. Jacob, and R. Grashey, *Ber.*, **92**, 2206 (1959).

[37] A. L. J. Beckwith and M. J. Thomson, *J. Chem. Soc.*, **1961**, 73.

appropriate charge-separated resonance structures. The trends are difficult to show with a localization MO model but follow readily from the π-extension model. The argument is an extension of that developed above

TABLE 13.6
RELATIVE RATES OF ARYLATION WITH ARYL RADICALS AT 80° (REF. 34*b*)

	Rate Relative to Benzene of Substituted Benzene		
Radical	Nitrobenzene	Chlorobenzene	Toluene
p-$NO_2C_6H_4^{\cdot}$	0.94	1.17	2.61
p-$ClC_6H_4^{\cdot}$	1.53	1.02	1.32
p-$CH_3OC_6H_4^{\cdot}$	2.92	1.56	—
$C_6H_5^{\cdot}$	4.0	1.49	1.68
p-$CH_3C_6H_4^{\cdot}$	5.13	2.05	1.03

for the effect of polar groups in hydrogen abstraction from substituted toluenes and may be worked out in more detail by the reader.

Even the π-extension model does not seem to be entirely satisfactory, at least in its simplest form. The reaction of benzyl radicals with azulene has been reported recently to provide comparable amounts of 1- and 2-benzylazulene.[38] Both the localization and the π-extension models

TABLE 13.7
RADICAL REACTIONS WITH AZULENE

Position	Localization Model $\Delta L_r^{\cdot}$	π-extension Model $\Delta M^{\cdot}$	Free Valence F_r
1-position	2.262	0.8254	0.480
2-position	2.362	0.7927	0.420
4-position	2.240	0.8417	0.482
5-position	2.341	0.8018	0.429
6-position	2.559	0.8161	0.454

predict that the 4-position should be the most active, followed by the 1-position (Table 13.7). The 2-position should be the least reactive according to both models. There seems to be no way within these models to interpret this surprising experimental observation. The same predictions follow also from the free-valence indices.

[38] J. F. Tilney-Bassett and W. A. Waters, *J. Chem. Soc.*, **1959**, 3123.

A final word is necessary. Relative rates of arylation are normally determined by analysis of the substitution products which are frequently formed in mediocre yield and accompanied by more or less by-product. An assumption implicit in all such work is that the relative amounts of different monosubstituted products reflect the relative rates of reaction with the attacking radical and are not distorted by subsequent reactions of the first-formed products. For example, if two positions have equal reactivity toward a given radical but the initial product of one undergoes more extensive subsequent reaction and loss, a simple product analysis would lead to the conclusion that the two positions have unequal reactivities.

Kice and co-workers[39] have recently determined the reactivities of some non-AH hydrocarbons toward polymethacrylate radicals. The results, summarized in Table 13.8 show no correlation at all with localization energies. Dibenzofulvene, for example, should involve greater loss in π-energy on combination with a radical at the methylene position than should dibenzoheptafulvene, yet the former is by far the more reactive. The discrepancies may be due at least in part to large charge-transfer interactions.

13.3 Copolymerization

The free radical polymerization of a monomer olefin starts with addition of an initiator radical to the double bond with formation of a new radical, which, in turn, can add to another olefin. Polymerization of styrene, for example, takes place as follows:

$$\text{In·} + \phi\text{—CH}=\text{CH}_2 \longrightarrow \text{InCH}_2\dot{\text{C}}\text{H}\phi$$

$$\text{InCH}_2\overset{\displaystyle\phi}{\underset{\displaystyle|}{\text{CH}}}\text{·} + \text{CH}_2=\overset{\displaystyle\phi}{\underset{\displaystyle|}{\text{CH}}} \longrightarrow \text{InCH}_2\overset{\displaystyle\phi}{\underset{\displaystyle|}{\text{CH}}}\text{CH}_2\overset{\displaystyle\phi}{\underset{\displaystyle|}{\text{CH}}}\text{·}$$

$$\cdot$$
$$\cdot$$
$$\cdot$$

$$\text{InCH}_2(\overset{\displaystyle\phi}{\underset{\displaystyle|}{\text{CH}}}\text{CH}_2)_x\overset{\displaystyle\phi}{\underset{\displaystyle|}{\text{CH}}}\text{·} + \text{CH}_2=\overset{\displaystyle\phi}{\underset{\displaystyle|}{\text{CH}}} \longrightarrow \text{InCH}_2(\overset{\displaystyle\phi}{\underset{\displaystyle|}{\text{CH}}}\text{CH}_2)_{x+1}\overset{\displaystyle\phi}{\underset{\displaystyle|}{\text{CH}}}\text{·}$$

After a few monomer units have joined together, the character and properties of the growing radical are virtually independent of the number of monomer units present; for example, when x is sufficiently large, the difference between x and $x + 1$ is small. We may symbolize the reaction

[39] J. L. Kice, *J. Am. Chem. Soc.*, **80**, 348 (1958); J. L. Kice and F. M. Parham, *J. Am. Chem. Soc.*, **80**, 3792 (1958); J. L. Kice and F. Taymoorian, *J. Am. Chem. Soc.*, **81**, 3405 (1959).

TABLE 13.8

RELATIVE REACTIVITIES OF SOME HYDROCARBONS TOWARD
POLYMETHACRYLATE RADICALS

Hydrocarbon	M Hydrocarbon	M Localized Radical (Terminal Position)	$L^{\cdot}$ Terminal Position	Relative Reactivity (Ref. 39)
Styrene	10.424	8.721	1.703	12
1,1-Diphenylethylene	18.815	17.301	1.514	7.2
Dibenzofulvene	19.224	17.725	1.499	2500
Dibenzoheptafulvene	21.684	20.262	1.422	1.5

at this stage by the following expression in which P· is the polymer radical and M is the monomer:

$$P· + M \longrightarrow P·$$

Note that in the *homopolymerization* of styrene the growing radical is always a benzyl radical. If two monomers, M_1 and M_2, are present, either can react with the growing polymer radical. When the polymer radical reacts with M_1, the radical end of the new polymer radical is determined only by M_1; this polymer radical may be symbolized as $P_1·$. Similarly, reaction of a polymer radical with M_2 yields $P_2·$. We assume that the polymer radical, $\cdots M_1 - M_1·$, behaves just like $\cdots M_2 - M_1·$.[40] Both are symbolized as $P_1·$.

Four reactions are possible within the system:

$$P_1· + M_1 \xrightarrow{k_{11}} P_1·$$

$$P_1· + M_2 \xrightarrow{k_{12}} P_2·$$

$$P_2· + M_1 \xrightarrow{k_{21}} P_1·$$

$$P_2· + M_2 \xrightarrow{k_{22}} P_2·$$

We define the reactivity ratios: $r_1 = k_{11}/k_{12}$; $r_2 = k_{22}/k_{21}$. Note that $r > 1$ means that a radical prefers its own type of molecule; $r < 1$ means a preference for the other molecule.[41]

Localization theory predicts that for any two molecules r is either greater than or less than unity, regardless of the nature of P·; a variation in r simply reflects a variation in the effective value of β for the reaction.[42] In many actual cases this type of behavior obtains; for example, for M_1 = styrene, M_2 = butadiene, $r_1 = 0.65$, $r_2 = 1.83$. For M_1 = styrene, M_2 = 2-vinylnaphthalene, $r_1 = 0.5$, $r_2 = 1.4$.[43] In such cases the polymer will tend to contain more units of the more reactive monomer.

For many pairs of monomers, however, *both* r's are either greater than or less than unity. When r_1 and r_2 are both greater than unity, the system tends to form a mixture of two homopolymers. When both r_1 and r_2

[40] This is clearly only a reasonable first approximation. Actually, there are established cases of small but measureable differences in behavior between $\cdots M_1 - M_1·$ and $\cdots M_2 - M_1·$; for an example, cf. F. R. Mayo, A. A. Miller, and G. A. Russell, *J. Am. Chem. Soc.*, **80**, 2500 (1958).

[41] For further details see ref. 9, Chap. 4.

[42] For one application of localization theory to radical polymerizations, see R. Bhattacharya and S. Basu, *Trans. Faraday Soc.*, **54**, 1223 (1958).

[43] Copolymerization data are available in ref. 41 and in a number of books devoted to polymerization; examples are T. Alfrey, Jr., J. J. Bohrer, and H. Mark, *Copolymerization*, Interscience Publishers, New York (1952), and F. W. Billmeyer, Jr., *Textbook of Polymer Chemistry*, Interscience Publishers, New York (1957).

are less than unity, the system tends to form an alternating polymer, $-M_1-M_2-M_1-M_2-M_1-M_2-$. The localization approximation fails to explain this phenomenon.

Fukui and his associates[44] have found that the π-extension theory provides a satisfying interpretation of the varying copolymerization ratios. In fact, their application seems to be the first use of the π-extension model in the literature. In this model the attacking polymer radical and the olefin substrate are joined by a p_σ-bond to form an extended π-system. The transition state is modeled after Fig. 13.4. These authors treated the bond

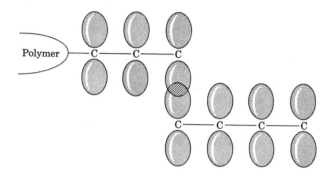

Model: $C\!\!=\!\!\!=\!\!\!C\!\!=\!\!\!=\!\!\!C----C\!\!=\!\!\!=\!\!\!C\!\!=\!\!\!=\!\!\!C\!\!=\!\!\!=\!\!\!C$

Fig. 13.4   Possible transition state for reaction of a polymer radical having an allyl radical end with butadiene.

between the two units as a perturbation and applied the results of perturbation theory to a host of copolymer systems. Alternatively, we can assign a value to the k for this bond and calculate in the usual way by solving the secular equations for the extended system. Even a value of unity for k gives satisfactory answers for many cases, although a somewhat smaller value, say $k = 0.7$–0.8, would seem to be more realistic.

With the choice of $k = 1$, however, calculations reduce to simple HMO computations and serve to demonstrate the procedure. As an example, we

[44] T. Yonezawa, K. Hayashi, C. Nagata, S. Okamura, and K. Fukui, *J. Polymer Sci.*, **14**, 312 (1954); K. Hayashi, T. Yonezawa, C. Nagata, S. Okamura, and K. Fukui, *J. Polymer Sci.*, **20**, 537 (1956); T. Yonezawa, T. Higashimura, K. Katagiri, S. Okamura, and K. Fukui, *Kobunshi Kagaku*, **14**, 533 (1957); T. Fueno, T. Tsuruta, and J. Furukawa, *Nippon Kagaku Zasshi*, **78**, 1075 (1957). Some of the *h*- and *k*-parameters used by these authors for heteroatoms are open to criticism; however, these papers are among the first to discuss the π-extension model.

may treat the copolymerization of styrene (M_1) and butadiene (M_2). P_1 then becomes benzyl radical and P_2 is an allyl radical. The four possible reactions give the results:

8.721   10.424   19.911
$P_1\cdot$ M_1 $\Delta E_{11}=0.766$

8.721   4.472   13.982
$P_1\cdot$ M_2 $\Delta E_{12}=0.790$

2.828   10.424   13.983
$P_2\cdot$ M_1 $\Delta E_{21}=0.731$

2.828   4.472   8.055
$P_2\cdot$ M_2 $\Delta E_{22}=0.755$

Butadiene is the more reactive monomer toward both radicals; hence, r_1 should be <1, r_2 should be >1, in complete agreement with the experimental findings (*vide supra*). Indeed, the magnitude of the difference is the same toward both radicals and we should expect $r_1r_2 \simeq 1$. The experimental product is 1.2. The detail of such agreement is remarkable for such a crude model and should not be expected generally.

Another example is provided by the copolymerization of styrene (M_1) and maleic anhydride (M_2). As our simplified model for maleic anhydride, we use the structure, VI, in which we take all β's equal and $h_X = 1$. Correspondingly, our model for P_2' is VII.

The four reactions reduce to:[45]

8.721	10.424	19.911
$P_1 \cdot$	M_1	$\Delta E_{11} = 0.766$

8.721	9.464	18.973
$P_1 \cdot$	M_2	$\Delta E_{12} = 0.788$

4.939	10.424	15.282
$P_2 \cdot$	M_1	$\Delta E_{21} = -0.081$

4·939	9.464	14.190
$P_2 \cdot$	M_2	$\Delta E_{22} = -0.213$

The results lead to the expectation that both r_1 and r_2 are less than unity. The experimental numbers are $r_1 = 0.01 - 0.04$, $r_2 = 0$. The significance of these examples is to show that simple MO methods can account for the gross aspects of such copolymerization behavior in terms of π-energies alone, although inductive effects and dipole-dipole interactions probably contribute to the finer details. The same technique clearly can be extended to ionic polymerization; the Japanese innovators have already published some of their results in this connection.[46]

[45] Most of the π-extension data in this chapter are unpublished calculations by the author in collaboration with J. B. Bush and J. I. Brauman.

[46] T. Yonezawa, T. Higashimura, K. Katagiri, K. Hayashi, S. Okamura, and K. Fukui, *J. Polymer Sci.*, **26**, 311 (1957); T. Fueno, T. Tsuruta, and J. Furukawa, *Nippon Kagaku Zasshi*, **78**, 1080 (1957).

14 Carbanions

14.1 Acidity of C—H Bonds

Hydrocarbons vary in acidity over a wide range. Aliphatic saturated hydrocarbon anions are known only as salts which are undoubtedly largely covalent in bonding; for example, ethyl lithium, methyl sodium. Conjugation of the anionic center to unsaturation functions increases the stability of the anion and the acidity of the corresponding hydrocarbon. Alkali metal salts of such anions undoubtedly possess increased ionic bonding; for example, benzyl sodium is colored.

Few hydrocarbons are sufficiently acidic to be measurable in the usual ways. One of these is fluoradene, I, which, with a pK_a of 13.5 in 97% aqueous methanol, is perhaps the most acidic hydrocarbon known.[1]

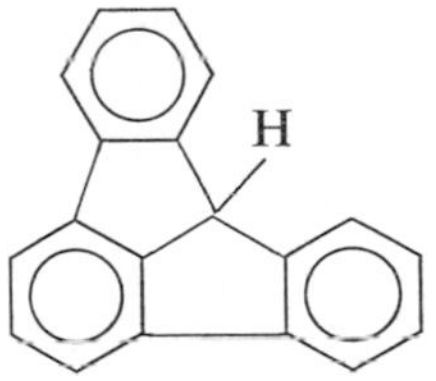

I

Actually, rather few determinations of hydrocarbon acidities have been reported. Most of the techniques compare relative acidities via equilibria between organometallic derivatives in inert solvents; for example,

$$R'Na + R''H \rightleftharpoons R'H + R''Na$$

The position of equilibrium has been adjudged by carbonation and analysis of the mixture of resulting carboxylic acids, although in other cases the direction of equilibrium has been determined simply from the color of the mixture.

[1] H. Rapoport and G. Smolinsky, *J. Am. Chem. Soc.*, **82**, 934 (1960).

413

TABLE 14.1
ACIDITY OF HYDROCARBONS

Compound		ΔM	pK Calculated (Ref. 8)	pK Experimental (Refs. 2, 4.)
Indene		1.747	(21)	21
9-Phenylfluorene		1.980	18	21
Fluorene		1.523	(25)	25
Diphenyl-p-biphenylylmethane		1.819	20	31
Diphenylmethane	$\phi\diagdown\diagup\phi$	1.301	28	35
Toluene		0.721	(37)	37
Cycloheptatriene		1.110	31	>25
Perinaphthene		1.697	22	16–25
Cyclopentadiene		2.000	17	16*
Fluoradene		2.115	15	11,† 13.5‡

TABLE 14.1 (continued)

Compound		ΔM	pK Calculated (Ref. 8)	pK Experimental (Refs. 2, 4.)
Cyclopropene		0	~48	high
1,2,3-Triphenylcyclopropene		1.301	28	high
4,5-Methylenephenanthrene		1.514	25	<31§

* R. E. Dessy, personal communication.

† In water (ref. 1).

‡ In 97% aqueous methanol (ref. 1).

§ A. Streitwieser, Jr., and J. I. Brauman, unpublished results.

The early work of Conant and Wheland[2] provided in this way an order of relative acidities of a number of hydrocarbons. Even at this early stage of the development of the simple MO theory of organic chemistry, Wheland[3] showed that this order was consistent with HMO theory; he made the now usual assumption that σ-bond energy changes were constant for a series and that the acidity differences were determined by the π-energy changes, $E_\pi(ArCH_2^-) - E_\pi(ArH)$, for the process, $ArCH_3 \rightarrow ArCH_2^-$.

Shortly thereafter McEwen[4] amplified the experimental techniques and derived a list of approximate pK-values for a series of hydrocarbons (Table 14.1). This list, although more than two decades old, has not yet been replaced by more accurate values. Relative orders of acidity of other hydrocarbons have been derived from metalation reactions, principally by Morton[5] and Gilman.[6]

That McEwen's pK-values are not unreasonable was shown by

[2] J. B. Conant and G. W. Wheland, *J. Am. Chem. Soc.*, **54**, 1212 (1932).

[3] G. W. Wheland, *J. Chem. Phys.*, **2**, 474 (1934).

[4] W. K. McEwen, *J. Am. Chem. Soc.*, **58**, 1124 (1936).

[5] A. A. Morton, *Chem. Rev.*, **35**, 1 (1944).

[6] H. Gilman in *Organic Reactions*, Vol. VIII, John Wiley and Sons, New York (1954), p. 258.

Shatenshtein,[7] who used the rate of base-catalyzed proton abstraction as a measure of acidity. A number of weak acids exchange the acidic hydrogens in liquid deuteroammonia; Shatenshtein found $\log k$ for such rates to correlate well with McEwen's pK-values. The occurrence or nonoccurrence of

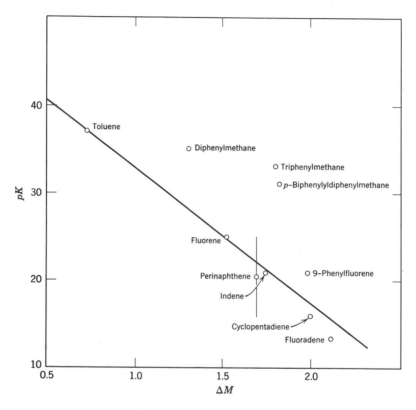

Fig. 14.1 Acidity of hydrocarbons.

other base-catalyzed reactions serve to provide limits for the pK-values of still further hydrocarbons.[8]

A plot of these estimated pK-values against the ΔE_π-values, as calculated by HMO theory (Table 14.1), shows a good linear correlation for the carbanions that are expected to be coplanar[8] (Fig. 14.1). In this treatment any strain energies are assumed to be the same in hydrocarbon and carbanion. Furthermore, any differences between the conversions, C_{sp^3}—C_{sp^3} → C_{sp^2}—C_{sp^3}, and C_{sp^3}—H → C_{sp^2}—H, are ignored (cf. Sec. 12.2).

[7] A. I. Shatenshtein, *Doklady Akad. Nauk. S.S.S.R.*, **60**, 1029 (1950).
[8] A. Streitwieser, Jr., *Tetrahedron Letters*, No. 6, 23 (1960).

Phenyl substituents do not decrease the pK as much as predicted by the simple correlation; the discrepancy is about $4\,pK$ units per phenyl group, an amount consistent with a reasonable degree of twisting of the phenyl group from coplanarity. The experimental relative acidities of cyclo-heptatriene and cyclopentadiene are in complete accord with the expecta-tions from Hückel's $4n + 2$ rule (Secs. 10.4 and 10.6).

Shatenshtein has extended the rate of exchange procedure to a number of hydrocarbons of rather low acidity by the use of the strong base KND_2 in liquid ND_3.[9] Related to this method is the protodedeuteration of toluene-α-d with lithium cyclohexylamide in cyclohexylamine. In this reaction substituent methyl groups are found to slow the rate of exchange; relative rates are H, 1.00; m-methyl, 0.68; p-methyl, 0.29; α-methyl, 0.11.[10]

In the inductive model of the methyl group the attached carbon is given a small negative h_r (Sec. 5.7). From the first approximation of the effect of a change of α on the π-energy, $\Delta E_\pi = q_r h_r \beta$, we derive the effect of a methyl group on the π-energy change for the ionization equilibrium as:

$$\Delta\Delta E_\pi = [q_r(\text{ArCH}_2^-) - q_r(\text{ArII})]h_r\beta \tag{1}$$

For methyl substitution at the *para*-position of toluene, the electron density difference within the brackets is positive $(1.143 - 1 = 0.143)$; since h_r is negative, the methyl substituent causes a decrease in the addi-tional conjugation energy attendant on ionization. The simple theory predicts that a *meta*-methyl substituent should have no effect (q_r for both ArH and ArCH_2^- are unity in the *meta*-position), but inductive effects will put some negative charge at this position. The effect of methyl substituents on acidity is in qualitative accord with the expectations of simple MO theory.

The effect of replacing a ring carbon by an electronegative atom such as nitrogen can be examined qualitatively by the same theoretical procedure.[11] Toluene is converted to γ-picoline, II, by assigning a positive h_r to the *para*-carbon. From (1) it is clear that γ-picoline should be a stronger acid than toluene, in accord with experiment. β-Picoline, III, is also a stronger acid than toluene, a fact that may also be attributed to inductive effects.[12]

[9] Reviewed by A. I. Shatenshtein, *Uspekhi Khim.*, **24**, 377 (1955); *Isotope Exchange and Replacement of Hydrogen in Organic Compounds*, (in Russian), Academy of Sciences of the U.S.S.R., Moscow (1960).

[10] A. Streitwieser, Jr., and D. E. van Sickle, unpublished results; cf. A. Streitwieser, Jr. Abstracts of 16th National Organic Chemistry Symposium, Seattle, June 15–17, 1959, p. 74.

[11] H. C. Longuet-Higgins, *Proc. Roy. Soc.*, **A207**, 121 (1951).

[12] H. C. Brown and W. A. Murphey, *J. Am. Chem. Soc.*, **73**, 3308 (1951); D. A. Brown and M. J. S. Dewar, *J. Chem. Soc.*, **1953**, 2406; **1954**, 2151.

In the foregoing the carbanion is used as the model of the transition state for proton removal. This treatment implies extensive C—H bond breaking at the transition state. A related problem is the preferred orientation for

recombination of a mesomeric anion with a proton, a reaction that is the microscopic reverse of proton removal. Since both reactions must have the same transition state, C—H bond making for the recombination reaction must be small and a perturbation treatment is convenient. The situation is much the same as the reaction of mesomeric cations with bases (Sec. 12.3); in precisely the analogous way we may derive that the proton

will tend to add to the carbon having the highest electron density. In cinnamyl anion, IV, the α- and γ-positions calculate to have the same charge density in the simple theory. Protonation of the lithium salt does give almost equal amounts of allylbenzene and β-methylstyrene.[13]

An alkyl group at either position may be treated as putting a negative h at the position thereby reducing the π-electron density at the substituted

position.[14] Hence a γ-alkylcinnamyl anion would be expected to protonate preferentially at the α-position. The Birch reduction of naphthalene probably involves the protonation of the anion, V (Sec. 14.3). The product

---

[13] H. F. Herbrandson and D. S. Mooney, *J. Am. Chem. Soc.*, **79**, 5809 (1957).
[14] N. Bouman and G. J. Hoijtink, *Rec. trav. chim.*, **76**, 841 (1957).

of protonation is mostly the unconjugated dihydronaphthalene, VI. This orientation agrees completely with the results of the HMO model. Similarly, the protonation of crotyl lithium in ether gives about a 5:1 ratio of 1-butene to 2-butene.[15] We should caution that the foregoing treatment applies to kinetically controlled products—in many cases such products are not those expected on the basis of thermodynamic stability.

14.2   Acidity of X—H Bonds

In many cases the acid-base equilibrium of proton bonds to heteroatoms, X, also involves a change in π-energy. We might expect such changes to be calculable by an MO method, at least for a related series of compounds. If suitable values for the heteroatom parameters are available, such correlations can be tested by direct calculations of the π-energies of the acid and base. The experimental acidities would be expected to parallel the corresponding calculated π-energy differences.

Longuet-Higgins[16] has shown how simple approximation techniques give a satisfactory accounting of observed acidities and basicities. One example is the protonation of heterocyclic bases of the pyridine and quinoline type:

VII   VIII

Both quinoline, VII, and quinolinium ion, VIII, as an example, are derived from naphthalene. If, as a first approximation, we consider that the only change is that of the Coulomb integral of one carbon, r, we may approximate the π-energy change for the equilibrium as

$$\Delta E_\pi \cong q_r(h_{N^+} - h_{\dot{N}})\beta \qquad (2)$$

q_r for all AH is unity; hence, to a first approximation, all such heterocyclic amines should have the same basicity. The data in Table 14.2 show that such amines differ in basicity by little more than a factor of ten. Nitrogens in different types of positions are included, and there is little doubt that much of the observed variation is due to variations in steric effects.

Amino derivatives of these heterocyclic systems are generally more basic but have pK's that vary over a wide range. This variation is accountable

[15] V. A. Kropachev, B. A. Dolgoplosk, and K. V. Danilovich, *Proc. Acad. Sci. U.S.S.R.* (English Translation), **111**, 763 (1957).

[16] H. C. Longuet-Higgins, *J. Chem. Phys.*, **18**, 275 (1950).

TABLE 14.2

BASICITIES OF SOME HETEROCYCLIC NITROGEN BASES*

Compound	pK_a Aqueous	50% Aqueous Alcohol
Pyridine	5.23	
Quinoline	4.94	
Acridine	5.60	4.11
Isoquinoline	5.14	
5,6-Benzoquinoline	5.15	3.90
6,7-Benzoquinoline	5.05	3.84
7,8-Benzoquinoline	4.25	3.15
Phenanthridine	—	3.30
3,4-Benzacridine	4.70	4.16
2,3-Benzacridine	—	4.52
1,2-Benzacridine	—	3.45

* From ref. 16.

by a simple approximation method because both acid and base are derived from a common odd-AH system.[16] An example is the protonation equilibrium of 4-aminoquinoline, IX. Both IX and X may be derived from XI;

IX   X   XI

to a first approximation the π-energy difference in the equilibrium is due only to the change of α at position 4 of XI and is given by (3) in which the q's are the electron densities of XI:

$$\Delta E_\pi \cong q_4(h_{N^+} - h_{\dot{N}})\beta \tag{3}$$

If this is a reasonable approximation, a plot of pK versus q_r, in which r is the ring carbon replaced by nitrogen, should give a linear correlation. Of course, q_r is easily calculated by Longuet-Higgins' simple technique (Sec. 2.8). Some of the available data are summarized in Table 14.3 and are plotted in Fig. 14.2. The correlation obtained is excellent for such a crude approximation; the two points farthest from the line are compounds

TABLE 14.3

BASICITIES OF SOME HETEROCYCLIC AMINES*

No. in Fig. 14.2	Amine	q_r	pK_a 50% Aqueous Ethanol, 20°
1	5-Aminoacridine	1.286	9.45
2	4-Aminoacridine	1.095	5.50
3	3-Aminoacridine	1.000	5.03
4	2-Aminoacridine	1.118	7.61
5	1-Aminoacridine	1.000	3.59†
6	4-Amino-1-azaphenanthrene	1.176	7.99
7	8-Amino-1-azaphenanthrene	1.019	4.10
8	7-Amino-1-azaphenanthrene	1.000	4.02
9	4-Amino-1-azaanthracene	1.214	8.75
10	3-Amino-1-azaanthracene	1.000	3.73
11	1-Amino-4-azaphenanthrene	1.167	7.68
12	9-Aminophenanthridine	1.286	6.75†
13	5-Amino-3,4-benzacridine	1.250	8.41
14	7-Amino-3,4-benzacridine	1.000	5.03
15	8-Amino-3,4-benzacridine	1.091	6.51
16	5-Amino-2,3-benzacridine	1.300	9.72
17	7-Amino-2,3-benzacridine	1.000	5.38
18	5-Amino-1,2-benzacridine	1.231	8.13
19	7-Amino-1,2-benzacridine	1.000	4.05
20	8-Amino-1,2-benzacridine	1.088	5.97

* From ref. 16.
† Ortho- or peri-.

in which the ring N and NH_2 group are *ortho-* or *peri-* to one another (filled circles in Fig. 14.2).

This method has been extended to heterocyclic phenols by Mason[17] and to aryldiamines by Hush.[18]

A related approach that can be applied to arylamines derives from Dewar and Sampson's demonstration of the relationship between the conjugation energy of an arylmethyl cation and the charge on the exocyclic carbon of the cation (Sec. 12.3).[19] Because AH's are involved, the precisely similar relationship should hold for arylmethyl anions. Such a relationship has also been demonstrated by Hush;[18] that is, there is a linear correlation between $E_\pi(ArCH_2^-) - E_\pi(ArH)$ and q_r of the exocyclic carbon of $ArCH_2^-$:

$$\Delta E_1 = E_\pi(ArCH_2^-) - E_\pi(ArH) \cong (C_1 - C_2 q_r)\beta \qquad (4)$$

[17] S. F. Mason, *J. Chem. Soc.*, **1958**, 674.
[18] N. S. Hush, *J. Chem. Soc.*, **1953**, 684.
[19] M. J. S. Dewar and R. J. Sampson, *J. Chem. Soc.*, **1956**, 2789.

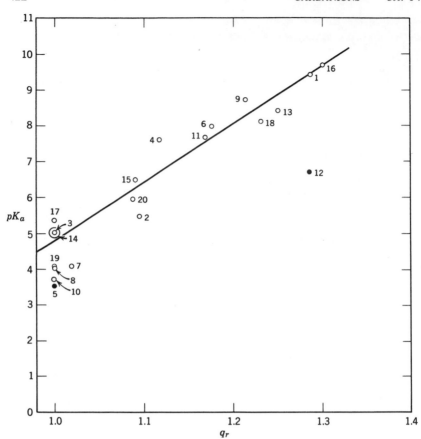

Fig. 14.2 Basicities of some aminoheterocycles. Numbers refer to compounds in Table 14.3.

The π-energy change for the equilibrium

$$ArNH_3^+ \rightleftharpoons ArNH_2 + H^+$$

may be approximated as

$$ArH \xrightarrow{\Delta E_1} ArCH_2^- \xrightarrow{\Delta E_2} ArNH_2$$

We neglect any effect of the $-NH_3^+$ group on the π-system of the hydrocarbon. ΔE_2 is given approximately as $q_r h_{\ddot{N}}\beta$; hence the total π-energy change, ΔE, is given as

$$\frac{\Delta E}{\beta} \cong C_1 + q_r(h_{\ddot{N}} - C_2) \qquad (5)$$

A plot of pK versus q_r should be linear. Elliott and Mason[20] have shown that such a correlation does indeed obtain (Table 14.4 and Fig. 14.3).

The effect of q_r on the pK_a of aminoarenes is small but definite. The

TABLE 14.4

BASICITIES OF ARYLAMINES*

No. in Fig. 14.3	Compound	q_r	pK_a 50% Aqueous Alcohol, 20°
1	Aniline	1.572	4.19
2	m-Aminobiphenyl	1.572	3.82
3	p-Aminobiphenyl	1.516	3.81
4	o-Aminobiphenyl	1.516	3.03
5	2-Aminonaphthalene	1.529	3.77
6	2-Aminophenanthrene	1.453	3.60
7	3-Aminophenanthrene	1.510	3.59
8	2-Aminoanthracene	1.471	3.40
9	1-Aminonaphthalene	1.450	3.40
10	1-Aminophenanthrene	1.463	3.23
11	9-Aminophenanthrene	1.446	3.19
12	1-Aminoanthracene	1.381	3.22
13	1-Aminopyrene	1.364	2.91
14	9-Aminoanthracene	1.286	2.7

* From ref. 20.

TABLE 14.5

ACIDITY OF PHENOLS

Compound	pK_a
Phenol	9.96
1-Naphthol	9.85
2-Naphthol	9.93
1-Anthrol	9.82
2-Anthrol	9.92

[Ref. K. Lauer, *Ber.*, **70**, 1288 (1937); H. Schenkel, *Experientia*, **4**, 383 (1948).]

term, $h_{\ddot{N}} - C_2$, in (5) is small and negative. The situation for phenols is comparable but more extreme. If the conjugation of the —OH group in the neutral phenol is neglected as being comparatively unimportant compared to that in the phenolate anion, the π-energy change for ionization can again be represented by (5) except that $h_{\ddot{N}}$ is replaced by $h_{\ddot{O}}$.[17] Because

[20] J. J. Elliott and S. F. Mason, *J. Chem. Soc.*, **1959**, 2353.

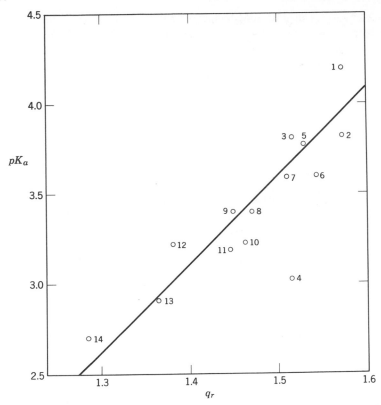

Fig. 14.3 Basicity of arylamines. Numbers refer to compounds in Table 14.4.

oxygen is more electronegative than nitrogen, the term, $h_{\ddot{O}} - C_2$, is less negative than the comparable term for amines; in fact, the two constants experimentally are now about the same and the entire dependence on q_r vanishes. Hydroxyarenes hardly vary in pK_a (Table 14.5).

For compounds derived from non-AH systems, the approximation methods may not hold and recourse must be had to more complete

$$\text{XII}$$

calculations. Such a change in total calculated π-energy, for example accounts for the relatively high basicity of 1-aminoazulene, XII.[21]

[21] J. Schulze and E. Heilbronner, *Helv. Chim. Acta.*, **41**, 1492 (1958).

14.3   Reductions of Hydrocarbons with Alkali Metals

By definition, reduction involves enrichment by electrons. Many chemical reductions in practice are complicated by surface phenomena; however, the reduction products of alkali metal reactions are apparently determined in homogeneous solution and are more amenable to theoretical treatments.

The equilibria between hydrocarbons and alkali metals and the salts of the corresponding radical anions and dianions are well known in ether and liquid ammonia solutions and have been discussed previously (Sec. 6.5).

$$ArH + Na \rightleftharpoons ArH^- + Na^+$$
$$ArH^- + Na \rightleftharpoons ArH^= + Na^+$$

Addition of a proton source such as alcohol to these solutions produces reduction products of various types which may be derived from the radical anion and dianion.[22] Among the possible reactions are the following:

1. Bond cleavage to liberate a relatively stable anion.

Example. $CH_2{=}CHCH_2OH \longrightarrow CH_2{=}CHCH_2^- + OH^-$

2. Dimerization of the radical anion.

Example.

$$(C_6H_5)_2CHCH_2CH_2CH(C_6H_5)_2$$

3. Addition of two hydrogens.

Example.

With polycyclic aromatic hydrocarbons, reaction 3 is the most common mode of behavior. The positions occupied by the two incoming hydrogens can be predicted to a remarkable degree by HMO theory. Hückel[23]

[22] Reviewed by A. J. Birch, *Quart. Revs.*, **4**, 69 (1950).
[23] E. Hückel, International Conference on Physics, London, 1934, Vol. II, p. 9.

suggested that the protons occupy positions of highest electron density in the dianion as calculated by MO theory. The suggestion has been applied to the reduction products of unsaturated hydrocarbons at the dropping mercury electrode by Hoijtink and van Schooten.[24] The process of this reduction is similar to the reduction by alkali metals (Sec. 7.1).

Protonation of either the mono- or dianion or both conceivably could be involved in the product-determining steps;[22,25] for example, the protonation steps could be either of the two reactions:

$$\mathrm{ArH} \Big\langle \begin{array}{c} \nearrow \mathrm{ArH^-} \xrightarrow{\ \mathrm{H^+}\ } \mathrm{ArH_2} \\[2ex] \searrow \mathrm{ArH^=} \xrightarrow{\ \mathrm{H^+}\ } \mathrm{ArH_2^-} \longrightarrow \mathrm{ArH_3} \end{array} $$

The detailed reaction may depend on the relative amounts of the ions present in equilibrium, and these in turn depend on the electron affinity of the hydrocarbon. A common experimental technique with polycyclic systems is to form the sodium or lithium adduct in liquid ammonia or in a suitable ether and then to add alcohol as a proton donor. The predominant reaction is probably with the dianion. This technique frequently leads to dihydro derivatives which are not reduced further. Benzene and its simple derivatives have such low electron affinity that little conversion occurs even to the monoanion; facile reduction requires the simultaneous presence of alkali metal and proton source in liquid ammonia. The reaction probably occurs via protonation of the radical-anion that is continuously generated.[26] Unconjugated dihydrobenzenes are still produced because isolated double bonds are not attacked under the conditions.

Fortunately, for AH at least, the detailed reduction process is unimportant for the MO application. The position of highest electron density, at which the first proton is considered to react, is determined by the coefficients of the lowest vacant MO; for this purpose it does not matter whether this orbital is singly or doubly occupied. The second proton reacts with an ordinary carbanion and again reacts at the position of highest electron density (Sec. 14.1).

Example. Naphthalene. c_r^2 for the lowest vacant orbital are given in Fig. 14.4. (Note that these values are the same as for the highest occupied orbital

[24] G. J. Hoijtink and J. van Schooten, *Rec. trav. chim.*, **71**, 1089 (1952); **72**, 691 (1953).

[25] A. J. Birch, *Australian J. Chem.*, **7**, 256, 261 (1954); **8**, 96 (1955).

[26] The point is still controversial; cf. A. P. Krapcho and A. A. Bothner-By, *J. Am. Chem. Soc.*, **81**, 3658 (1959); **82**, 751 (1960); J. F. Eastham and D. R. Larkin, *J. Am. Chem. Soc.*, **81**, 3652 (1959); J. F. Eastham, C. W. Keenan and H. V. Secor, *J. Am. Chem. Soc.*, **81**, 6523 (1959).

for AH.) The first proton reacts at carbon 1. The anion with which the second proton reacts is V. Without the methylene group the anion is a cinnamyl anion, IV, in which two positions have equal highest electron densities. The

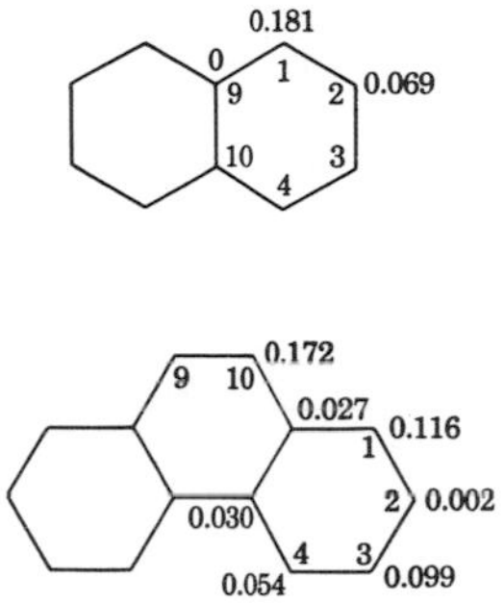

Fig. 14.4 c_r^2 for lowest vacant orbital of naphthalene and phenanthrene.

methylene group increases the electron density at carbon 4 (Sec. 14.1) and the product should be IV. The entire process is

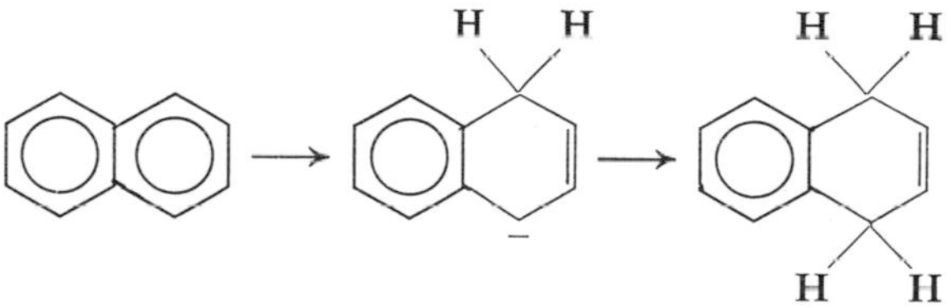

Example. Phenanthrene. c_r^2 for the lowest vacant orbital is also given in Fig. 14.4. The first proton reacts at carbon 9. The next anion is an *o*-phenyl-benzyl anion; the highest electron density is unambiguously the exocyclic carbon (XIII). The perturbing effect of the methylene group need not be considered. The product should be 9,10-dihydrophenanthrene.

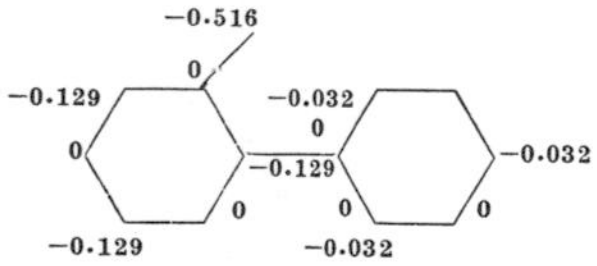

XIII

The results of such calculations are compared with experiment in Table 14.6. The agreement is almost perfect; even when the dihydro compound is not the isolated reduction product, MO theory helps to interpret the course of the reaction. The reduction of fluoranthene is a conspicuous example. The principal product from the addition of alcohol to a solution

TABLE 14.6

ALKALI METAL REDUCTIONS OF AROMATIC HYDROCARBONS

Hydrocarbon	Calculated Dihydro Reduction Product	Experimental Result	Refs.
Benzene	1,4		*, †
Toluene	2,5	CH₃	*
Naphthalene	1,4		‡, §
Anthracene	9,10		‖
Phenanthrene	9,10		§, ¶
1,2-Benzanthracene	7,12		**, ††
1,2,3,4-Dibenzanthracene	9,14		††
1,2,5,6-Dibenzanthracene	7,14		**

TABLE 14.6 (*continued*)

Hydrocarbon	Calculated Dihydro Reduction Product	Experimental Result	Refs.
Chrysene	5,6	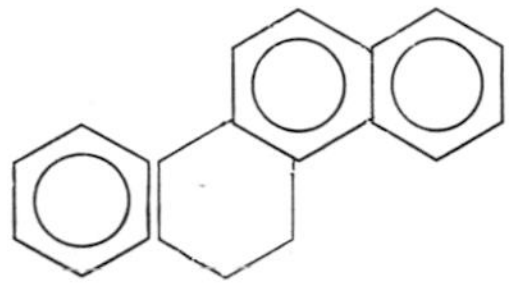	‡‡
Pyrene	1,6; 1,8	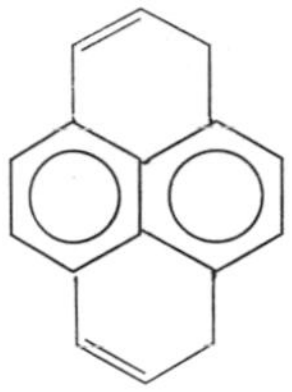 or 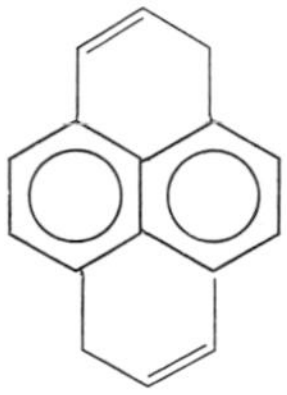 or both	§§
Fluoranthene	3,12	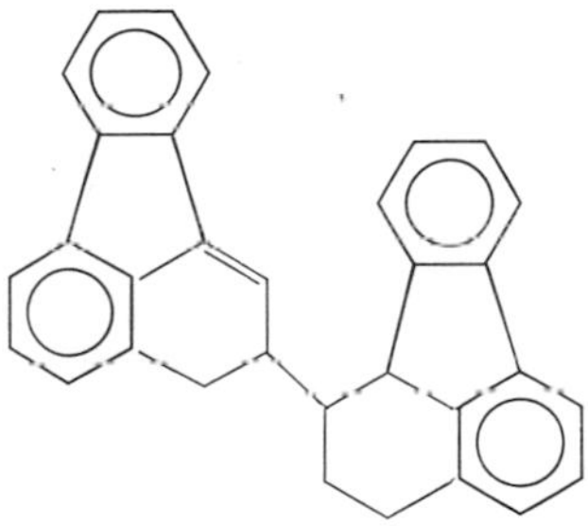	‖‖
Biphenyl	1,4	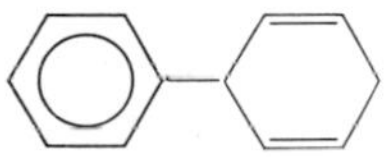	¶¶
Picene	5,6	—	
Perylene	3,14	—	
Pentaphene	5,14	—	
3,4-Benzophenanthrene	5,12*b*	—	
Biphenylene	2,10	—	

Footnotes to Table 14.6

  * A. P. Krapcho and A. A. Bothner-By, *J. Am. Chem. Soc.*, **81**, 3658 (1959).
  † C. B. Wooster and K. L. Godfrey, *J. Am. Chem. Soc.*, **59**, 586 (1937).
  ‡ W. Hückel and H. Bretschneider, *Ann.*, **540**, 157 (1937).
  § W. Schlenk and E. Bergmann, *Ann.*, **463**, 83 (1928).
  ‖ W. Schlenk, J. Appenrodt, A. Michael, and A. Thal, *Ber.*, **47**, 473 (1914);
W. Schlenk and E. Bergmann, *Ann.*, **463**, 134 (1928).

of fluoranthene and sodium in liquid ammonia is 1,2,3,12,2',3'-hexa-hydro-1,2'-bifluoranthyl, XIV.[27] The expected product, 3,12-dihydro-fluoranthene, XV, is sufficiently acidic (calculated pK by the MO method

of Sec. 14.1 is 16)[28] to be converted rapidly to the anion, XVI, and 2,3-dihydrofluoranthene, XVII. A condensation of the Michael type between XVI and XVII leads to the observed product, XIV. Such condensations with fulvene derivatives such as XVII are well known (Sec. 10.4).

Under normal conditions, the dihydroaromatics are generally sufficiently stable to be isolated before isomerization; however, further action can result in additional reduction. Further reaction of naphthalene, for

[27] A. Streitwieser, Jr., and S. Suzuki, *Tetrahedron*, in press.
[28] Unpublished calculations of A. Streitwieser, Jr., and J. I. Brauman.

¶ A. Jeanes and R. Adams, *J. Am. Chem. Soc.*, **59**, 2008 (1937).
** W. E. Bachmann, *J. Org. Chem.*, **1**, 347 (1936).
†† W. E. Bachmann and L. H. Pence, *J. Am. Chem. Soc.*, **59**, 2339 (1937).
‡‡ S. E. Hunt and A. S. Lindsey, *J. Chem. Soc.*, **1958**, 2227.
§§ O. Neunhoeffer, H. Woggon, and S. Dähne, *Ann.*, **612**, 98 (1958).
|| || A. Streitwieser, Jr., and S. Suzuki, unpublished results; see text.
¶¶ W. Hückel and R. Schwen, *Ber.*, **89**, 150 (1956). However, for a different structure, see I. P. Egorov, E. P. Kaplan, Z. I. Letina, V. A. Shliapochnikov, and A. D. Petrov, *J. Gen. Chem. U.S.S.R.* (*English Translation*), **28**, 3284 (1958).

example, produces tetralin.[29]   Reductions with alkali metals in alcohols probably proceed by much the same course as the foregoing metalations. Now, however, the basic action of the sodium alkoxide may cause iso-merization of the initial dihydro compounds to systems in which the double bond is conjugated with the remaining aromatic rings.   Further reduction is then facile.   The sodium-amyl alcohol reduction of naphthalene to tetralin is an example:

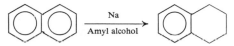

A general theoretical approach to many such reductions has been worked out by Hoijtink.[30]

[29] C. B. Wooster and F. B. Smith, *J. Am. Chem. Soc.*, **53**, 179 (1931);   W. Hückel and H. Bretschneider, *Ann.*, **540**, 157 (1937).
   [30] G. J. Hoijtink, *Rec. trav. chim.*, **76**, 885.

15 Four-center reactions

15.1 Diels-Alder Reactions

The Diels-Alder reaction, or diene synthesis, is a union of a diene and an olefin to form a new six-membered ring:[1]

For the purposes of this reaction, many aromatic hydrocarbons behave as dienes and yield polycyclic adducts:

The reaction with polycyclic aromatics is reversible; the relative positions of equilibrium have been determined for the reaction of several hydrocarbons with maleic anhydride.[2] For a series of related compounds, say anthracene derivatives, the principal difference lies in the change in π-energies during reaction.

Brown[3] has proposed a reactivity index, *para*-localization energy, L_p, for use in such cases. L_p is defined as the loss in π-bonding energy that

---

[1] For reviews, see M. C. Kloetzel, *Organic Reactions*, Vol. IV, John Wiley and Sons, New York (1948) Chap. 1; H. L. Holmes, *ibid.*, Chap. 2; L. W. Butz and A. W. Rytina, *ibid.*, Vol. V, Chap. 3 (1949); C. Walling, *The Chemistry of the Petroleum Hydrocarbons*, Vol. 3, Reinhold Publishing Corp., New York (1955) Chap. 47.

[2] W. E. Bachmann and M. C. Kloetzel, *J. Am. Chem. Soc.*, **60**, 481 (1938).

[3] R. D. Brown, *Australian J. Sci. Research*, **2A**, 564 (1949); *J. Chem. Soc.*, **1950**, 3249; **1951**, 1950.

results when two carbons having a mutual *para*-orientation are removed from a π-network:[4]

$$L_{p_{r,s}} = M - M^{*r,s} \tag{1}$$

$L_{p_{9,10}}$ for anthracene is $M_{\text{anthracene}} - 2M_{\text{benzene}}$, or 3.314. The relative equilibrium constants towards maleic anhydride for 1,2,5,6-dibenzanthracene, 1,2-benzanthracene, and anthracene are approximately 1, 60,

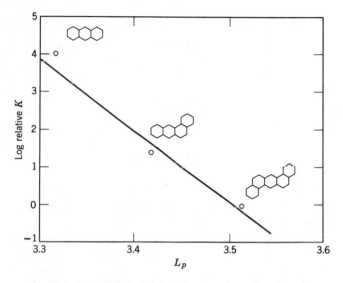

Fig. 15.1   Equilibria in Diels-Alder reactions of anthracenes.

and 20,000, respectively. When plotted against L_p, a reasonable linear correlation is obtained, the slope of which corresponds to $\beta \cong -36$ kcal. (Fig. 15.1).[5]

In many diene syntheses the reaction can be run substantially to completion; the relative rate of reaction is then of synthetic interest. Two general types of reaction mechanisms have been proposed in the past for the Diels-Alder reaction—the two-step and one-step mechanisms.[6] In the two-step mechanism, the new bonds are formed successively, see p. 434.

[4] As here defined, the localization energy is the negative of that used in the literature, but it is more convenient in practice because L_p are usually positive numbers. Note that our L_p are dimensionless numbers because they are given in terms of the β-coefficients of the π-energies.

[5] This treatment is a slight modification of that of R. D. Brown, *J. Chem. Soc.*, **1951**, 1612.

[6] For reviews and further references, see refs. 1 and 7.

[7] R. B. Woodward and T. J. Katz, *Tetrahedron*, **5**, 70 (1959); *Tetrahedron Letters*, No. 5, 19 (1959); M. J. S. Dewar, *Tetrahedron Letters*, No. 4, 16 (1959).

In the one-step mechanism, the two new bonds are formed simultaneously:

On the basis of the one-step mechanism, in which two carbons of the diene are starting to become localized, Brown[8] suggested the application of *para*-localization energies as a MO model of the transition state. The use of such a model should not be disturbed by a possible prior intervention of a complex between diene and dienophile, say of the charge-transfer type.[9]

Quantitative rate data are limited, but Brown has shown how simple MO theory follows rather qualitative observations; in particular, aromatic hydrocarbons apparently give an adduct with maleic anhydride under usual experimental conditions only if L_p is $< 3.6^8$ (Fig. 15.2). Brown[10] has also examined the effects of substituents on *para*-localization energies.

For large molecules, calculation of localization energies is tedious without a computer. For AH, however, Hopff and Schweizer[11] have shown that Dewar's simple localization approximation (Sec. 4.3) may be applied to give satisfactory results. In this method, the *para*-localization energy is approximated as the sum of two atom-localization energies. The method was applied to a number of large polycyclic aromatic hydrocarbons and gave results in good qualitative agreement with experiment.

In principle, unsymmetrical dienes and dienophiles can give isomeric adducts:

[8] R. D. Brown, *J. Chem. Soc.*, **1950**, 691, 2730; *Quart. Revs.*, **6**, 63 (1952).

[9] For example, see J. A. Berson and R. D. Reynolds, *J. Am. Chem. Soc.*, **77**, 4434 (1955); L. J. Andrews and R. M. Keefer, *J. Am. Chem. Soc.*, **77**, 6284 (1955).

[10] R. D. Brown, *J. Chem. Soc.*, **1951**, 1955; **1952**, 2229.

[11] H. Hopff and H. R. Schweizer, *Helv. Chim. Acta*, **42**, 2315 (1959).

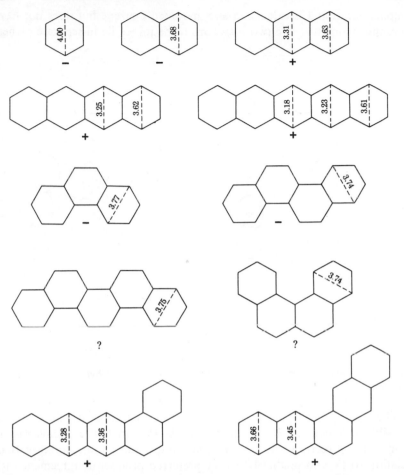

Fig. 15.2 Formation of adducts with maleic anhydride. L_p-values are indicated. Experimental formation of adduct is indicated by $+$; no adduct formed under usual conditions is indicated by $-$.

Frequently, only one isomer is obtained. Localization energies clearly do not account for such specificity or allow a prediction of the course of reaction. If the transition state for the diene synthesis is represented as in Fig. 15.3a, the treatment as an extended π-network follows. The transition state for the reaction between butadiene and ethylene then resembles a modified benzene, I, in which the p_σ-p_σ bonds may be assigned a different value of β. Such a model was originally suggested by Evans.[12] It can be applied to the problem of unsymmetrical reagents. It should be

[12] M. G. Evans, *Trans. Faraday Soc.*, **35**, 824 (1939).

emphasized that I is calculated as a modified benzene but does not have benzene symmetry; the two π-systems to be joined lie in separate planes.

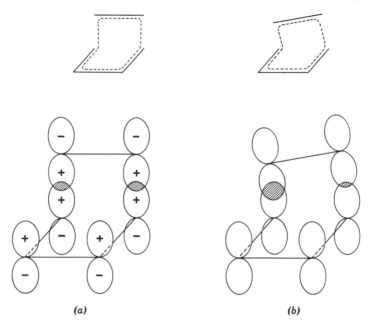

Fig. 15.3   Possible transition states for Diene Syntheses. (a) Symmetrical. (b) Unsymmetrical.

The reaction of 1-phenylbutadiene with acrolein provides an example. Possible products are III via II and V via IV. In II the phenyl ring is conjugated directly with the aldehyde group; this should stabilize II in relation to IV and predict III as the preferred product in agreement with experiment.[13]

I

Woodward and Katz[7] have recently focused attention on the one-step mechanism with an unsymmetrical transition state, a scheme that combines the best features of the one-step and two-step mechanisms. In the MO equivalent the model of the transition state is still an extended conjugation model, but the two p_σ-p_σ bonds are now given different values of β; that is, the model used derives from Fig. 15.3b. Application of this model

[13] E. Lehmann and W. Paasche, *Ber.*, **68**, 1146 (1935).

to the dimerization of acrolein leads to VI as the most stable transition state and predicts the correct product, VII. This model should be applicable generally to Diels-Alder reactions, but no detailed calculations or quantitative comparisons with experiment have yet been reported.

A further point should be noted: if localization theory can be applied to the 1,4-positions of a diene on reaction with a dienophile, the effect of the diene reaction on the dienophile is to localize the 1,2-positions of the

double bond. This localization energy may be called an *ortho-* or bond-localization energy, L_b.[3] We would anticipate that the reactivity of a dienophile would be greater the smaller the magnitude of L_b. Unfortunately, this prediction is not borne out by experiment; for example, butadiene is more reactive than ethylene toward dienes, despite its higher L_b: 2.472 and 2.000, respectively.

Although no extensive calculations with a variation of parameters have been reported, we can easily show by a simple example that the extended conjugation model can give at least qualitative agreement with experimental chemistry. For simplicity we take the primary bond to be 1β and the secondary bond to be $\frac{1}{2}\beta$. The transition state for the reaction of ethylene with butadiene then takes on the appearance of a benzene with one bond having a reduced β—the *π-systems* unite as

VIII

We derive the π-energy of VIII from benzene by the first approximation method for change in β (4.16). $\Delta E_\pi/\beta$ for the reaction becomes 8.000 + 2(0.667) (−0.5) − 4.472 − 2.000 = 0.861. Similarly, the reaction of butadiene with butadiene becomes

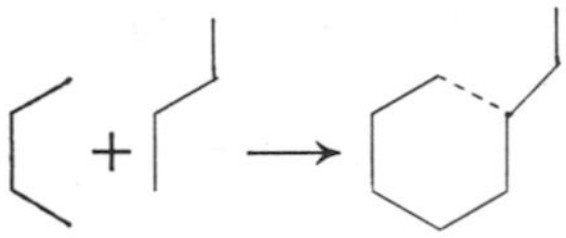

IX

We derive the π-energy of IX from styrene and obtain $\Delta E_\pi/\beta$ for the reaction as 10.424 + 2(0.610) (−0.5) − 2(4.472) = 0.870. Butadiene as dienophile gives a greater increase in π-bonding energy and is predicted to be more reactive. Conceivably, a more judicious choice of the two β-values could give extensive agreement with experiment for a variety of dieno-philes.

15.2 Other Double-Bond Reagents

Dienes may be considered as double-bond reagents toward dienophiles. Other double-bond reagents are known which apparently also react via 4-center mechanisms. Some of these cases have been treated by MO methods, particularly bond-localization theory.

The reaction of osmium tetroxide with double bonds gives an osmate

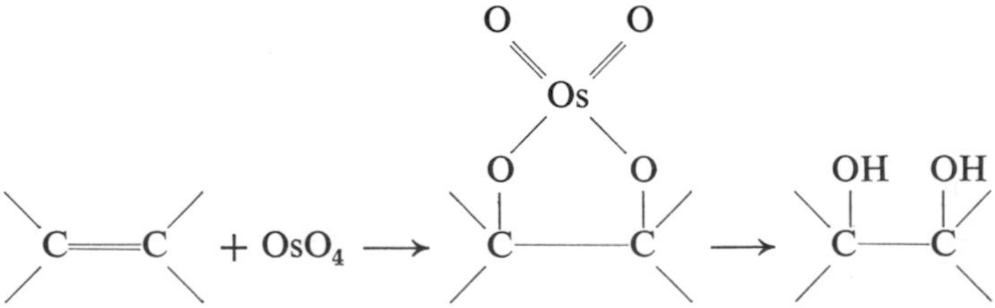

ester that on hydrolysis yields a *cis*-glycol.[14] Polycyclic aromatic hydro-carbons give the same reaction; Brown[15] has shown that the relative rates

[14] G. M. Badger, *Quart. Revs.*, **5**, 147 (1951).
[15] R. D. Brown, *Quart. Revs.*, **6**, 63 (1952).

of reaction of several such hydrocarbons follow semiquantitatively the bond-localization energies. Because of the relationship between bond order and the effect of a change in β on π-energy (4.16), L_b is approximately inversely proportional to the bond order of the bond.[3] Coulson showed that the experimental reactivities of the dinaphthylethylenes, 2,2 > 1,2 > 1,1, agrees with the expectations from the ethylenic bond orders: 1.814, 1.803 and 1.792, respectively.[16]

Ozone may be another double-bond reagent in the 4-center sense. Wibaut has accumulated evidence that ozone acts as an electrophilic reagent and has proposed that the initial step is[17]

Meinwald[18] has suggested that a terminal oxygen is the actual attacking species, for example,

This view of the reaction is difficult to reconcile with the observation that pyrene reacts at the 4-5 double bond; the 1-position is the most reactive position by far in electrophilic reactions (Chap. 11). In the one-step,

---

[16] C. A. Coulson, *J. Chem. Soc.*, **1950**, 2252; G. M. Badger, *Nature* **165**, 647 (1950).
[17] For a review see J. P. Wibaut, *Chimia*, **11**, 298, 321 (1957).
[18] J. Meinwald, *Ber.*, **88**, 1889 (1955).

4-center mechanism the initial adduct is formed directly:

The actual products are then thought to derive from further reactions and rearrangements of the first intermediate.[19] According to this mechanism, ozone would be expected to attack preferentially the bond of lowest L_b within a molecule. The behavior of pyrene is thus explained, but anthracene now becomes anomalous for anthraquinone is one the products.

Bailey has attempted to reconcile these differences with the proposal of an intermediate π-complex which he formulates as a localized bond complex, X. The complex could perhaps also be formulated as a delocalized charge-transfer complex, XI.

Wallenberger[20] has suggested that actual bond formation between aromatic hydrocarbons and ozone, perhaps with prior complex formation such as X or XI, takes place 1,2 or 1,4 such as to give the smallest decrease in resonance energy, In MO terms, this proposal means that reaction will occur 1,2 (to give bond cleavage) or 1,4 (to give quinone), depending on which L_b or L_p is the smallest. The proposal has also been made that the positions of reaction are those of lowest redox potential of the corresponding quinones;[21] however, these redox potentials have been shown to correlate with *para-* and bond-localization energies (Sec. 9.6). The summary of results in Table 15.1 shows that reaction of aromatic hydrocarbons with ozone does occur at the 1,2 and 1,4 positions with the lowest localization energies, provided that the *para-*positions are weighted by about 0.1

---

[19] For further details see the extensive review by P. S. Bailey, *Chem. Rev.*, **58,** 925 (1958).

[20] F. T. Wallenberger, *Tetrahedron Letters*, No. 9, 5 (1959).

[21] E. J. Moriconi, W. F. O'Connor, and F. T. Wallenberger, *Chem. and Ind.*, **1959,** 22; *J. Am. Chem. Sloc.*, **81,** 6466 (1959).

TABLE 15.1

REACTION OF AROMATIC HYDROCARBONS WITH OZONE

Hydrocarbon	Positions	L_b	L_p	Preferred Ozonization Positions (Ref. 20)
Naphthalene	1,2	3.259		1,2
	2,3	3.729		
	1,4		3.68	
Anthracene	1,2	3.204		9,10
	2,3	3.786		1,2
	1,4		3.63	
	9,10		3.31	
Phenanthrene	1,2	3.321		9,10
	2,3	3.655		
	3,4	3.338		
	9,10	3.065		
	1,4		3.77	
Chrysene	1,2	3.318		5,6
	2,3	3.712		
	3,4	3.303		
	5,6	3.121		
	1,4		3.74	
Tetracene	1,2	3.188		5,12
	3,4	3.798		
	1,4		3.62	
	5,12		3.25	
1,2-Benzanthracene	1,2	3.358		5,6
	2,3	3.604		7,12
	3,4	3.340		
	5,6	3.030		
	1,4		3.78	
	7,12		3.41	

TABLE 15.1 (*continued*)

Hydrocarbon	Positions	L_b	L_p	Preferred Ozonization Positions (Ref. 20)
1,2,5,6-Dibenzanthracene	1,2	3.349		5,6
	2,3	3.626		
	3,4	3.331		
	1,4		3.79	
	5,6	3.045		
	7,14		3.51	
Triphenylene	1,2	3.383		1,2
	2,3	3.595		
	1,4		3.79	
Pyrene	4,5	3.057		4,5

units. In using this technique for predictive purposes the reader should be cautioned that the mode of attack seems to be solvent-dependent.[22]

Other double-bond reagents involve reactive intermediates of the carbene type, $R_2C:$; such intermediates may be generated by loss of nitrogen from diazo compounds or by base dehydrohalogenation of suitable alkyl halides. Another typical double-bond reaction is the formation of epoxides with peracids. Some examples of each of these reactions follow, see p. 443. No extensive comparisons have been made between experimental reactivities in these cases and MO calculations, although we might anticipate an application of bond-localization theory.[14,15]

[22] F. Dobinson and P. S. Bailey, *Tetrahedron Letters*, No. 13, 14 (1960).

Finally, we may mention the formation of cyclobutanes by cyclo-addition of olefins.[23,24] Some olefins tend to form 4-membered rings on

$$CHCl_3 \mid (CH_3)_3COK \longrightarrow [:CCl_2] + KCl + (CH_3)_3COH$$

addition rather than the six-membered rings characteristic of Diels-Alder reactions (Sec. 15.1). Highly fluorinated double bonds, allenes, and ketenes are particularly prone to this kind of cycloaddition. Some examples are[24]

The mode of addition is generally explicable in terms of a two-stage addition with the prior formation of the most stable "diradical."

[23] E. Vogel, *Angew. Chem.*, **72**, 4 (1960).
[24] J. D. Roberts and C. M. Sharts, "Cyclobutane Derivatives from Thermal Cyclo-addition Reactions," in *Organic Reactions*, Vol. 12, 1961.

Instead of a true diradical (parallel spins), the intermediate or transition state may be one in which the electrons are of opposed spins and perhaps

$$2CH_2{=}CHCN \longrightarrow \begin{bmatrix} CH_2{-}\dot{C}H{-}CN \\ | \\ CH_2{-}\dot{C}H{-}CN \end{bmatrix} \longrightarrow \square \begin{array}{c} {-}CN \\ \\ {-}CN \end{array}$$

weakly bonded;[24] the transition state may be represented as in Fig. 15.4. In simple MO terms it becomes a distorted cyclobutadiene in which the two new β's are unequal, XII.

$$\beta_1 \; \square \; \beta_2 \qquad |\beta_1| > |\beta_2|$$

XII

This unsymmetrical addition is clearly analogous to that which may be involved in the Diels-Alder addition. By keeping the new β's unequal, the system partly avoids the onus of an unstable cyclobutadiene structure

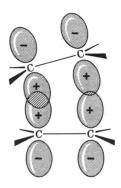

Fig. 15.4   Possible transition state for cycloaddition of two olefins to give a cyclobutane.

(Sec. 10.3). In this connection it is interesting that cycloadditions via quasi-eight-membered rings are rare. In part, this may be due to the entropy requirements of formation of medium-sized rings, but it may also be due to the electronic instability of the conjugated cyclooctatetraene structure (Sec. 10.7). A possible exception has been the catalyzed cyclo-dimerization of butadiene which yields in part *trans*-1,2-divinylcyclobutane, XIII, and 1,5-cyclooctadiene, XIV. Vogel[25] has shown recently that XIV

[25] E. Vogel, *Ann.*, **615**, 1 (1958).

undoubtedly comes not from a direct 4-center reaction via an eight-membered ring but by rearrangement of a first-formed *cis*-1,2-divinylcyclo-
butane, XV. A synthetic sample of XV rearranges to XIV at a lower
temperature than required for reaction.

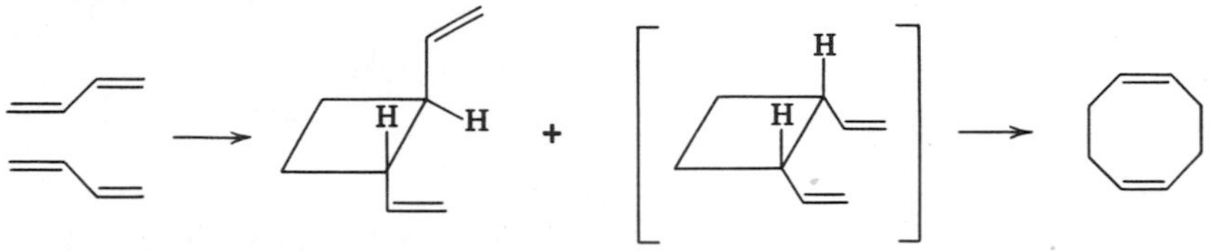

XIII   XV   XIV

The reader should be reminded at this point that the treatment of the
π-system in the transition state of cycloadditions as a quasi-aromatic
structure is at present still a *postulate* or *model* of potentially useful significance. Theoretical treatments are rare and further research in this area
should be encouraged.

15.3   Claisen and Cope Rearrangements

In 1912 Claisen[26] discovered the thermal rearrangement of ethyl
O-allylacetoacetate, XVI, to ethyl α-allylacetoacetate, XVII.

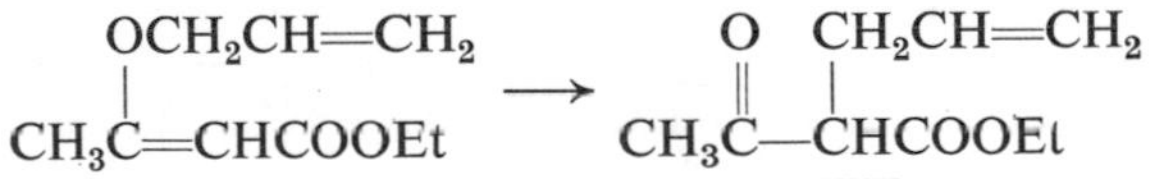

The Claisen rearrangement is usually associated with thermal rearrangement of aryl allyl ethers to phenols:

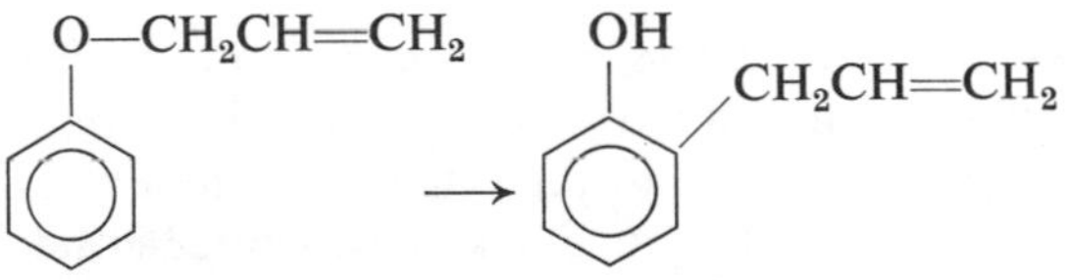

[26] L. Claisen, *Ber.*, **45**, 3157 (1912).
[27] C. D. Hurd and M. A. Pollack, *J. Org. Chem.*, **3**, 550 (1939).
[28] A detailed discussion of the evidence is beyond the scope of this book; for reviews
see D. S. Tarbell, *Organic Reactions*, Vol. II, John Wiley and Sons, New York (1944),
Chap. 1; G. W. Wheland, *Resonance in Organic Chemistry*, John Wiley and Sons, New
York (1955), p. 537; J. Hine, *Physical Organic Chemistry*, McGraw-Hill Book Company,
(1956), p. 453; E. S. Gould, *Mechanism and Structure in Organic Chemistry*, Henry Holt
and Company, New York (1959), p. 644.

There is no longer any doubt that the mechanism is substantially that suggested by Hurd and Pollack:[27,28]

XVIII　　　　　　　　　XIX

This arrangement of systems isoelectronic with 1,5-hexadienes seems to be rather general

A number of such systems has been studied by Cope[29] and are known as Cope rearrangements. Two examples are

Although no MO treatments of the reaction have yet been reported, our previous discussions lead to some models that may be considered. In the Claisen rearrangement, to the extent that the energy of the intermediate, XIX, determines the rate of reaction, a possible index of reactivity may be the bond-localization energy of the aromatic bond. The relative values of L_b for the 1,2- and 2,3-positions of naphthalene, 3.259 and 3.729, respectively, may explain why rearrangements that involve the first bond are facile, whereas those that involve the second bond are unknown:[26]

[29] A. C. Cope and E. M. Hardy, *J. Am. Chem. Soc.*, **62**, 441 (1940) and subsequent papers; see ref. 28.

$$\text{CH}_2\text{CH}{=}\text{CH}_2$$

OCH₂CH=CH₂ ... OH → ...

$$\text{OCH}_2\text{CH}{=}\text{CH}_2$$

R ... OCH₂CH=CH₂ ⤫→ R ... OH

$$\text{CH}_2\text{CH}{=}\text{CH}_2$$

For other purposes, however, an extended conjugation model may apply, especially if, as seems likely, the transition state for rearrangement resembles Fig. 15.5 or XX.[29a] This structure is a combination of p_σ- and

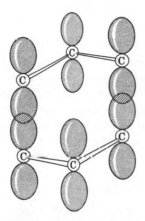

Fig. 15.5   Possible transition state for Cope rearrangement of 1,5-hexadiene.

p_π- bonds much as in the proposed transition state for the Diels-Alder reaction (Sec. 15.1). Data for an MO application of this model are not yet available, although some measurements of the relative rates of rearrangement of p-substituted phenyl allyl ethers are suggestive. The effect of the

XX

[29a] W. E. Doering (personal communication) has shown that for acyclic systems the transition state resembles the chairlike structure assumed in Fig. 15.5 rather than the alternative boatlike structure.

substituents follows a σ^+ correlation in a Hammett-type treatment,[30] as if some carbonium ion character were being generated at the phenoxy carbon in the transition state. Just such character can be derived from the MO model, XX, or Fig. 15.5. In the corresponding transition state, XXI, the phenyl-O bond resembles electronically that of a carbonyl group in the same way as in the carbonyl group, XXII, we may expect positive character at the phenyl carbon. Electron donating groups at X are expected to facilitate the rate. Testing of these ideas will require further experiments and detailed calculations of suitable systems.

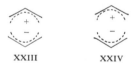

$$\diagdown \hspace{-0.3em} C\!=\!O \leftrightarrow \overset{+}{\underset{\diagup}{C}}\!-\!\bar{O}$$

XXI   XXII

The value of β for the p_σ-bonds in XX or Fig. 15.5 will depend on the corresponding bond distances in the transition state; the structure may well be unsymmetrical such that the two β's are unequal. With the substitution of suitable heteroatoms or odd-rings, the structure is no longer that of an AH and charge separation may occur. As the magnitude of the p_σ-β's is decreased or the difference in electronegativities of the two "halves" is increased, the structure will become increasingly that of a

XXIII   XXIV

charge-transfer complex, XXIII, or an ion pair, XXIV. Such considerations suggest further that the structure of the transition state (values of β, for example) may depend on the solvent and other reaction variables.

[30] W. N. White, D. Gwynn, R. Schlitt, C. Girard, and W. Fife, *J. Am. Chem. Soc.*, **80**, 3271 (1958); H. L. Goering and R. R. Jacobson, *J. Am. Chem. Soc.*, **80**, 3277 (1958).

16 Advanced MO methods

16.1 Antisymmetrization

A basic assumption of the simple MO methods (Sec. 4.1) is that the electrons are independent, and with assignments of opposite spin they doubly occupy each molecular orbital, ψ_j. In advanced MO methods, explicit consideration is given to electronic repulsions. For many purposes it is convenient to associate the electron spin with the molecular orbital and to define *spin orbitals*, ψ_j, for a spin of, say, $+\frac{1}{2}$ and $\bar{\psi}_j$ for a spin of $-\frac{1}{2}$. Each spin orbital is occupied by a single electron. The total wave function for the π-system is then a product wave function of occupied spin orbitals:

$$\Psi = \psi_1(1)\bar{\psi}_1(2)\psi_2(3)\bar{\psi}_2(4) \cdots \psi_m(2m-1)\bar{\psi}_m(2m) \tag{1}$$

In this product wave function electron (1) is in spin orbital, ψ_1, electron (2) is in $\bar{\psi}_1$, etc. Because the electrons cannot be labeled, we must allow alternative assignments of electrons to spin orbitals, such as $\psi_1(2)\bar{\psi}_1(1)\cdots$, etc. In characterizing such alternative assignments, we must be cognizant of a basic postulate of quantum mechanics, the Pauli principle, which states that a wave function must change sign with each interchange of electron labels.

A convenient way of formulating the product wave function that embodies the Pauli requirements is as a Slater determinant:[1]

$$\Psi = \frac{1}{(2m!)^{1/2}} \begin{vmatrix} \psi_1(1) & \bar{\psi}_1(1) & \psi_2(1) & \cdots \psi_m(1) \\ \psi_1(2) & \bar{\psi}_1(2) & \psi_2(2) & \cdots \psi_{2m}(2) \\ \cdots\cdots\cdots\cdots\cdots\cdots\cdots\cdots\cdots\cdots\cdots \\ \psi_1(2m) & \bar{\psi}_1(2m) & \psi_2(2m) & \cdots \psi_m(2m) \end{vmatrix} \tag{2}$$

Equation 2 is usually written in the short notation of (3):

$$\Psi = |\psi_1\bar{\psi}_1 \cdots \psi_j\bar{\psi}_j \cdots \psi_m\bar{\psi}_m| \tag{3}$$

---

[1] J. C. Slater, *Phys. Rev.*, **34**, 1293 (1929).

From the properties of determinants, this wave function may be shown to obey the Pauli principle because the determinant vanishes when two columns are identical (two electrons in one spin orbital) and changes sign when two columns are interchanged (interchanging electron labels.)

The product wave function in (2) or (3) represents one possible *configuration* or assignment of electrons to spin orbitals; (2) or (3) obviously represents the *ground configuration*. In excited configurations electrons are assigned to higher spin orbitals. An example of a *singly excited configuration* is

$$|\psi_1\bar\psi_1 \cdots \psi_m\psi_{m+1}|$$

(Recall that m is our index for the highest occupied MO in the ground configuration.) A possible *doubly excited configuration* is

$$|\psi_1\bar\psi_1 \cdots \psi_{m-1}\bar\psi_m\psi_{m+1}\bar\psi_{m+1}|$$

The ground and excited *states* may be approximated by appropriate *configurations*; however, a better approximation is that the various states are represented as linear combinations of configurations. This description of a state in terms of more than one configuration is known as *configuration interaction*. Note that in such descriptions only those configurations that belong to the same irreducible representation need be included: others cannot interact (p. 76).

In an LCAO MO procedure the total wave function is treated as a linear combination of antisymmetrized configurations of component molecular orbitals, which in turn are formulated as linear combinations of atomic orbitals. By the usual variation procedure, the function,

$$\frac{\int \Psi H \Psi \, d\tau}{\int \Psi^2 \, d\tau} \tag{4}$$

is minimized with respect to the coefficients. The Hamiltonian may be given the more explicit definition,

$$H = \sum_\mu^n H_\mu^c + \sum\sum_{\mu<\nu} \frac{e^2}{r_{\mu\nu}} \tag{5}$$

in which the indices, μ and ν, refer to electrons. H includes the attraction energy of electrons to the "core," H_μ^c, and the repulsion energy between electrons, $e^2/r_{\mu\nu}$. In the HMO procedure the latter term is neglected—the electrons are treated as being independent; hence there is no point in giving explicit attention to antisymmetrization in the HMO method.

Various levels of sophistication arise in the formulation of the core Hamiltonian. We may consider, for example, interactions only with appropriately shielded carbon nuclei in a π-network; we may include interactions with peripheral groups such as attached hydrogen atoms ("hydrogen penetration integrals"), or, at least in principle, we may treat interactions with the bare nuclei by including all of the electrons in the molecule, σ- and π- alike. Treatment of organic compounds, however, has generally been confined to π-electrons and corresponding cores in which the carbon nuclei are shielded by the σ- and inner-shell electrons.[2]

Because the molecular orbitals are taken as linear combinations of atomic orbitals, the integrals required for evaluation of the energy reduce to two types of integrals in terms of the atomic orbitals. The core integrals are[3]

$$\mathbf{H}_{rs}^c = \int \varphi_r \mathbf{H}^c \varphi_s \, d\tau \tag{6}$$

and are one-electron functions. The electron repulsion integrals or *Coulomb repulsion integrals* (not to be confused with the Coulomb integrals of the simple HMO theory, which are actually more closely associated with core integrals) are usually abbreviated as $(rs \mid tu)$ and defined as

$$(rs \mid tu) = \int \varphi_r(1)\varphi_s(1) \frac{e^2}{r_{12}} \varphi_t(2)\varphi_u(2) \, d\tau \tag{7}$$

This integral can be interpreted as a classical electrostatic repulsion between the two charge distributions, $\varphi_r\varphi_s$ and $\varphi_t\varphi_u$.

When more than one configuration is included, the wave function may be written as

$$\Phi = C'\Psi' + C''\Psi'' + \cdots \tag{8}$$

The coefficients, C, are determined by a variation method that requires overlap and Hamiltonian integrals between the various antisymmetrized wave functions, Ψ; however, these integrals still reduce to those involving only atomic wave functions as in (6) and (7).

Several types of approach have been popular in the literature. The LCAO MO's may be relatively arbitrary, say HMO's, but then a substantial amount of configuration interaction may have to be included to get good results (CI method); or self-consistent field MO's may be

[2] R. S. Mulliken, *J. chim. phys.*, **46**, 497, 675 (1949).

[3] Integrals in this chapter are formulated in terms of real functions. For complex functions the complex conjugate is used to the left of the operator.

evaluated, in which case configuration interaction may be minimal (SCF method). Furthermore, we may calculate or estimate the required integrals, using assumed forms of the atomic orbitals (usually taken as Slater atomic orbital functions) (nonempirical methods), or we may assign to some of the integrals values consistent with some experimental data (empirical or semiempirical methods). Thus the principal advanced MO applications to organic compounds may be divided approximately into four classes:

1. Configuration interaction
   (a) Nonempirical
   (b) Empirical
2. Self-consistent field
   (a) Nonempirical
   (b) Empirical

These methods are not reviewed in detail, but a few of the prime points are mentioned in succeeding sections.

16.2   Antisymmetrized MO Method with Configuration Interaction (ASMO CI)

The principal application of the ASMO CI procedure has been to the calculation of the relative energy levels of excited states for comparison with experimental spectra. One of the earliest applications was Goeppert-Mayer and Sklar's[4] ASMO treatment of benzene. The core Hamiltonian was given an explicit form and some Coulomb repulsion integrals were estimated or evaluated, although many others were neglected. A number of errors crept into these calculations and were corrected in later calculations, which also included evaluation of further integrals.[5] In general, the agreement with experimentally assigned levels is only fair.

In its complete form the ASMO CI method is difficult to apply; calculations on benzene are facilitated by the high symmetry of the molecule. Ethylene has also been examined in detail;[6] even with extensive configuration interaction, the nonempirical calculations give disappointing results.

---

[4] M. Goeppert-Mayer and A. L. Sklar, *J. Chem. Phys.*, **6**, 645 (1938).

[5] The most recent calculations are those of R. G. Parr, D. P. Craig, and I. G. Ross, *J. Chem. Phys.*, **18**, 1561 (1950), and F. A. Gray, I. G. Ross, and J. Yates, *Australian J. Chem.*, **12**, 347 (1959). The results of a number of ASMO CI calculations of benzene are summarized by W. Kauzmann, *Quantum Chemistry*, Academic Press, New York (1957), p. 472.

[6] R. Daudel, R. Lefebvre, and C. Moser, *Quantum Chemistry*, Interscience Publishers, New York (1959), Chap. 20.

Empirical quantities have been used in place of calculated energies in several different ASMO calculations,[7] but the most complete semiempirical ASMO CI theory is that of Pariser and Parr.[8] In this theory the core Hamiltonians are given empirical values: $H_{rr}^c \equiv \alpha_r$ is given a value based on an appropriate valence-state ionization potential, $H_{rs}^c \equiv \beta_{rs}$ is neglected between nonbonded atoms and is retained as a parameter for bonded atoms. Values used for β_{rs} are -2.92 ev for ethylene and -2.39 ev for benzene.[8] An important simplification is the *zero differential overlap* approximation, that is, that overlap integrals between atomic orbitals may be neglected[9]—the same approximation used in HMO Theory. In ASMO theory, however, one further effect of this neglect is to cause all of the $(rs \mid tu)$ integrals to vanish except those of the form $(rr \mid rr)$ and $(rr \mid ss)$.

For the p-orbital on an sp^2-hybridized carbon Slater functions give 16.93 ev as the value of the $(11 \mid 11)$ integral. This value seems to be too high—when two electrons are in one orbital they will tend to stay apart because of Coulombic repulsion. This type of electron correlation is not considered in the Slater functions; hence the nonempirical calculated value is substantially higher than it should be. Pariser and Parr assume that a better value is given empirically as the difference between the valence state ionization potential and electron affinity:

$$(11 \mid 11) \cong I - A \tag{9}$$

For sp^2-carbon the derived value is 10.53 ev. Part of the poor results of the nonempirical ASMO method seems clearly to be associated with the use of too high a value for this integral; the success of the Pariser-Parr theory is partly because it accounts to a certain degree for the correlation of electrons.[10,11] The remaining integrals, $(rr \mid ss)$, are assigned values tending to e^2/r at long distances.

[7] For some examples of the use of ionization potentials in place of core energies, cf. O. Chalvet and R. Daudel, *Compt. rend.*, **235**, 960 (1952); *J. chim. phys.*, **49**, 629 (1952); H. C. Lefkovits, J. Fain, and F. A. Matsen, *J. Chem. Phys.*, **23**, 1690 (1955).

[8] R. Pariser and R. G. Parr, *J. Chem. Phys.*, **21**, 466, 767 (1953); R. Pariser, *J. Chem. Phys.*, **21**, 568 (1953); **24**, 250, 1112 (1956); R. G. Parr and R. Pariser, *J. Chem. Phys.*, **23**, 711 (1955).

[9] R. G. Parr, *J. Chem. Phys.*, **33**, 1184 (1960), has recently discussed the significance and justification of this approximation.

[10] W. Kolos, *Roczniki Chem.*, **32**, 315 (1958).

[11] It is interesting that the integral calculated for two electrons restricted to opposite lobes of the p-orbital is close to the value adopted by Pariser and Parr [M. J. S. Dewar and C. E. Wulfman, *J. Chem. Phys.*, **29**, 158 (1958); M. J. S. Dewar and N. L. Nojvat, *J. Chem. Phys.*, **34**, 1232 (1961); L. C. Snyder and R. G. Parr, *J. Chem. Phys.*, **34**, 1661 (1961)].

The Pariser-Parr method using HMO's and including all appropriate singly excited configurations accounts very well for the spectra of benzene and the linear acenes. Applied to azulene, it predicts the observed spectrum, including the low intensity α-bands in the visible (Sec. 8.3), and it gives a reasonable value for the dipole moment (1.88 D; experimental, 1 D. Cf. Sec. 6.1).[8] A number of additional applications has been published.[12]

16.3 Self-Consistent Field MO Theory (SCF MO)

In atoms minimization of the energy in a variation procedure with self-consistent field orbitals leads to a series of simultaneous nonlinear equations called the Hartree-Fock equations. Applied to LCAO MO's these equations were shown by Roothaan[13] to reduce to the form

$$\sum_s F_{rs}c_{js} = \sum_s S_{rs}c_{js}E_j \tag{10}$$

in which the index j refers to an MO. S_{rs} is the familiar overlap integral, and F_{rs} is defined by

$$F_{rs} = \mathbf{H}^c_{rs} + \sum_t \sum_u p_{tu}[(rs \mid tu) - \tfrac{1}{2}(rt \mid su)] \tag{11}$$

$$p_{tu} = 2 \sum_{j=1}^{m} c_{jt}c_{ju} \tag{12}$$

p_{tu} has the same significance as the HMO bond order, but here it is defined between all pairs of atomic orbitals. The terms with the double summations are Coulombic repulsion integrals weighted by a type of "bond density" between the orbitals involved.

Because the F_{rs} are functions of the coefficients, (10) is nonlinear and is solved, in practice, by successive iterations. Starting with an assumed set of MO's (say, HMO's) and the corresponding c_{jr}'s, we may evaluate the F_{rs} quantities. We derive, in turn, a new set of MO's that may be used to calculate a corresponding set of new F_{rs}-values. When the derived MO's do not differ significantly from the old, self-consistency has been reached.

In more qualitative terms, this SCF procedure starts with a set of molecular orbitals to which electrons are assigned in a usual *Aufbau*

---

[12] Some of the examples are butadiene: C. M. Moser, *J. Chem. Soc.*, **1954**, 3455; pyrene: J. Baudet, *Compt. rend.*, **245**, 1730 (1957); benzoquinone: T. Anno, I. Matsubara, and A. Sado, *Bull. Chem. Soc. Japan*, **30**, 168 (1957); T. Anno, A. Sado, and I. Matsubara, *J. Chem. Phys.*, **26**, 967 (1957); amidinium dyes: S. P. McGlynn and W. T. Simpson, *J. Chem. Phys.*, **28**, 297 (1958); hydrocarbon radical ions: G. J. Hoijtink, *Mol. Phys.*, **2**, 85 (1959); nitrogen heterocycles: L. Paoloni, *Gazz. chim. ital.*, **87**, 313 (1957); M. J. S. Dewar and L. Paoloni, *Trans. Faraday Soc.*, **53**, 261 (1957).

[13] C. C. J. Roothaan, *Revs. Modern Phys.*, **23**, 69 (1951).

manner. Energy minimization with cognizance of the electronic repulsion between the electrons in these MO's and of antisymmetrization results in a new set of MO's. The electrons may be assigned to these new MO's and the process may be repeated until self-consistency is reached. In principle, by assigning electrons to higher energy MO's, configuration interaction can also be included.

Mulliken[2] has detailed how empirical quantities may be used to derive values for the various core integrals. This procedure used with nonempirical calculated Coulomb repulsion integrals has been applied to a number of examples.[14]

Pople[15] has introduced into SCF MO theory a set of simplifying approximations closely related to the Pariser-Parr approximations of ASMO CI theory (Sec. 16.2). The approximations are principally the neglect of overlap, which carries with it the elimination of all Coulomb repulsion integrals except those of the type $(rr \mid rr)$ and $(rr \mid ss)$, and the use of empirical quantities for the integrals. With these simplifications, Roothaan's F_{rs} terms become, for hydrocarbons

$$F_{rr} = \alpha_r + \tfrac{1}{2}q_r\gamma_{rr} + \sum_{s \neq r}(q_s - 1)\gamma_{rs} \tag{13}$$

$$F_{rs} = \beta_{rs} - \tfrac{1}{2}p_{rs}\gamma_{rs} \quad (r \neq s) \tag{14}$$

in which γ_{rr} and γ_{rs} are symbols for $(rr \mid rr)$ and $(rr \mid ss)$, respectively, and q_s is the electron density at atom s. Starting with HMO's, for example, p_{rs} and q_r may be evaluated; together with γ_{rs}-values,[16] the F_{rs}-values are calculated. α_r may be retained as a parameter and assigned a value from, say, an ionization potential. β_{rs} may be given the Pariser-Parr value, -2.39 ev for a benzenoid bond (vide supra). Diagonization of the matrix of F_{rs}-elements gives a new set of coefficients which may be used to repeat the procedure to self-consistency.

Energies calculated by this simplified procedure have been shown to give a good account of ionization potentials of a number of radicals and

[14] A few examples are butadiene: R. G. Parr and R. S. Mulliken, *J. Chem. Phys.*, **18**, 1338 (1950); C. M. Moser, *Compt. rend.*, **238**, 1585 (1954); methylenecyclopropene: A. Julg, *J. chim. phys.*, **50**, 652 (1953); fulvene: G. Berthier, *J. chim. phys.*, **50**, 344 (1953); *J. Chem. Phys.*, **21**, 953 (1953); heptafulvene: A. Julg, *J. chim. phys.*, **52**, 50 (1955).

[15] J. A. Pople, *Trans. Faraday Soc.*, **49**, 1375 (1953); A. Brickstock and J. A. Pople, *Trans. Faraday Soc.*, **50**, 901 (1954); J. A. Pople, *Proc. Roy. Soc.*, **A233**, 233 (1955); J. A. Pople and P. Schofield, *Proc. Roy. Soc.*, **A233**, 241 (1955); J. A. Pople, *J. Phys. Chem.*, **61**, 6 (1957).

[16] The usual values are $\gamma_{rs} = 10.53$, 7.30, 5.46, and 4.90 ev for internuclear distances 0, 1, $\sqrt{3}$, and 2 times the C—C distance in benzene. For larger separations γ_{rs} is approximated by the interaction energy of unit point charges.[13]

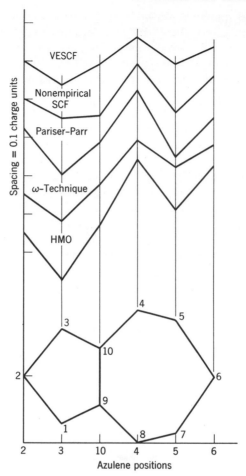

Fig. 16.1 Charge distribution pattern in azulene as calculated by various MO techniques. Curves are spaced 0.1 charge unit apart.

hydrocarbons[15,17] and to agree well with the stabilities and spectra of a number of carbonium ions.[18]

A recent elaboration of the simple procedure has been reported by Brown and Heffernan[19] to yield improved results. The principal modification in this "variable electronegativity" or VESCF method is the inclusion

[17] N. S. Hush, *J. Chem. Phys.*, **27**, 612 (1957).

[18] G. Dallinga, A. A. V. Stuart, P. J. Smit, and E. L. Mackor, *Z. Elektrochem.*, **61**, 1019 (1957); A. A. V. Stuart, and E. L. Mackor, *J. Chem. Phys.*, **27**, 826 (1957); G. Dallinga, E. L. Mackor, and A. A. V. Stuart, *Mol. Phys.*, **1**, 123 (1958).

[19] R. D. Brown and M. L. Heffernan, *Trans. Faraday Soc.*, **54**, 757 (1958); *Australian J. Chem.*, **12**, 319, 330, 543, 554 (1959); **13**, 38, 49 (1960).

of non-neighbor β's and the variation of α_r with q_r in a manner reminiscent of but differing in detail from the ω-technique of the simple LCAO procedure (Sec. 4.5).

Several of the advanced MO procedures have been applied to azulene and yield the charge densities in Table 16.1. The graphical comparison

TABLE 16.1

CHARGE DENSITIES OF AZULENE

Position	HMO	Nonempirical SCF*	Pariser-Parr†	VESCF‡	ω-Technique§
2	−0.047	+0.003	+0.021	−0.004	−0.048
3	−0.173	−0.049	0.096	−0.061	−0.118
10	−0.027	−0.042	−0.013	−0.009	−0.020
4	+0.145	+0.092	+0.121	+0.063	+0.095
5	+0.014	−0.034	−0.049	−0.009	+0.025
6	+0.130	+0.062	+0.052	+0.039	+0.084

* A. Julg, *J. chim. phys.*, **52**, 377 (1955).
† R. Pariser, *J. Chem. Phys.*, **25**, 1112 (1956).
‡ R. D. Brown and M. L. Heffernan, *Australian J. Chem.*, **13**, 38 (1960).
§ A. Streitwieser, Jr., J. B. Bush, and J. I. Brauman, unpublished calculations.

in Fig. 16.1 illustrates the degree to which the various procedures produce the same trends, although the absolute values differ. It is interesting to note that the advanced MO methods tend to reduce the magnitudes of the HMO charge densities, although the general pattern is unchanged.

Supplemental Reading

H. C. Longuet-Higgins in J. Prigogine's *Advances in Chemical Physics*, Vol. I, Interscience Publishers, New York (1958), p. 239.

R. Daudel, R. Lefebvre, and C. Moser, *Quantum Chemistry*, Interscience Publishers, New York (1959), especially Part II.

Appendix: Point group character tables

C_1	E
A	1

C_2	E	C_2
A	1	1
B	1	-1

C_3	E	C_3	C_3^2
A	1	1	1
E	1	ϵ^*	ϵ
	1	ϵ	ϵ^*

$\epsilon = e^{2\pi i/3}$

C_4	E	C_2	C_4	C_4^3
A	1	1	1	1
B	1	1	-1	-1
E	1	-1	$-i$	i
	1	-1	i	$-i$

C_{2v}	E	C_2	σ_v	σ_v'
A_1	1	1	1	1
B_2	1	-1	-1	1
A_2	1	1	-1	-1
B_1	1	-1	1	-1

C_{3v}	E	$2C_3$	$3\sigma_v$
A_1	1	1	1
A_2	1	1	-1
E	2	-1	0

459

$\mathbf{C}_{4v}$	E	C_2	$2C_4$	$2\sigma_v$	$2\sigma_d$
A_1	1	1	1	1	1
A_2	1	1	1	-1	-1
B_1	1	1	-1	1	-1
B_2	1	1	-1	-1	1
E	2	-2	0	0	0

$\mathbf{C}_{5v}$	E	$2C_5$	$2C_5^2$	$5\sigma_v$
A_1	1	1	1	1
A_2	1	1	1	-1
E_1	2	$2\cos\phi$	$2\cos\phi$	0
E_2	2	$2\cos 2\phi$	$2\cos 4\phi$	0

$$\phi = \frac{2\pi}{5}$$

$\mathbf{C}_{6v}$	E	C_2	$2C_3$	$2C_6$	$3\sigma_d$	$3\sigma_v$
A_1	1	1	1	1	1	1
A_2	1	1	1	1	-1	-1
B_1	1	-1	1	-1	-1	1
B_2	1	-1	1	-1	1	-1
E_1	2	-2	-1	1	0	0
E_2	2	2	-1	-1	0	0

Author index

461

Subject index

477